PRINCIPLES OF
ENVIRONMENTAL PHYSICS

PRINCIPLES OF ENVIRONMENTAL PHYSICS

3rd Edition

JOHN MONTEITH
Edinburgh, United Kingdom

MIKE UNSWORTH
Corvallis, Oregon, United States

AMSTERDAM • BOSTON • HEIDELBERG • LONDON
NEW YORK • OXFORD • PARIS • SAN DIEGO
SAN FRANCISCO • SINGAPORE • SYDNEY • TOKYO

Academic Press is an imprint of Elsevier

ELSEVIER

Publishing Editor: Jeanne Lawson
Production Director: Lori A. Buck
Project Manager: Lori A. Buck
Manufacturing Manager: André A. Cuello

Academic Press is an imprint of Elsevier
30 Corporate Drive, Suite 400, Burlington, MA 01803, USA
525 B Street, Suite 1900, San Diego, California 92101-4495, USA
84 Theobald's Road, London WC1X 8RR, UK

This book is printed on acid-free paper. ∞

Library of Congress Cataloging-in-Publication Data
Application Submitted

British Library Cataloguing-in-Publication Data
A catalogue record for this book is available from the British Library.

ISBN: 978-0-12-505103-3

For information on all Academic Press publications
visit our Web site at www.books.elsevier.com

Printed in the United States of America
08 09 10 9 8 7 6 5 4 3 2 1

QH
505
.M5P
2008

B2345356

Contents

Preface   ix

Acknowledgments   xiii

Symbols   xvi

1   The Scope of Environmental Physics   1

2   Properties of Gases and Liquids   4
Gases and Water Vapor   4
Liquid   20
Stable Isotopes   24
Problems   25

3   Transport of Heat, Mass, and Momentum   26
General Transfer Equation   26
Molecular Transfer Processes   28
Diffusion Coefficients   32
Problems   37

4   Transport of Radiant Energy   38
The Origin and Nature of Radiation   38
Spatial Relations   44
Problems   51

5   Radiation Environment   52
Solar Radiation   52
Attenuation of Solar Radiation in the Atmosphere   57
Solar Radiation at the Ground   62
Terrestrial Radiation   73

Net Radiation 79
Problems 84

6   Microclimatology of Radiation (i)   **86**
Radiative Properties of Natural Materials 86
Problems 99

7   Microclimatology of Radiation (ii)   **100**
Geometric Principles 101
Diffuse Radiation 110
Problems 115

8   Microclimatology of Radiation (iii)   **116**
Interception of Radiation by Plant Canopies 116
Interception of Radiation by Animal Coats 133
Net Radiation 137
Problems 139

9   Momentum Transfer   **142**
Boundary Layers 142
Momentum Transfer to Natural Surfaces 147
Lodging and Windthrow 155
Problems 159

10 Heat Transfer   **160**
Convection 160
Measurements of Convection 166
Conduction 179
Insulation 181
Problems 188

11 Mass Transfer (Gases and Water Vapor)   **190**
Non-Dimensional Groups 191
Measurements of Mass Transfer 193
Ventilation 197
Mass Transfer through Pores 200
Coats and Clothing 208
Problems 210

12 Mass Transfer (Particles)   **212**
Steady Motion 212

Non-Steady Motion 214
Particle Deposition 215
Problems 227

13 Steady State Heat Balance (i)   **229**
Heat Balance Equation 230
Heat Balance of Thermometers 232
Heat Balance of Surfaces 238
Developments from the Penman Equation 250
Problems 256

14 Steady State Heat Balance (ii)   **258**
Heat Balance Components 259
The Thermo-Neutral Diagram 266
Specification of the Environment 270
Case Studies 272
Sheep 273
Problems 281

15 Transient Heat Balance   **283**
Time Constant 283
General Cases 285
Heat Flow in Soil 290
Problems 298

16 Micrometeorology (i)   **300**
Turbulent Transfer 301
Flux-Gradient Methods 309
Methods for Indirect Measurements of Flux above Canopies . . 323
Relative Merits of Methods of Flux Measurement 328
Turbulent Transfer in Canopies 329
Density Corrections to Flux Measurements 331
Problems 332

17 Micrometeorology (ii)   **335**
Resistance Analogues 335
Case Studies 343
Transport within Canopies 359
Problems 365

References   **367**

Bibliography   390

Appendix A   394

Solutions to Selected Problems   401

Index   414

Preface to the Third Edition

In the time since the first edition of *Principles of Environmental Physics* was published in 1973, the subject has developed substantially; indeed, many users of the first and second editions have contributed to the body of research that makes this third edition larger than the first. From the start, this text has been aimed at two audiences: undergraduate and graduate students seeking to learn how the principles of physics can be applied to study the interactions between plants and animals and their environments; and the research community, particularly those involved in multidisciplinary environmental research. In many ways, environmental physics has become more thoroughly embedded in environmental research over the decades. For example, in ecology and hydrology, concepts of atmospheric exchange of gases and energy between organisms and the atmosphere, and the resistances (or conductances) controlling them are commonly applied. And in atmospheric science, soil-vegetation-atmosphere transfer schemes (SVATS) are an integral part of general circulation, mesoscale, and climate models. This "union of ideas" across the disciplines has made it challenging to define the scope of this third edition and to keep the size manageable. In doing so we have been guided by the word *Principles* in the title, so have focused, as in previous editions, on describing the critical principles of energy, mass, and momentum transfer, and illustrating them with a number of examples of their applications, taken from a range of classic and more recent publications.

Several themes have waxed and waned over the editions. At the time of the first edition, agricultural crop micrometeorology was a dominant application of environmental physics, beginning with the desire to quantify the water use and irrigation requirements of crops, and extending, as new instrumentation became available, to the analysis of carbon dioxide exchange in efforts to identify the environmental controls of crop productivity. There has been much less new work on agricultural crop micrometeorology in the last decade or two, but applications of environmental physics to the study of managed and natural forests and other ecosystems gathered pace through the 1970s and 1980s, and these topics, rather than agricultural applications, probably currently account for the major fraction of the annual reviewed publications in environmental physics. Also in the 1970s and 1980s, concern grew over human influences on air quality (particularly acid rain and ozone), leading to the application of environmental physics to study fluxes of pollutant gases, acidic particles, and mist to crops and forests. Additionally, the technology for remote sensing from satellites developed considerably.

The second edition of *Principles of Environmental Physics*, published in 1990, reflected these developments by adding a new chapter on particle transfer, new material on radiative transfer, and expanding the sections on micrometeorological methods. It also expanded treatment of the environmental physics of animals and their environments, influenced by the work of a number of researchers studying the heat balance of livestock and wild animals, who began to use the terminology of environmental physics, thus establishing parallels with the integration of environmental physics into plant science. The identification of the ozone hole above Antarctica in 1985, and its influence on ultraviolet radiation at the surface, received a short mention in the second edition, but emerging research on deposition of nitrogen-containing gases to vegetation was not covered; both topics receive more attention in this edition.

Through the 1980s, extending to the present time, concern over rising concentrations of carbon dioxide and other greenhouse gases in the atmosphere and consequent likely effects on climate has been a dominant topic, leading to an explosion of measurement and modeling research programs that make use of principles described in this book. Two developments have been particularly important: improved instrumentation allowing the eddy covariance technique in micrometeorology to be applied for studies of land-atmosphere exchange of carbon dioxide, water vapor and some other trace gases over seasonal and multi-annual periods; and theoretical advances to enable models of plant-atmosphere exchange to be scaled up from the leaf scale to landscape, regional, and even global scales, creating links between the principles described in this book at organism and canopy levels with the type of regional and global modeling necessary to address climate-change concerns. This edition contains two substantially revised chapters on micrometeorology with expanded treatment of the eddy covariance method, and with several new case studies to illustrate the application of micrometeorological methods over forests and natural landscapes. We have also expanded the material on solar and terrestrial radiation with new discussion of the roles of radiatively active greenhouse gases and aerosols.

Although eddy covariance has become the method of choice for micrometeorology in many situations, we have retained the material describing profile (similarity) techniques for deducing fluxes, because an understanding of similarity methods is essential for large-scale models and because profile methods have advantages in terms of simplicity of instrumentation when designing student projects or working with limited resources.

Other changes in this edition resulted from our own experience and feedback from those using this book as a teaching text: several sections that were particularly condensed in earlier editions have been expanded to aid clarity, and some sections have been rearranged to improve the flow; more worked examples have been included in the text; some specialized material (for example, details of the physics of radiative emission, and of radiation interaction with aerosols) has been added in text boxes that can be omitted by readers seeking a briefer treatment of the subject; and numerical problems have been added at the end of each chapter. Many of the numerical problems are more extensive than typically found in text books. This reflects requests we have had over the years from teachers who would like to explore realistic applications of the subject; many of the problems have been used in our own undergraduate and graduate teaching at Nottingham and Oregon State Universities, and we thank many students for their feedback and suggestions for improvements to the problems.

In planning this third edition, we debated whether to change nomenclature in flux equations from resistances to conductances, and whether to express quantities in "mole" units rather than "m-kg-s" units. Biologists increasingly use conductances and moles in their analyses, and there are some good theoretical and didactic arguments for this. However, we preferred to retain the "resistance" terms and "s m^{-1}" units used in earlier editions: the analogy with Ohm's Law emphasizes the underlying physics of many analyses used in this book, and units of s m^{-1} for resistances are more intuitive for heat and mass transfer calculations in energy balance and hydrological applications. Nevertheless, recognizing that many readers will be familiar with conductance and mol units, we have discussed conversions of units at several appropriate points in the text. There are many advantages in environmental physicists becoming familiar with both systems of units to facilitate communication across the disciplines.

We intend this text to be useful for teaching undergraduates and graduate students specializing in physics, biology, and the environmental sciences. The mathematical treatment is deliberately kept relatively simple, with little use of calculus; the biology is also strictly limited, consisting principally of material essential for understanding the physical applications. There is a bibliography directing readers to more detailed texts if necessary. For our other category of readers, research scientists, we have continued the approach of previous editions by including a large number of references to peer-reviewed literature, identifying a mix of papers that we consider classics and ground-breaking research applications; about 33% of the references in this edition have been published since 1990.

In the Preface to the second edition we expressed the hope that our book would encourage more university physics departments to expose their students to environmental physics. Our impression is that progress has been slow. This surely cannot be because of a lack of career opportunities—current environmental concerns open many possibilities for environmental physicists in the atmospheric sciences, hydrology, ecology, and biology, particularly if they enjoy the challenges of multidisciplinary work. Nor does it seem to be because physics students lack interest in environmental subjects. Perhaps it is inevitable that the crowded physics curriculum leaves little room for options such as environmental physics, but it would be satisfying if, by the time the fourth edition of this book appears, environmental physics became as common as astronomy or meteorology as an optional course in physics departments.

John Monteith and Mike Unsworth, 2007

Acknowledgments

We thank the following for allowing us to use diagrams, photographs, and original data:

Dr C.D. Keeling and the database provided by the U.S. Department of Energy through its Carbon Dioxide Information Analysis Center at Oak Ridge National Laboratory (Figure 2.2);

Dr. J. Lean for providing the data used in Figure 5.1; Dr. S.T. Henderson and Adam Hilger Publishers (5.2);

Dr. J.A Coakley (5.3); NASA Goddard SpaceFlight Center for the TOMS data in Figure 5.4; The Solar Energy Research Institute for the computer models used to construct Figs. 5.5;

Dr. M.D. Steven and the Royal Meteorological Society (5.6) from the *Quarterly Journal of the Royal Meteorological Society*;

Dr. F. Vignola for providing the data used in Figure 5.8;

Dr. J.V. Lake for providing the data used in Fig 5.9;

Mr. F.E. Lumb and the Royal Meteorological Society (5.10) from the *Quarterly Journal of the Royal Meteorological Society*;

Dr. R. von Fleischer and the Deutschern Wetterdienstes (5.15);

Dr R. Nakamura for providing the data used in Fig 5.17; Dr. K. Bible for providing the data used in Fig 5.18;

Dr. E.L. Deacon and Elsevier Publishing Co. (6.1 and 15.7);

Dr. S.A. Bowers and Lippincott Williams and Wilkins Co. (6.3) from *Soil Science*;

Dr. K.J. McCree and Elsevier Publishing Co. (6.4) from *Agricultural Meteorology*;

Dr. G. Stanhill and Pergamon Press (6.6) from *Solar Energy*; Professor L.E. Mount and Edward Arnold (6.7, 14.3 and 14.4); Dr. J.C.D. Hutchinson and Pergamon Press (6.8) from Comparative Biochemistry and Physiology; Dr. W. Porter for providing data used in Figure 6.9; Dr. C.R. Underwood and Taylor and Francis Ltd (7.5) from Ergonomics; Dr. G.S. Campbell and Nottingham University Press (8.3); Dr. K. Cena and the Royal Society of London (8.6) from the Proceedings of the Royal Society; Dr. J. Grace and Oxford University Press (9.3) from *Journal of Experimental Botany*; The Royal Meteorological Society (9.4, 9.5, 13.6, 13.7, 13.8, 13.9, 15.3, 17.13 and 17.14) from the *Quarterly Journal of the Royal Meteorological Society*; Dr. W.C. Hinds and John Wiley & Sons Inc. (9.6, 12.7, 12.8, 12.9, 12.10); Dr. D. Aylor and the American Society of Plant Physiologists (9.7) from *Plant Physiology*; Dr. A Stokes and Cambridge University Press (9.8); Drs. C.J. Wood and R. Belcher and D. Reidel Publishing Co. (9.10) from Boundary Layer Meteorology; Dr. J.A. Clark and D. Reidel Publishing Co. (10.3) from *Boundary Layer Meteorology*; Dr. B.J. Bailey and the International Society for Horticultural Science (10.4) from *Acta Horticulturae*; Dr. S. Vogel and Clarendon Press (10.5); Dr. A.J. McArthur and the Royal Society of London (10.7, 10.8) from the *Proceedings of the Royal Society*; Dr. R.P. Clark and *The Lancet* (10.9, 10.11) and Cambridge University Press (10.8) from *Journal of Physiology*; Dr. P.F. Scholander and the Marine Biological Laboratory (10.12); Dr. I. Impens for allowing us to use unpublished measurements in Figure 11.1. Dr. T. Haseba and the Society of Agricultural Meteorology in Japan (11.3) from *Journal of Agricultural Meteorology*; Dr. H.G. Jones and Cambridge University Press (11.7); Dr. D. Aylor and Pergamon Press (12.5) from *Atmospheric Environment*; Professor N.A. Fuchs and Pergamon Press (12.1); Dr. A.C. Chamberlain and Academic Press (12.3), and D. Reidel Publishing Co. (17.1) from *Boundary Layer Meteorology*;

Dr. P. Little and Pergamon Press (12.6) from Atmospheric Environment;

Dr. K. Raschke and Springer–Verlag (13.5) from *Planta*;

R. Milstein (13.10); Elsevier Publishing Co (13.11, 13.12) from the *Journal of Hydrology* and (17.11) from *Agricultural and Forest Meteorology*;

Dr. A.M. Hemmingsen (14.2);

Dr. D.M. Gates and Springer-Verlag (15.2);

Dr. J. van Eimern and the Deutschern Wetterdienstes (15.5);

Dr. W.R. van Wijk and North Holland Publishing Co. (15.6);

D. Vickers and Dr. L. Mahrt (16.3);

Dr. J. Finnigan and D. Reidel Publishing Co. (16.4) from *Boundary Layer Meteorology* and (17.16, 17.17);

Dr. R.H. Shaw and Elsevier Publishing Co. (16.8) from *Agricultural Meteorology*; Academic Press (16.9 and 16.10);

Dr. M.R. Raupach and Annual Reviews Inc. (16.11) from *Annual Review of Fluid Mechanics*;

Dr. T.A. Black and Blackwell Scientific (17.9, 17.10) from Global Change Biology;

Dr. D. Baldocchi (17.7) and D. Reidel Publishing Co. (17.19) from *Boundary Layer Meteorology*;

Dr. M. Sutton and the Royal Society (17.15) from the *Philosophical Transactions of the Royal Society of London*.

We particularly thank the staff at ChapterTwo for their skill and patience in converting our text, prepared in Scientific Word, into this book.

►

Symbols

The main symbols used in this book are arranged here in a table containing brief definitions of each quantity. A few of the symbols are universally accepted (e.g. R, g), some have been chosen because they appear very frequently in the literature of environmental physics (e.g., r_s, z_0, K_M), and some have been devised for the sake of consistency. In particular, the symbols $\mathbf{S}$ and $\mathbf{L}$ are used for flux densities of short-and long-wave radiation with subscripts to identify the geometrical character of the flux, e.g., $\mathbf{S}_d$ for the flux of diffuse short-wave radiation from the sky.

Flux densities of momentum, heat and mass are printed in bold case throughout the book (e.g., $\boldsymbol{\tau}$, $\mathbf{E}$) and so is the latent heat of vaporization of water $\boldsymbol{\lambda}$, partly to distinguish it from wavelength λ and partly because it is often associated with $\mathbf{E}$. Upper case subscripts are used to refer to momentum, heat, vapor, carbon dioxide, etc., e.g., r_V, K_M; most other subscripts are lower case, e.g., c_p for the specific heat of air at constant pressure.

The complete set of symbols represents the best compromise that could be found between consistency, clarity and familiarity.

ROMAN ALPHABET

A	area; azimuth angle with respect to south
A_b	area of solid object projected on a horizontal plane
A_p	area of solid object projected on plane perpendicular to solar beam
$A(z)$	amplitude of soil temperature wave at depth z
$\mathbf{B}$	total energy emitted by unit area of full radiator or black body

B	wet-bulb depression
$\mathbf{B}(\lambda)$	energy per unit wavelength in spectrum of full radiator or black body
c	volume fraction at CO_2 (e.g., v.p.m.); fraction of sky covered by cloud; velocity of light; mean velocity of gas molecules
c_d	drag coefficient for form drag and skin friction combined
c_f	drag coefficient for form drag
c_l	specific heat of a liquid
c_p	specific heat of air at constant pressure; efficiency of impaction of particles
c_s	specific heat of solid fraction of a soil
c_v	specific heat at constant volume
c'	bulk specific heat of soil
C	flux of heat per unit area by convection in air
$\mathcal{C}$	heat capacity of an organism per unit surface area
d	zero plane displacement
D	saturation vapor pressure deficit; diffusion coefficient for a gas in air (subscripts V for water vapor; C for CO2); damping depth $(2\kappa'/\omega)^{1/2}$
e	partial pressure of water vapor in air
$e_s(T)$	saturation vapor pressure of water vapor at temperature T
δe	saturation deficit, i.e., $e_s(T) - e$
E_q	energy of a single quantum
$\mathbf{E}$	flux of water vapor per unit area; evaporation rate
$\mathbf{E}_r$	respiratory evaporation rate of animal
$\mathbf{E}_s$	rate of evaporation from skin
$\mathbf{E}_t$	rate of evaporation from vegetation
F	generalized stability factor $(\phi_v\phi_m)^{-1}$
F	drag force on a particle; retention factor
$\mathbf{F}$	mass flux of a gas per unit area; flux of radiant energy
g	acceleration by gravity (9.81 m s^{-2})
$\mathbf{G}$	flux of heat by conduction, per unit area
h	Planck's constant $(6.63 \times 10^{-34} \text{ J s})$; relative humidity of air; height of cylinder, crop, etc.
$\mathbf{H}$	total flux of sensible and latent heat, per unit area
i	intensity of turbulence, i.e., root mean square velocity/mean velocity
$\mathbf{I}$	intensity of radiation (flux per unit solid angle)

J	rate of change of stored heat per unit area
k	von Karman's constant (0.41); thermal conductivity of air; attenuation coefficient; Boltzmann constant $(1.38 \times 10^{-23}\ \text{J K}^{-1})$
k'	attenuation coefficient; thermal conductivity of a solid
K	diffusion coefficient for turbulent transfer in air (subscripts H for heat, M for momentum, V for water vapor, C for CO_2)
$\mathcal{H}$	canopy attenuation coefficient
$\mathcal{H}_s$	ratio of horizontally projected area of an object to its plane or total surface area
l	mixing length, stopping distance; length of plate in direction of airstream
L	leaf area index; Monin–Obukhov length
L	flux of long-wave radiation per unit area (subscript u for upwards; d for downwards; e from environment; b from body)
m	mass of a molecule or particle; air mass number
M	rate of heat production by metabolism per unit area of body surface
M	grammolecular mass (subscripts a for dry air, v for water vapor)
n	represents a number or dimensionless empirical constant in several equations
N	Avogadro constant (6.02×10^{23}); number of hours of daylight
N	radiance (radiant flux per unit area per unit sold angle)
P	latent heat equivalent of sweat rate per unit body area
p	total air pressure; interception probability in hair coats
q	specific humidity of air (mass of water vapor per unit mass of moist air)
Q	rate of mass transfer
r	radius; resistance to transfer (subscripts M momentum, H heat, V water vapor, C for CO_2); usually applied to boundary layer transfer
r_a	resistance to transfer in the atmosphere (subscripts M, H, V, C as above)
r_b	boundary layer resistance of crop for mass transfer
r_c	canopy resistance
r_d	thermal resistance of human body
r_f	thermal resistance of hair, clothing; resistance for forced ventilation in open-top chambers
r_p	resistance of pore for mass transfer
r_s	resistance of a set of stomata

r_h	resistance of hole (one side) for mass transfer
r_i	incursion resistance for open-top chambers
r_t	total resistance of single stoma
r_H	resistance for heat transfer by convection, i.e., sensible heat
r_R	resistance for radiative heat transfer $(\rho c_p / 4\sigma T^3)$
r_{HR}	resistance for simultaneous sensible and radiative heat exchange, i.e., r_H and r_R in parallel
r_V	resistance for water vapor transfer
R	Gas Constant (8.31 J mol^{-1} K^{-1})
$\mathbf{R}_n$	net radiation flux density
$\mathbf{R}_{ni}$	net radiation absorbed by a surface at the temperature of the ambient air
s	amount of entity per unit mass of air
S	gas concentration
S_d	diffuse solar irradiance on horizontal surface
S_e	solar radiation received by a body, per unit area, as a result of reflection from the environment
S_p	direct solar irradiance on surface perpendicular to solar beam
S_b	direct solar irradiance on horizontal surface
S_t	total solar irradiance (usually) on horizontal surface
t	diffusion pathlength
T	temperature
T_a	air temperature
T_b	body temperature
T_c	cloud-base temperature
T_d	dew point temperature
T_e	equivalent temperature of air $(T + (e/\gamma))$
T_{e*}	apparent equivalent temperature of air $(T + (e/\gamma^*))$
T_f	effective temperature of ambient air
T_s, T_o	temperature of surface losing heat to environment
T_v	virtual temperature
T'	thermodynamic wet bulb temperature
T^*	standard temperature for vapor pressure specification
u	optical pathlength of water vapor in the atmosphere
$u(z)$	velocity of air at height z above earth's surface
u_*	friction velocity

υ molecular velocity

υ_d deposition velocity

υ_s sedimentation velocity

V_m molar volume at STP (22.4 L)

$\dot{V}$ volume

w vertical velocity of air; depth of precipitable water

W body weight of animal

x volume fraction (subscripts s for soil; l for liquid; g for gas); ratio of cylinder height to radius

z distance; height above earth's surface

z_0 roughness length

Z height of equilibrium boundary layer

GREEK ALPHABET

α absorption coefficient (subscripts p for photosynthetically active; T for total radiation; r for red; i for infrared)

$\alpha(\lambda)$ absorptivity at wavelength λ

β solar elevation; ratio of observed Nusselt number to that for a smooth plate

γ psychrometer constant ($= c_p p/\lambda\varepsilon$)

γ^* apparent value of psychrometer constant ($= \gamma r_v/r_H$)

Γ dry adiabatic lapse rate, DALR ($9.8 \times 10\ \mathrm{k\ m^{-1}}$)

δ depth of a boundary layer

Δ rate of change of saturation vapor pressure with temperature, i.e., $\partial e_s(T)\partial T$

ε ratio of molecular weights of water vapor and air (0.622)

ε_a apparent emissivity of the atmosphere

$\varepsilon(\lambda)$ emissivity at wavelength λ

θ angle with respect to solar beam; potential temperature

κ thermal diffusivity of still air

κ' thermal diffusivity of a solid, e.g., soil

λ wavelength of electromagnetic radiation

$\boldsymbol{\lambda}$ latent heat of vaporization of water

μ coefficient of dynamic viscosity of air

ν coefficient of kinematic viscosity of air; frequency of electromagnetic radiation

ρ	reflection coefficient (subscripts p for photosynthetically active; c for canopy; s for soil; r for red; i for infrared; T for total radiation); density of a gas, e.g., air including water vapor component
ρ_a	density of dry air
ρ_c	density of CO_2
ρ_l	density of a liquid
ρ_s	density of a solid component of soil
ρ'	bulk density of soil
ρ	reflection coefficient
$\rho(\lambda)$	reflectivity of a surface at wavelength λ
σ	Stefan–Boltzmann constant (5.67×10^{-8} W m^{-2} K^{-4})
Σ	the sum of a series
τ	flux of momentum per unit area; shearing stress
τ	fraction of incident radiation transmitted, e.g., by a leaf; relaxation time, time constant, turbidity coefficient
ϕ	mass concentration of CO_2, e.g., g m^{-3}; angle between a plate and airstream
Φ	radiant flux density
χ	absolute humidity of air
$\chi_s(T)$	saturated absolute humidity at temperature $T(^\circ C)$
ψ	angle of incidence
ω	angular frequency; solid angle

NON-DIMENSIONAL GROUPS

Le	Lewis number (κ/D)
Gr	Grashof number
Nu	Nusselt number
Pr	Prandtl number (v/κ)
Re$_*$	Roughness Reynolds number ($u_* z_0/v$)
Re	Reynolds number
Ri	Richardson number
Sc	Schmidt number (v/D)
Sh	Sherwood number
Stk	Stokes number

LOGARITHMS

ln	logarithm to the base e
log	logarithm to the base 10

Chapter 1

The Scope
of Environmental Physics

Physics has always been concerned with understanding the natural environment, and, in its early days was often referred to as "Natural Philosophy." Environmental Physics, as we choose to define it, is the measurement and analysis of interactions between organisms and their environments.

To grow and reproduce successfully, organisms must come to terms with the state of their environment. Some microorganisms can grow at temperatures between −6 and 100°C and, when they are desiccated, can survive down to −272°C. Higher forms of life, on the other hand have adapted to a relatively narrow range of environments by evolving sensitive physiological responses to external physical stimuli. When environments change—for example, because of natural variation or because of human activity, organisms may, or may not, have sufficiently flexible responses to survive.

The physical environment of plants and animals has five main components that determine the survival of the species:

(i) The environment is a source of radiant energy that is trapped by the process of photosynthesis in green cells and stored in the form of carbohydrates, proteins, and fats. These materials are the primary source of metabolic energy for all forms of life on land and in the oceans.

(ii) The environment is a source of the water, carbon, nitrogen, other minerals, and trace elements needed to form the components of living cells.

1

(iii) Factors such as temperature and daylength determine the rates at which plants grow and develop, the demand of animals for food, and the onset of reproductive cycles in both plants and animals.

(iv) The environment provides stimuli, notably in the form of light or gravity, which are perceived by plants and animals and provide frames of reference both in time and in space. These stimuli are essential for resetting biological clocks, providing a sense of balance, etc.

(v) The environment determines the distribution and viability of pathogens and parasites that attack living organisms, and the susceptibility of organisms to attack.

To understand and explore relationships between organisms and their environment, the biologist should be familiar with the main concepts of the environmental sciences. He or she must search for links between physiology, biochemistry, and molecular biology on the one hand, and atmospheric science, soil science, and oceanography on the other. One of these links is environmental physics. The presence of an organism modifies the environment to which it is exposed, so that the physical stimulus received *from* the environment is partly determined by the physiological response *to* the environment.

When an organism interacts with its environment, the physical processes involved are rarely simple and the physiological mechanisms are often imperfectly understood. Fortunately, physicists are trained to use Occam's Razor when they interpret natural phenomena in terms of cause and effect; i.e., they observe the behavior of a system and then seek the simplest way to describe it in terms of governing variables. Boyle's Law and Newton's Laws of Motion are classic examples of this attitude. More complex relations are avoided until the weight of experimental evidence shows they are essential. Many of the equations discussed in this book are approximations to reality that have been found useful to establish and explore ideas. The art of environmental physics lies in choosing robust approximations that maintain the principles of conservation for mass, momentum, and energy.

Such approximations are often described as *models*. These models may be either theoretical or experimental, and both types are found in this book. We have not considered models of plant or animal systems based on computer simulations. They can rarely be *tested* in the sense that physicists use the word because so many variables and assumptions are deployed in their derivation. Consequently, although they can be useful for identifying the sensitivity of systems to environmental variables, they seldom seem to us to contribute to an understanding of the *principles* of environmental physics.

Several volumes would be needed to cover all relevant principles of environmental physics, and the definite article was deliberately omitted from the title of this book because it makes no claim to be comprehensive. However, the topics that it covers are central to the subject: the exchange of radiation, heat, mass, and momentum between organisms and their environment. Within these topics, similar analysis can be applied to a number of closely related problems in plant, animal, and human ecology. The short bibliography at the end of the book should be consulted for more specialized treatments; for example, of subjects such as the physics of water, heat, and solute transfer in soils.

The lack of a common language is often a barrier to progress in interdisciplinary subjects, and it is not easy for a physicist or atmospheric scientist with no biological training to communicate with a physiologist or ecologist who is fearful of formulae. Throughout the book, therefore, simple electrical analogues are used to describe rates of transfer and exchange between organisms and their environment, and calculus has been kept to a minimum. The concept of *resistance* (and its reciprocal, *conductance*) has been familiar to plant physiologists for many years, mainly as a way of expressing the physical factors that control rates of transpiration and photosynthesis, and animal physiologists have used the term to describe the insulation provided by clothing, coats, or by a layer of air. In micrometeorology, aerodynamic resistances derived from turbulent transfer coefficients can be used to calculate fluxes from a knowledge of the appropriate gradients, and resistances that govern the loss of water from vegetation are now incorporated in models of the atmosphere that include the behavior of the earth's surface. Ohm's Law has therefore become an important unifying principle of environmental physics; the basis of a common language for biologists and physicists.

The choice of units was dictated by the structure of the Système International, modified by retaining the centimeter. For example, the dimensions of leaves are quoted in mm and cm. To adhere strictly to the metre or the millimeter as units of length often needs powers of 10 to avoid superfluous zeros and sometimes gives a false impression of precision. As most measurements in environmental physics have an accuracy between ± 1 and $\pm 10\%$, they should be quoted to 2 or at most 3 significant figures, preferably in a unit chosen to give quantities between 10^{-1} and 10^3. The area of a leaf would therefore be quoted as 23.5 cm^2 rather than 2.35 $\times$ 10^{-3} m^2 or 2350 mm^2.

Conversions from **SI** to c.g.s. are given in Appendix A.1.

Chapter 2

Properties of Gases and Liquids

The physical properties of gases influence many of the exchanges that take place between organisms and their environment. The relevant equations for air therefore form an appropriate starting point for an environmental physics text. They also provide a basis for discussing the behavior of water vapor, a gas whose significance in meteorology, hydrology, and ecology is out of all proportion to its relatively small concentration in the atmosphere. Because the evaporation of water from soils, plants, and animals is also an important process in environmental physics, this chapter reviews the principles by which the state of liquid water can be described in organisms and soil, and by which exchange occurs between liquid and vapor phases of water.

2.1 Gases and Water Vapor

PRESSURE, VOLUME, AND TEMPERATURE

The observable properties of a gas such as temperature and pressure can be related to the mass and velocity of its constituent molecules by the *Kinetic Theory of Gases*, which is based on Newton's Laws of Motion. Newton established the principle that when force is applied to a body, its momentum, the product of mass and velocity, changes at a rate proportional to the magnitude of the force. Appropriately, the unit of force in the Système

Internationale is the Newton; and the unit of pressure (force per unit area) is the Pascal—from the name of another famous natural philosopher.

The pressure p that a gas exerts on the surface of a liquid or solid is a measure of the rate at which momentum is transferred to the surface from molecules which strike it and rebound. Assuming that the kinetic energy of all the molecules in an enclosed space is constant and by making further assumptions about the nature of a *perfect* gas, a simple relation can be established between pressure and kinetic energy per unit volume. When the density of the gas is ρ and the mean square molecular velocity is $\bar{v}^2$, the kinetic energy per unit volume is $\rho\bar{v}^2/2$ and

$$p = \rho\bar{v}^2/3 \qquad (2.1)$$

implying that pressure is two-thirds of the kinetic energy per unit volume.

Although Equation 2.1 is central to the Kinetic Theory of Gases, it has little practical value. A number of congruent but more useful relations can be derived from the observations of Boyle and Charles whose gas laws can be combined to give

$$pV \propto T \qquad (2.2)$$

where V is the volume of a gas at an absolute temperature T (K). To establish a constant of proportionality, a standard amount of gas is defined by the volume V_m, occupied by a mole at standard pressure and temperature (STP, i.e., 101.325 kPa and 273.15 K), which is 0.0224 m^3 (22.4 liters). Then

$$pV_m = RT \qquad (2.3)$$

where $R = 8.314$ J mol^{-1} K^{-1}, the *molar gas constant*, has the dimensions of a molecular specific heat.

Since the pressure exerted by a gas is a measure of its kinetic energy per unit volume, pV_m is proportional to the kinetic energy of a mole. A mole of any substance contains N molecules where $N = 6.02 \times 10^{23}$ is the *Avogadro constant*. It follows that the mean energy per molecule is proportional to

$$pV_m/N = (R/N)\,T = kT \qquad (2.4)$$

where k is the *Boltzmann constant*.

Equation 2.3, the *Ideal Gas Equation*, is sometimes used in the form

$$p = \rho RT/M \qquad (2.5)$$

obtained by writing the density of a gas as its molecular mass divided by its molecular volume; i.e.,

$$\rho = M/V_m \qquad (2.6)$$

TABLE 2.1   Composition of dry air

Gas	Molecular weight (g)	Density at STP (kg m^{-3})	Percent by volume	Mass concentration (kg m^{-3})
Nitrogen	28.01	1.250	78.09	0.975
Oxygen	32.00	1.429	20.95	0.300
Argon	38.98	1.783	0.93	0.016
Carbon dioxide	44.01	1.977	0.03	0.001
Air	29.00	1.292	100.00	1.292

For unit mass of any gas with volume V, $\rho = 1/V$, so Equation 2.5 can also be written in the form

$$pV = RT/M \qquad (2.7)$$

Equation 2.7 provides a general basis for exploring the relation between pressure, volume, and temperature in unit mass of gas and is particularly useful in four cases:

1. constant volume—p proportional to T

2. constant pressure (isobaric)—V proportional to T

3. constant temperature (isothermal)—V inversely proportional to p

4. constant energy (adiabatic)—p, V, and T may all change.

When the molecular weight of a gas is known, its density at STP can be calculated from Equation 2.6, and its density at any other temperature and pressure from Equation 2.5. Table 2.1 contains the molecular weights and densities at STP of the main constituents of dry air. Multiplying each density by the appropriate volume fraction gives the concentration of each component, and the sum of these concentrations is the density of dry air. From a density of 1.292 kg m^{-3} and from Equation 2.5, the effective molecular weight of dry air (in g) is 28.96, or 29 within 0.1%.

Since air is a mixture of gases, it obeys *Dalton's Law*, which states that the total pressure of a mixture of gases that do not react with each other is given by the sum of the partial pressures. *Partial pressure* is the pressure a gas would exert at the same temperature as the mixture if it alone occupied the volume the mixture occupies.

THE HYDROSTATIC EQUATION

Although the atmosphere is usually in motion, the upward force acting on a thin slab of air due to the decrease of pressure with height is generally balanced by the downward force imposed by gravity. Hence

$$-dp = g\rho \, dz$$

or

$$\frac{dp}{dz} = -g\rho \tag{2.8}$$

Equation 2.8, the *hydrostatic equation*, describes how pressure decreases with increasing height.

THE FIRST LAW OF THERMODYNAMICS, AND SPECIFIC HEATS

The First Law of Thermodynamics states that energy in a system is conserved if heat is taken into account. When a unit mass of gas is heated but not allowed to expand, the increase in total heat content per unit increase of temperature is known as the *specific heat at constant volume*, usually given the symbol c_v. Conversely, if the gas is allowed to expand in such a way that its *pressure* stays constant, additional energy is needed for expansion, so that the *specific heat at constant pressure* c_p is larger than c_v.

To evaluate the *difference* between c_p and c_v, the work done by expansion can be calculated by considering the special case of a cylinder with cross-section A fitted with a piston that applies a pressure p on gas in the cylinder, thereby exerting a force pA. If the gas is heated and expands to push the cylinder a distance x, the work done is the product of force pA and distance x or pAx, which is also the product of the pressure and the change of volume Ax. The same relation is valid for any system in which gas expands at a constant pressure.

The work done for a small expansion dV of unit mass of gas at constant pressure p is found by differentiating Equation 2.7 to give

$$p \, dV = (R/M) \, dT \tag{2.9}$$

As the *difference* between the two specific heats is the work done in expansion per unit increase of temperature, it follows that

$$c_p - c_v = p dV/dT = R/M \tag{2.10}$$

The *ratio* of c_p to c_v for gases depends on the energy associated with the vibration and rotation of its molecules and therefore depends on the number of atoms forming the molecule. For diatomic molecules such as nitrogen and oxygen, which are the major constituents of air, the theoretical value of c_p/c_v is $7/5$, in excellent agreement with experiment. Because $c_p - c_v = R/M$, it follows that $c_p = (7/2)R/M$ and $c_v = (5/2)R/M$. In the natural environment, most processes involving the exchange of heat in air occur at a pressure (atmospheric) that is effectively constant, and since the molecular weight of air is 28.96,

$$c_p = (7/2) \times (8.314/28.96) = 1.01 \text{ J g}^{-1} \text{ K}^{-1}$$

LATENT HEAT

When heat is supplied to a substance without its temperature changing, it is described as *latent heat*. The increase in internal energy is associated with a change of phase, involving changes in molecular configurations. For example, evaporation and condensation of water are common phenomena in environmental physics. The *latent heat of vaporization* λ is the the heat that must be supplied to convert unit mass of water from liquid to vapor without a change in temperature. For water at STP, λ is 2500 J g^{-1}; the *latent heat of condensation* has the same value. Values of λ vary with temperature, as given in Appendix A.3. Similarly, if heat is supplied to ice at STP, the temperature remains at 0°C until all the ice has melted. The *latent heat of melting* is 334 J g^{-1} and is equal to the latent heat of fusion.

LAPSE RATE

Meteorologists apply thermodynamic principles to the atmosphere by imagining that discrete infinitesimal "parcels" of air are transported either vertically or horizontally by the action of wind and turbulence. Relations between temperature, pressure, and height can be deduced by assuming that

▶ processes within a parcel are adiabatic; i.e., that the parcel neither gains energy from its environment (e.g., by heating) nor loses energy,

▶ the parcel is always at the same pressure as the air surrounding it, which is assumed to be in hydrostatic equilibrium,

▶ the parcel is moving sufficiently slowly for its kinetic energy to be negligible

Suppose a parcel containing unit mass of air makes a small ascent so that it expands as external pressure falls by dp. If there is no external supply of heat, the energy for expansion must come from cooling of the parcel by an amount dT. It is convenient to treat this as a two-stage process:

1. parcel ascends and cools at constant pressure and volume providing energy $c_v \, dT$;

2. parcel expands against external pressure p requiring energy $p \, dV$.

For an adiabatic process, the sum of these quantities must be zero; i.e.,

$$c_v \, dT + p \, dV = 0 \qquad (2.11)$$

Differentiating Equation 2.7 and putting $R/M = c_p - c_v$, gives

$$V \, dP + p \, dV = (c_p - c_v) \, dT \qquad (2.12)$$

Eliminating $p \, dV$ from the last two equations gives

$$c_p \, dT = V \, dP \qquad (2.13)$$

Substituting $V = RT/Mp$ from Equation 2.7 gives

$$\frac{dT}{T} = \left(\frac{R}{Mc_p}\right)\frac{dp}{p} \qquad (2.14)$$

Two important relationships can be derived from Equation 2.14. First, rearranging the hydrostatic equation (Equation 2.8) and substituting for ρ from Equation 2.5 gives

$$\frac{dp}{p} = -\left(\frac{gM}{RT}\right)dz \qquad (2.15)$$

Substituting Equation 2.15 in Equation 2.14 then gives

$$-\frac{dT}{dz} = \frac{g}{c_p} \qquad (2.16)$$

The quantity g/c_p is known as the *Dry Adiabatic Lapse Rate* or DALR, usually given the symbol Γ. When both g and c_p are expressed in Sl units, the DALR is

$$\Gamma = \frac{9.8 \left(\text{m s}^{-2}\right)}{1.01 \times 10^3 \left(\text{J kg}^{-1}\text{K}^{-1}\right)} \approx 1 \text{ K per 100 m}$$

"Dry" in this context implies that no condensation or evaporation occurs within the parcel. A *Saturated Adiabatic Lapse Rate* operates within cloud where the lapse rate is smaller than the DALR because of the release of latent heat by condensation. Unlike the DALR, the Saturated Adiabatic Lapse Rate depends strongly on pressure and temperature.

The difference between the actual lapse rate of air and the DALR is a measure of the vertical stability of the atmosphere. During the day, the lapse rate up to a height of at least 1 km is usually larger than the DALR and is many times larger immediately over dry sunlit surfaces. Consequently, ascending parcels (which cool at the DALR) rapidly become warmer than their surroundings and experience buoyancy that accelerates their ascent and promotes turbulent mixing, so that the atmosphere is said to be *unstable*. Conversely, temperature usually increases with height at night ("inversion" of the day-time lapse) so that rising parcels become cooler than their environment and further ascent is inhibited by buoyancy. Turbulence is suppressed and the atmosphere is said to be *stable*.

POTENTIAL TEMPERATURE

A second quantity that can be derived from Equation 2.14 is *potential temperature*. Equation 2.14 shows that the temperature of an air parcel that is subject to adiabatic expansion or contraction is a function of pressure alone. This fact makes it possible to "label" an air parcel, which is at an arbitrary pressure, by the temperature it would reach if brought adiabatically to a standard pressure p_0, usually taken as 100 kPa. This is the potential temperature θ of the air parcel. If an air parcel is subject only to adiabatic processes as it moves in the atmosphere, its potential temperature remains constant. Quantities that remain constant during such transformations are called *conservative quantities*.

To derive an expression relating θ to pressure, temperature, and p_0, Equation 2.14 can be integrated upward from p_0 (where $T = \theta$) to p, giving

$$\frac{Mc_p}{R} \int_{\theta}^{T} \frac{dT}{T} = \int_{p_0}^{p} \frac{dp}{p} \tag{2.17}$$

or

$$\frac{Mc_p}{R} \ln \frac{T}{\theta} = \ln \frac{p}{p_0}$$

which may be written

$$\theta = T \left(\frac{p_0}{p} \right)^{R/Mc_p} \tag{2.18}$$

The expression $R/\left(Mc_p\right) = \left(c_p - c_v\right)/c_p = 0.29$ for dry air. Most problems in environmental physics are concerned with the lowest few tens of meters in the atmosphere, so p_0/p is within 1 or 2 percent of unity. Consequently, it is seldom necessary to use potential temperature (Equation 2.18) when calculating temperature gradients, but it is an important concept in larger-scale atmospheric physics.

In an atmosphere where θ is constant with height, the temperature T (z) at any level z is exactly that which leads by adiabatic ascent or descent to the value θ at 100 kPa. Hence

$$T(z) = \theta - \Gamma z \qquad (2.19)$$

The atmosphere is then in a state of adiabatic (or convective) equilibrium, also called *neutral stability*.

WATER VAPOR AND ITS SPECIFICATION

The evaporation of water at the earth's surface to form water vapor in the atmosphere is a process of major physical and biological importance because the latent heat of vaporization is large in relation to the specific heat of air. The heat released by condensing 1 g of water vapor is enough to raise the temperature of 1 kg of air by 2.5 K. Water vapor has been called the "working substance" of the atmospheric heat engine because of its role in global heat transport. The total mass of water vapor in the air at any moment is enough to supply only one week of the world's precipitation, so the process of evaporation must be very efficient in replenishing the atmospheric reservoir. On a much smaller scale, it is the amount of latent heat removed by the evaporation of sweat that allows man and many other mammals to survive in hot climates. Sections that follow describe the physical significance of different ways of specifying the amount of vapor in a sample of air and relations between them.

Vapor Pressure

When both air and liquid water are present in a closed container, molecules of water continually escape from the surface into the air to form water vapor, but there is a counter-flow of molecules recaptured by the surface. If the air is dry initially, there is a net loss of molecules from the liquid, recognized as "evaporation," but as the partial pressure (e) of the vapor increases, the evaporation rate decreases, reaching zero when the rate of loss is exactly balanced by the rate of return. The air is then said to be "saturated" with vapor and the partial pressure is the *saturation vapor pressure of water* (SVP), often written $e_s(T)$ because it depends strongly on temperature. When a surface is maintained at a lower temperature than the air above it, it is possible for molecules to be captured faster than they are lost and this net gain is recognized as "condensation."

The second law of thermodynamics can be used to derive an equation for the rate of change of saturated vapor pressure above a liquid with temperature, namely the *Clausius–Clapeyron equation*

$$\frac{de_s}{dT} = \frac{L}{T(\alpha_2 - \alpha_1)} \tag{2.20}$$

where α_1 and α_2 are the specific volumes of liquid and vapor, respectively, at temperature T.

A rigorous expression for the dependence of $e_s(T)$ on T can be obtained by integrating the Clausius–Clapeyron equation, but as the procedure is cumbersome, a simpler (and unorthodox) method will be used here, with the advantage that it relates vapor pressure to the concepts of latent heat and free energy.

Suppose that the evaporation of unit mass of water can be represented by the isothermal expansion of vapor at a fictitious and large pressure e_0 to form a much larger volume of saturated water vapor at a smaller pressure $e_s(T)$. Water vapor can be treated as an ideal gas in this example. If the work done during this expansion is identified as the latent heat of vaporization, λ, and V is the volume of the gas at any point during expansion,

$$\lambda = \int_{e_0}^{e_s(T)} e \, dV \tag{2.21}$$

Differentiating Equation 2.7 for an isothermal system and putting $p = e$ gives

$$e \, dV + V \, de = 0$$

so that, from Equation 2.7,

$$e \, dV = -(R/M_w) \, T \, de/e$$

where M_w is the molecular weight of water.

Substitution in Equation 2.21 then gives

$$\lambda = -\frac{RT}{M_w} \int_{e_0}^{e_s(T)} \frac{de}{e} = \frac{RT}{M_w} \ln\left[e_0/e_s(T)\right] \tag{2.22}$$

Given values of T and $e_s(T)$ as in Appendix A.4, the initial pressure e_0 can now be calculated, by rearranging Equation 2.22 to give

$$e_0 = e_s(T) \exp\left(\lambda M_w/RT\right) \tag{2.23}$$

Adopting a constant value of $\lambda = 2.48$ kJ g^{-1} at 10°C, e_o is found to change very little with temperature: it is 2.076×10^5 MPa at 0°C and 2.077×10^5 MPa at 20°C. (In fact, the value of λ decreases with temperature by about 2.4 J g^{-1} K^{-1}, and e_o decreases with temperature if this is taken into account.)

Equation 2.23 therefore provides a simple expression for the dependence of $e_s(T)$ on T with the form

$$e_s(T) = e_o \exp\left(-\lambda M_w / RT\right) \qquad (2.24)$$

where e_o can be regarded as constant, at least over a restricted temperature range. However, it is more convenient to eliminate e_o by normalizing the equation to express $e_s(T)$ as a proportion of the saturated vapor pressure at some standard temperature T^* so that

$$e_s(T^*) = e_o \exp\left(-\lambda M_w / RT^*\right) \qquad (2.25)$$

Dividing Equation 2.24 by Equation 2.25 then gives

$$e_s(T) = e_s(T^*) \exp\left\{A\left(T - T^*\right) / T\right\} \qquad (2.26)$$

where $A = \lambda M_w / RT^*$.

For $T^* = 273$ K (0°C), $\lambda = 2470$ J g^{-1} K^{-1} so that $A = 19.59$; but over the range 273 to 293 K, more exact values of $e_s(T)$ (to within 1 Pa) are obtained by giving A an arbitrary value of 19.65 (see Appendix A.4). Similarly, with $T^* = 293$, the calculated value of A is 18.3, but $A = 18.00$ gives values of $e_s(T)$ from 293 to 313 K. (The need to adjust the value of A arises from the slight dependence of e_o on temperature, coupled with the sensitivity of $e_s(T)$ to the value of A).

An empirical equation introduced by Tetens [1930] has almost the same form as Equation 2.26 and is more exact over a much wider temperature range. As given by Murray [1967], it is

$$e_s(T) = e_s(T^*) \exp\left\{A\left(T - T^*\right) / \left(T - T'\right)\right\} \qquad (2.27)$$

where

$$A = 17.27$$
$$T^* = 273 \text{ K } (e_s(T^*) = 0.611 \text{ kPa})$$
$$T' = 36 \text{ K}$$

Values of saturation vapor pressure from the Tetens formula are within 1 Pa of the exact values given in Appendix A.4 up to 35°C.

The rate of increase of $e_s(T)$ with temperature is an important quantity in micrometeorology (see Chapter 11) and is usually given the symbol Δ or s. Between 0 and $30°C$, $e_s(T)$ increases by about 6.5 % per $°C$, whereas the pressure of unsaturated vapor of any ideal gas increases by only 0.4 % $(1/273)$ per $°C$. By differentiating Equation 2.24 with respect to T it can be shown that

$$\Delta = \lambda M_w e_s(T) / \left(R T^2 \right) \tag{2.28}$$

and, up to $40°C$, this expression is exact enough for all practical purposes.

Dew Point

The *dew point* (T_d) of a sample of air with vapor pressure e is the temperature to which it must be cooled to become saturated; i.e., it is defined by the equation $e = e_s(T_d)$. When the vapor pressure is known, the dew point can be found approximately from tables of SVP, or more exactly by inverting a formula such as Equation 2.26 to obtain dew point temperature as a function of vapor pressure; i.e.,

$$T_d = \frac{T^*}{1 - A^{-1} \ln e/e_s(T^*)} \tag{2.29}$$

The specification of a dew point is most useful in problems of dew formation, which occur when the temperature of a surface is below the dew point of the ambient air.

Saturation Vapor Pressure Deficit

The *saturation vapor pressure deficit* of an air sample (sometimes "*vapor pressure deficit*, vpd" or just "*saturation deficit*" for short) is the difference between the saturation vapor pressure and the actual vapor pressure i.e. $e_s(T) - e$. In ecological problems, vpd is often regarded as a measure of the "drying power" of air, because it plays an important part in determining the relative rates of growth and transpiration in plants. In micrometeorology, the vertical gradient of saturation deficit is a measure of the lack of equilibrium between a wet surface and the air passing over it (Chapter 13).

Mixing ratio

The *mixing ratio* (w) of water vapor in a volume of air is defined as the ratio of the mass of water vapor to the mass of dry air in the volume. In the

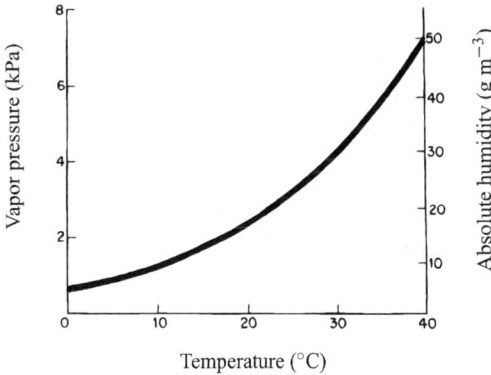

FIGURE 2.1 The relation between saturation vapor pressure, absolute humidity, and temperature.

atmosphere, the magnitude of w is typically a few grams per kilogram in middle latitudes, but may reach 20 g kg^{-1} in humid environments. When neither evaporation nor condensation occurs, the mixing ratio is a conservative quantity.

Specific and absolute humidity

Specific humidity (q) is the mass of water vapor per unit mass of moist air and is useful in problems of vapor transport in the lower atmosphere because it is independent of temperature (but see p. 327). It is closely related to *absolute humidity* (χ), also called *vapor density*, which is the mass of water vapor per unit *volume* of moist air. If the density of the moist air is ρ, $\chi = \rho q$. Figure 2.1 shows the relation between saturation vapor pressure, absolute humidity, and temperature.

Both absolute and specific humidity can be expressed as functions of total air pressure ρ and vapor pressure e. For the vapor component only, Equation 2.3 can be written in the form

$$e = \chi \, (R/M_{\mathrm{w}}) \, T$$

so that with $M_{\mathrm{w}} = 18$ g mol^{-1},

$$\chi \left(\mathrm{g \, m}^{-3} \right) = \frac{M_{\mathrm{w}} e}{R T} = \frac{2165 e \ (\mathrm{kPa})}{T \ (\mathrm{K})} \tag{2.30}$$

Similarly, the dry air component has a pressure $p - e$ and therefore has a density of

$$\rho_A = M_A (p - e) / RT$$

where the molecular weight of air is $M_A = 29 \text{ g mol}^{-1}$.

The density of the moist air is the sum of the density of its components; i.e.,

$$\rho = \{M_w e + M_A (p - e)\} / RT \qquad (2.31)$$

and the specific humidity is therefore

$$q = \chi / \rho = \frac{\varepsilon e}{(p - e) + \varepsilon e} \approx \frac{\varepsilon e}{p} \qquad (2.32)$$

where $\varepsilon = M_w / M_A = 0.622$. The approximation is usually valid in microclimatic problems because e is two orders of magnitude smaller than p.

Virtual Temperature

The density of *dry* air with the same temperature and pressure as *moist* air is given by

$$\rho' = p M_A / RT \qquad (2.33)$$

and substitution in Equation 2.31 gives the density of moist air as

$$\rho = \rho' \{1 - e (1 - \varepsilon) / p\} \qquad (2.34)$$

Moist air is therefore less dense than dry air at the same temperature and is relatively buoyant. In problems involving the transfer of heat as a consequence of buoyancy (see Chapter 11), it is convenient to express the difference in density produced by water vapor in terms of a *virtual temperature* at which dry air would have the same density as a sample of moist air at an actual temperature T. Combination of Equations 2.31 and 2.33 gives the virtual temperature as

$$T_V = T / \{1 - (1 - \varepsilon) e / p\} \approx T (1 + (1 - \varepsilon) e / p) \qquad (2.35)$$

where the approximation is valid when e is much smaller than p.

Relative humidity

The *relative* humidity of moist air (h) is defined as the ratio of its actual vapor pressure to the saturation vapor pressure at the same temperature;

i.e., $h = e/e_s(T)$. Although this quantity is frequently quoted in climatic statistics, and is widely regarded as a measure of the drying power of air (like saturation deficit), its fundamental significance lies in the specification of thermodynamic equilibrium between liquid water and water vapor, discussed in Section 2.2.

Wet-bulb temperature

An unsaturated parcel of air containing a small quantity of liquid water can be saturated adiabatically if all the energy used for evaporation is obtained by cooling the air. The minimum temperature achieved in an ideal process of this kind is called the *thermodynamic wet-bulb temperature* because it is closely related to the observed temperature of a thermometer covered with a wet sleeve. The derivation of the wet-bulb temperature and discussion of its environmental significance is deferred to Chapter 13.

Summary of methods for specifying water vapor amount

To review several methods of specifying the amount of water vapor in air, Figure 2.2 contains a point X representing a sample of air with temperature T and vapor pressure e. The change of saturation vapor pressure with temperature is given by the exponential curve SS'. Dew point temperature T_d is given by the point Z at which the curve intersects a line of constant

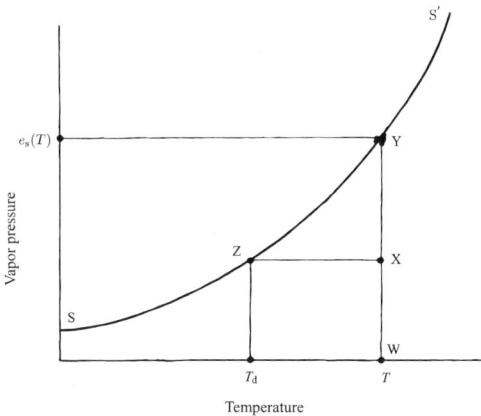

FIGURE 2.2 Relation between dew point temperature, saturation vapor pressure deficit, and relative humidity of an air sample (see text and Figure 13.2).

vapor pressure through X, and saturation vapor pressure $e_s(T)$ is given by the intersection of the curve with a vertical line through X meeting the temperature axis at W. Saturation vapor pressure deficit $e_s(T) - e$ is YW − XW = YX, and relative humidity is XW/YW.

OTHER GASES

Although water vapor plays a major role in the energy balance and survival of living organisms, several trace gases in the atmosphere interact also with organisms and soil. The concentration of many of these gases has increased as a result of human activity, and in the past few decades, several trace gases have achieved prominence: gases associated with the sulphur and nitrogen biogeochemical cycles because of their contributions to acidic deposition and aerosol formation; tropospheric ozone because it affects plant growth, human health, visibility, and climate; carbon dioxide because of its roles in the carbon cycle and as a greenhouse gas modifying climate; and other greenhouse gases such as methane and nitrous oxide because of their likely contribution to global warming.

Acidic deposition results principally from emissions of sulphur dioxide (SO_2) and nitrogen oxides (NO_x). These gases are removed relatively rapidly from the atmosphere (lifetimes of hours to days) either by deposition to the surface or by conversion to aerosols. Consequently, acid deposition tends to be a regional problem [Colls 1997]. Tropospheric ozone (O_3) is another shortlived photochemical pollutant formed in the atmosphere during complex chemical reactions between motor vehicle and other emissions in combination with sunlight. Important human-related sources of carbon dioxide (CO_2), methane (CH_4), and nitrous oxide (N_2O) are fossil fuel combustion, rice and animal production, and fertilizer use, respectively.

Trace Gas Concentrations

Both *gravimetric* units (e.g., g m^{-3}) and *volumetric* units (e.g., parts per billion by volume, ppb) are used to describe trace gas concentrations. Conversion between the two makes use of principles described earlier. For example, sulphur dioxide (SO_2) has a molecular weight of 64 g mol^{-1}. One mol occupies 0.0224 m^3 at STP, so that the density of pure SO_2 is $64/0.0224 = 2857$ g m^{-3} at STP. Consequently, since volumetric concentration of the pure gas is, by definition, 10^9 ppb,

$$1 \text{ ppb } SO_2 = 2.86 \ \mu\text{g m}^{-3} \tag{2.36}$$

TABLE 2.2 Past and recent (2001–2002) concentrations and "lifetimes" of some important greenhouse gases

Species	Formula	Concentration (ppb in 1998)	Trend (ppb/yr in 1990s)	Lifetime (yr)	Radiative forcing (W m^{-2} ppb^{-1})
Carbon dioxide	CO_2	365×10^3	1500	50–200	1.5×10^{-5}
Methane	CH_4	1745	7.0	8.4	3.7×10^{-4}
Nitrous oxide	N_2O	314	0.8	120	3.7×10^{-3}
Tropospheric ozone	O_3	50	unknown	0.01–0.05	2.0×10^{-2}
CFC-12	CCl_2F_2	0.533	4.4	100	2.8×10^{-1}

Gravimetric concentrations at other pressure and temperatures can then be found using the gas law, Equation 2.7 in the form

$$\frac{P_1 V_1}{T_1} = \frac{P_2 V_2}{T_2}$$

Table 2.2 (from IPCC 2001) summarizes information about past and recent (2001–2002) concentrations and "lifetimes" of some important greenhouse gases. *Lifetime* is defined as the total atmospheric burden divided by the global mean rate of removal of the gas; for CO_2, this concept is difficult to define because it is not strictly removed, but is circulated between terrestrial, oceanic, and atmospheric reservoirs. Table 2.2 also shows the increased radiative forcing attributed to the increase in each gas since pre-industrial times (see Chapter 5).

Figure 2.3 illustrates the increase in atmospheric concentration of CO_2 since 1958 at a remote site (Mauna Loa, Hawaii).

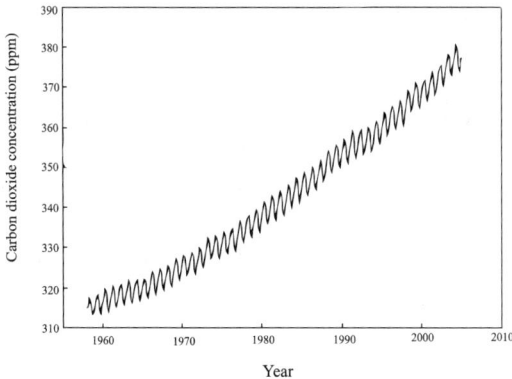

FIGURE 2.3 The increase in atmospheric concentration of CO_2 measured at Mauna Loa, Hawaii since 1958 (data courtesy of CDICAC).

The average rate of increase in the global concentration of CO_2 has been about 0.4% per year since 1980, largely as a result of fossil fuel combustion and land use change. At this rate, the concentration will exceed twice the pre-industrial value before the end of the twenty-first century. Other greenhouse gases are also increasing steadily. Implications for global thermodynamics and therefore for all forms of life are the subject of vigorous controversy [IPCC, 2001].

2.2   Liquid

WATER CONTENT AND POTENTIAL

In environmental physics, the rate of evaporation of water from the tissue of plants and animals, and from soil surfaces, is paramount for the energy balance and for physiological responses of organisms. This section describes the variables that determine the state of liquid water in organisms and soil, and the physical principles relating liquid-phase water to the vapor phase. Campbell and Norman [1998] give a more complete treatment.

Two variables are used to define the state of water in a system: water content and water potential. The water content can be expressed as the ratio of the volume of water in a substance to its total volume (*volumetric water content, θ*), or as the ratio of the mass of water to the dry mass of the substance (*mass water content, w*).

Water potential is a concept recognizing that water in living tissue or soil may be bound to the material, diluted by solutes, and under pressure or tension. Consequently, its energy state may be different from that of pure unconfined water. Water potential is the potential energy of water per unit of the substance (i.e., per mole, kilogram, etc.). The water potential of pure unconfined water is defined as zero. Gradients of water potential are the driving forces for liquid movement in biological systems.

The units of the most fundamental expression of potential, water potential per unit mass, are $J\ kg^{-1}$. However, since water is almost incompressible at pressures encountered in environmental physics, its density is independent of potential. Hence there is direct proportionality between water potential per unit mass and water potential per unit volume. The units of the latter expression are pressure ($J\ m^{-3}$ or $N\ m^{-2}$), and are given the S.I. name *pascals*. Historically, water potential in biological systems has been expressed in pressure units. Numerically, taking the density of water as $10^3\ kg\ m^{-3}$, it follows that $1\ J\ kg^{-1} = 1\ kPa$.

The total water potential ψ in a material can be expressed as the sum of several components

$$\psi = \psi_g + \psi_m + \psi_p + \psi_o \qquad (2.37)$$

where the subscripts g, m, p, and o stand for gravitational, matric, pressure, and osmotic potential, respectively.

Gravitational potential is the potential energy of water as a result of its position; e.g.,

$$\psi_g = gh \qquad (2.38)$$

where h is the height above or below a reference level (positive *above* the level). For example, water in tissue at the top of a tall tree would have a larger gravitational potential than water at the ground.

Matric potential is a consequence of the binding of water to materials such as soil particles, cellulose, and proteins, and the enclosure of water in capillaries or other fine pores; both mechanisms reduce the potential energy below that of free water. For simple pores, such as capillaries, the reduction in potential energy is inversely proportional to pore diameter. The relationship between the matric potential in a substance and the water content is called the *moisture characteristic*, and can often be described by an empirical equation of the form

$$\psi_m = aw^{-b} \qquad (2.39)$$

where a and b are empirical constants and w is the water content. Matric potential is always negative or zero.

Pressure potential arises from hydrostatic pressure, such as the turgor pressure in plant cells or the blood pressure in animals. Pressure potential can be positive or negative.

Osmotic potential occurs when solutes are dissolved in water and constrained by a semipermeable membrane; for example, in plant or animal cells. For a perfect semipermeable membrane, ψ_o is given by

$$\psi_o = -C\phi v R T \qquad (2.40)$$

where C is the solute concentration (mol kg^{-1}), ϕ is the osmotic coefficient (unity for an ideal solute, and usually within 10% of unity for solutions encountered in biological systems and their environments), v is the number of ions per mole (e.g., 2 for sodium chloride, 1 for sucrose), R is the molar Gas Constant, and T is temperature (K).

When soils are saturated with pure water, their total water potential is near zero, but they quickly drain by gravity to potentials between -30

and -10 kPa, reaching a water content called *field capacity*. As water is lost from soils by plant uptake and evaporation, their water potential decreases to a point where roots cannot extract further water (i.e., the root water potential cannot drop below the soil water potential). Typically, this occurs when $\psi \simeq -1500$ kPa; the corresponding soil water content is called the *permanent wilting point*.

The osmotic potential in cells of plant leaves is typically between -500 and -7000 kPa, with most mesophytic plants operating in the range -1000 to -2000 kPa. During the day, if plants are transpiring rapidly, the pressure potential (turgor) in leaf cells is close to zero, so Equation 2.37 shows that leaf water potential is close to osmotic potential. At night, when transpiration rates are very slow, the water potential of leaves approaches that of the soil, close to zero if the soil is wet. For comparison, the osmotic potentials of blood and other bodily fluids in mammals are usually in the range -200 to -700 kPa.

LIQUID–AIR INTERFACES

For a plane surface of pure water, equilibrium is established when the air above it is saturated, so that the relative humidity of air at the interface is unity. But when water contains dissolved salts (as in plant cells) or is held by capillary and absorptive forces in a porous medium (such as soil), the water potential is negative and the equilibrium relative humidity over the water surface is less than unity.

To take a homely example, samples of cake or biscuits (cookies) contain water in equilibrium with a unique relative humidity h. If a sample is exposed to air with a higher value of h it will absorb water, but it will lose water to air with a lower value of h. Because biscuits have much smaller pores than cakes, their equilibrium relative humidity is lower. The relative humidity of air in most kitchens is intermediate between the equilibrium value for biscuits and cakes, so biscuits left on the table usually absorb water, becoming soft, whereas cakes lose water, becoming crisp.

The physics of drying and wetting in soils and other porous materials depends on the water potential or "free energy" of water in the sample, which may be treated as the amount of work needed to move unit mass of water from its particular environment to a pool of pure water at atmospheric pressure and at the same temperature. The relation between water potential and relative humidity is more useful than this abstract definition. To fix ideas, consider a salt solution at equilibrium in a closed, isothermal container above which is air with a vapor pressure $e < e_s(T)$. The energy needed to change unit mass of water from its liquid state to saturated va-

TABLE 2.3 Equilibrium relative humidity (at 20°C) and equivalent free energy of some common solutions and substances

	Relative humidity (%)	Free energy (Eq. 2.32) (MPa)
Plant cellular fluid	100 to 98	0 to −3
Ocean water	98	−3
Fresh salami	80	−3
Saturated NaCl	75	−39
Crisp cornflakes	20	−218
Saturated LiCl.H$_2$O	11	−299

por is simply the latent heat of vaporization, already evaluated in terms of expansion from an initial pressure e_0 to $e_s(T)$ in Equation 2.22. Because the system is closed (an essential point), further work is needed to expand the vapor from its volume at $e_s(T)$ to its larger volume at e. By use of Equation 2.22 again, this energy is

$$E = (R/M_w) \, T \ln \{e_s(T)/e\} \qquad (2.41)$$

This expression gives the work, in addition to latent heat, needed to change liquid water from a state of equilibrium with air at $e_s(T)$ to vapor at a pressure e where $e/e_s(T)$ is the relative humidity h. In other words, the water behaves as if it had a *negative* water potential of

$$-E = (R/M_w) \, T \ln h \qquad (2.42)$$

Table 2.3 illustrates the range of water potentials and equilibrium relative humidities for a number of solutions and common substances.

The argument used in Equation 2.41 to establish E as a function of h is not valid in an open system where pure water evaporates into unsaturated air, because all parts of the system are exposed to the same pressure—atmospheric. The energy needed to evaporate unit mass of water is simply the latent heat of vaporization and is not augmented by a "gas expansion" term as physiologists believed for many years [Monteith, 1972]. For similar reasons, the rate at which water evaporates depends on the difference between the saturation vapor pressure at the appropriate temperature and the vapor pressure of air moving over the water. The rate is not a direct function of the corresponding difference in free energy [Monteith and Campbell, 1980].

| 2.3 | ## Stable Isotopes

Many elements occur in forms that differ in atomic mass because they possess extra neutrons in their nuclei, but have nearly identical chemical activities. Isotopes that do not decay radioactively are known as *stable isotopes*. The stable isotopes of carbon, hydrogen, and oxygen provide important ways for tracking the past history of water vapor and carbon dioxide in the environment. Table 2.4 shows the percentage abundance of each isotope. Because the ratios of the heavier to the lighter forms of stable isotopes (*abundance ratios*) are so small, isotopic compostion is usually expressed as deviations (δ) in parts per thousand ($^0/_{00}$) from the abundance ratio in a standard. For example, for ^{13}C, the deviation is given by

$$\delta^{13}C^0/_{00} = \{[(^{13}C/^{12}C)_{sample}/(^{13}C/^{12}C)_{standard}] - 1\} \times 1000 \quad (2.43)$$

The reference standard for carbon is a carbonate rock with a $^{13}C/^{12}C$ ratio of 0.01124.

Reactions that depend on the kinetic energy of molecules, such as diffusion, and biochemical reactions such as photosynthesis, discriminate against the heavier isotopes, resulting in small differences in concentrations of the stable isotope relative to the standard isotope of an element that has taken part in a dscriminatory process. For example, the source of most water vapor in the atmosphere is tropical oceans. Water molecules contain stable isotopes of hydrogen (deuterium, D) and oxygen (^{18}O). The heavier isotopes of water evaporate less rapidly than the lighter, so water vapor has a negative $\delta^{18}O$ and δD with respect to sea water (the standard). As water vapor cools (e.g., by moving to higher latitudes or by ascent), rain or snow forms by condensation. The process of condensation discriminates in favor of the heavier isotopes, leaving the lighter isotopes in the vapor. Consequently, as water vapor is transported to cooler regions, and some of the heavier isotopes are preferentially removed by precipitation, its isotopic signatures become steadily lighter; there is thus a good correlation between both $\delta^{18}O$ and δD in precipitation and the ambient

TABLE 2.4   Abundance (in percent) of stable isotopes of carbon, oxygen and hydrogen

Carbon	Oxygen	Hydrogen
^{12}C 98.89	^{16}O 99.763	H 99.9844
^{13}C 1.11	^{17}O 0.0375	D 0.0156
	^{18}O 0.1995	

air temperature. This correlation has been used to infer air temperatures in the past from the analysis of $\delta^{18}O$ and δD in ice cores [IPCC 2001]. Similarly, precipitation that falls in the winter months has a more negative δD than summer rain. Since winter rain is largely responsible for replenishing groundwater in semi-arid environments, it is possible to determine whether plants are using deep groundwater or recent rainfall by measuring δD in transpired water [Dawson 1993]. Further examples of isotopic discrimination in the water and carbon cycles are given in Waring and Running [1998].

2.4 Problems

1. Using Equation 2.8, construct a graph to show how atmospheric pressure decreases with height above the earth's surface from 0 to 5 km.

2. Using Equation 2.16, construct a graph of atmospheric temperature in an atmosphere with neutral stability as a function of height above the earth's surface to a height of 1 km. In the real world, what factors might distort this graph?

3. Why does the saturation vapor pressure of a series of gases at the same temperature decrease as their molecular weight increases? Illustrate this dependence graphically.

4. Using Equation 2.26 or Equation 2.27, plot graphs to demonstrate:

    a) how saturation vapor pressure depends on temperature between 0 and 40°C.

    b) how relative humidity depends on temperature between 10 and 30°C when vapor pressure is fixed at 1.0., 1.5, or 2.0 kPa.

5. Use Equation 2.26 or Equation 2.27 to plot the dependence of dewpoint temperature on relative humidity in the range 10–100%.

6. What would the Dry Adiabatic Lapse Rate be on a planet with an atmosphere similar to that on earth in terms of chemical composition but with a molecular weight of only 10 g mol^{-1}?

Chapter 3

Transport of Heat, Mass, and Momentum

The last chapter was concerned primarily with ways of specifying the state of the atmosphere in terms of properties such as pressure, temperature, and gas concentration. To continue this introduction to some of the major concepts and principles on which environmental physics depends, we now consider how the transport of entities such as heat, mass, and momentum is determined by the state of the atmosphere and the corresponding state of the surface involved in the exchange, whether soil, vegetation, the coat of an animal, or the integument of an insect or seed.

3.1 General Transfer Equation

A simple general equation can be derived for transport within a gas by "carriers," which may be molecules or particles or eddies, capable of transporting units of a property P such as heat, water vapor, or gas concentration. Even when the carriers are moving randomly, net transport may occur in any direction provided that P decreases with distance in that direction. The carrier can then unload its excess of P at a point where the local value is less than at the starting point.

To evaluate the net flow of P in one dimension, consider a volume of gas with unit horizontal cross-section and a vertical height l assumed to be the mean distance for unloading a property of the carrier (Figure 3.1).

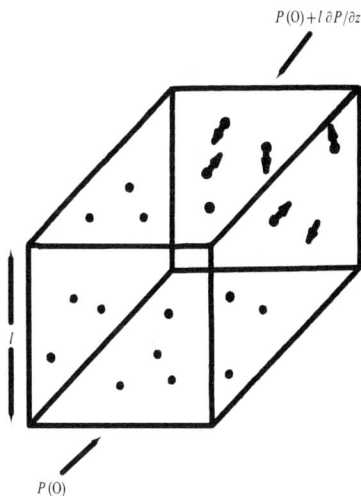

FIGURE 3.1   Volume of air with unit cross-section and height l containing n carriers per unit volume moving with random velocity c (see text).

Over the plane defining the base of the volume, P has a uniform value $P(0)$, and if the vertical gradient (change with height) of P is dP/dz, the value at height l will be $P(0) + l\,dP/dz$. Carriers that originate from a height l will therefore have a load corresponding to $P(0) + l\,dP/dz$, and if they move a distance l vertically downward to the plane where the standard load corresponds to $P(0)$, they will be able to unload an excess of $l\,dP/dz$. To find the rate of transport equivalent to this excess, assume that n carriers per unit volume move with a mean *random velocity* c so that the number moving toward one face of a cube at any instant is $nc/6$ per unit area. The downward flux of P (quantity per unit area and per unit time) is therefore given by $(ncl/6)dP/dz$. However, there is a corresponding upward flux of carriers reaching the same plane from below after setting off with a load given by $P - l\,dP/dz$. Mathematically, the upward flow of a deficit is equivalent to the downward flow of an excess, so the net *downward* flux of P is

$$\mathbf{F} = (ncl/3)\,dP/dz \tag{3.1}$$

For the application of Equation 3.1 to the transport of entities by molecular movement, the mean velocity w is often related to the root mean square molecular velocity c by assuming that, at any instant, one-third of the molecules in the system are moving in the z direction so that $w = c/3$.

On the other hand, micrometeorologists studying transfer by turbulent eddies are concerned with a form of Equation 3.1 in which P is replaced by the amount of an entity per unit mass of air (the *specific concentration*, q) rather than per unit volume (since the latter depends on temperature); the two quantities are related by $nP = \rho q$, where ρ is air density. The flux equation therefore becomes

$$\mathbf{F} = -\overline{\rho w l} \,(d\bar{q}/dz) \qquad (3.2)$$

The minus sign is needed to indicate that the flux is downward if q increases upward; averaging bars are a reminder that both w and q fluctuate over a wide range of time-scales as a consequence of turbulence. In this context, the quantity l is known as the 'mixing length' for turbulent transport.

It is also possible to write the instantaneous values of q and w as the sum of mean values $\bar{q}$ or $\bar{w}$ and corresponding deviations from the mean q' and w'. The net flux across a plane then becomes

$$\overline{(\rho\bar{w} + \rho w')\,(\bar{q} + q')} = \overline{\rho w' q'} \qquad (3.3)$$

where $\bar{q}'$ and $\bar{w}'$ are zero by definition and $\bar{w}$ is assumed to be zero near the ground when averaging is performed over a period long compared with the lifetime of the largest eddy (say, 10 minutes). This relation provides the "eddy correlation" method of measuring vertical fluxes discussed in Chapter 16.

3.2   Molecular Transfer Processes

According to Equation 2.1, the mean square velocity of molecular motion in an ideal gas is $\overline{v^2} = 3p/\rho$. Substituting $p = 10^5$ N m^{-2} and $\rho = 1.29$ kg m^{-3} for air at $0°$C gives the root mean square velocity as $(\overline{v^2})^{1/2} = 480$ m s^{-1}. Molecular motion in air is therefore extremely rapid over the whole range of temperatures found in nature and this motion is responsible for a number of processes fundamental to micrometeorology: the transfer of momentum in moving air responsible for the phenomenon of viscosity; the transfer of heat by the process of conduction; and the transfer of mass by the diffusion of water vapor, carbon dioxide and other gases. Because all three forms of transfer are a direct consequence of molecular motion, they are described by similar relationships that will be considered for the simplest possible case of diffusion in one dimension only.

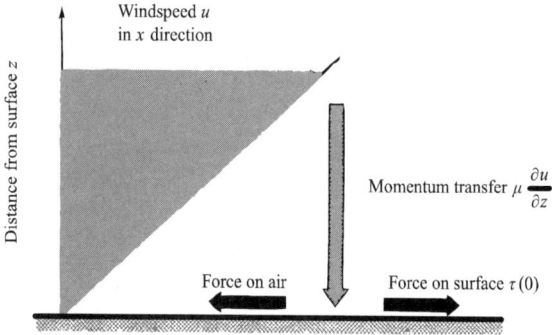

FIGURE 3.2 Transfer of momentum from moving air to a stationary surface, showing related forces.

MOMENTUM AND VISCOSITY

When a stream of air flows over a solid surface, its velocity increases with distance from the surface. For a simple discussion of viscosity, the velocity gradient $\partial u/\partial z$ will be assumed linear as shown in Figure 3.2. (A more realistic velocity profile will be considered in Chapter 9). Provided the air is isothermal, the velocity of molecular agitation will be the same at all distances from the surface, but the horizontal component of velocity in the x direction increases with vertical distance z. As a direct consequence of molecular agitation, there is a constant interchange of molecules between adjacent horizontal layers with a corresponding vertical exchange of horizontal momentum.

The horizontal momentum of a molecule attributable to motion of the gas as distinct from random motion is mu, so from Equation 3.1 the rate of transfer of momentum, otherwise known as the *shearing stress*, can be written

$$\tau = (ncl/3)\, d\,(mu)\,/dz = (cl/3)\, d\,(\rho u)\,/dz \qquad (3.4)$$

as the density of the gas is $\rho = mn$. This is formally identical to the empirical equation defining the *kinematic viscosity* of a gas ν, viz.

$$\tau = \nu d\,(\rho u)\,/dz \qquad (3.5)$$

showing that ν is a function of molecular velocity and mean free path. Where the change of ρ with distance is small, it is more convenient to write

$$\tau = \mu du/dz \qquad (3.6)$$

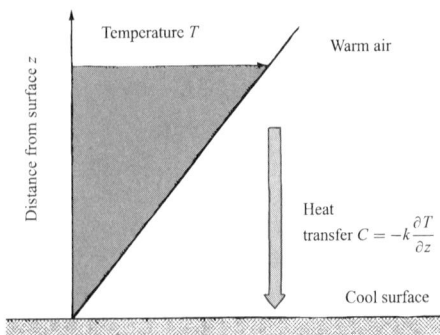

FIGURE 3.3   Transfer of heat from still, warm air to a cool surface.

where $\mu = \bar{\rho}\nu$ is the coefficient of *dynamic viscosity* and $\bar{\rho}$ is a mean density. By convention, the flux of momentum is taken as positive when it is directed *toward* a surface and it therefore has the same sign as the velocity gradient (see Figure 3.2).

The momentum transferred layer by layer through the gas is finally absorbed by the surface, which therefore experiences a frictional force acting in the direction of the flow. The reaction to this force required by Newton's Third Law is the frictional drag exerted on the gas by the surface in a direction opposite to the flow.

HEAT AND THERMAL CONDUCTIVITY

The conduction of heat in still air is analogous to the transfer of momentum. In Figure 3.3, a layer of warm air makes contact with a cooler surface. The velocity of molecules therefore increases with distance from the surface and the exchange of molecules between adjacent layers of air is responsible for a net transfer of molecular energy and hence of heat. The rate of transfer of heat is proportional to the gradient of heat content per unit volume of the air and may therefore be written

$$C = -\kappa d\left(\rho c_{\mathrm{p}} T\right)/dz \tag{3.7}$$

where κ, the *thermal diffusivity* of air, has the same dimensions $(\mathrm{L}^2\,\mathrm{T}^{-1})$ as the kinematic viscosity, and ρc_{p} is the heat content per unit volume of air relative to the value at $T = 0$ ($^\circ$C or K). As in the treatment of momentum, it is convenient to assume that ρ has a constant value of $\bar{\rho}$ over the distance

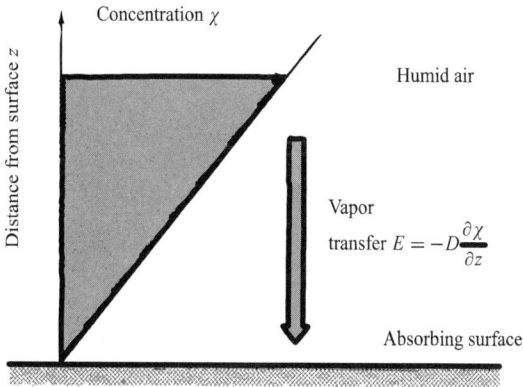

FIGURE 3.4 Transfer of vapor from humid air to an absorbing surface.

considered and to define a *thermal conductivity* as $k = \bar{\rho}c_p\kappa$ so that

$$\mathbf{C} = -k\,dT/dz \tag{3.8}$$

identical to the equation for the conduction of heat in solids.

In contrast to the convention for momentum, $\mathbf{C}$ is taken as positive when the flux of heat is *away* from the surface in which case dT/dz is negative. The equation therefore contains a minus sign.

MASS TRANSFER AND DIFFUSIVITY

In the presence of a gradient of gas concentration, molecular agitation is responsible for a transfer of mass, generally referred to as "diffusion" although this word can also be applied to momentum and heat. In Figure 3.4, a layer of still air containing water vapor makes contact with a hygroscopic surface where water is absorbed.

The number of molecules of vapor per unit volume increases with distance from the surface, and the exchange of molecules between adjacent layers produces a net movement toward the surface. The transfer of molecules expressed as a mass flux per unit area ($\mathbf{E}$) is proportional to the gradient of concentration, and the transport equation analogous to Equations 3.6 and 3.8 is

$$\mathbf{E} = -D\,d\chi/dz = -Dd(\rho q)/dz = -\bar{\rho}D\,dq/dz \tag{3.9}$$

where $\bar{\rho}$ is an appropriate mean density, D (dimensions $L^2\,T^{-1}$) is the molecular diffusion coefficient for water vapor, and other symbols are defined on p. 16. The sign convention in this equation is the same as for heat.

fined on p. 16.

3.3 Diffusion Coefficients

Because the same process of molecular agitation is responsible for all three types of transfer, the diffusion coefficients for momentum, heat, water vapor and other gases are similar in size and in their dependence on temperature. Values of the coefficients at different temperatures calculated from the Chapman–Eskog kinetic theory of gases agree well with measurements and are given in Appendix A.3. The temperature and pressure dependence of the diffusion coefficients is usually expressed by a power law; e.g.,

$$D(T)/D(0) = \{T/T(0)\}^n \{P(0)/P\} \qquad (3.10)$$

where $D(0)$ is the coefficient at base temperature and pressure $T(0)(K)$ and $P(0)$ (kPa) respectively, and n is an index between 1.5 and 2.0. Within the limited range of temperature relevant to environmental physics, say -10 to $50°C$, a simple temperature coefficient of 0.007 is accurate enough for practical purposes; i.e.,

$$D(T)/D(0) = \kappa(T)/\kappa(0) = v(T)/v(0) = (1 + 0.007T)$$

where T is the temperature in $°C$ and the coefficients in units of $m^2 \, s^{-1}$ are

$$v(0) = 13.3 \times 10^{-6} \text{ (momentum)}$$
$$\kappa(0) = 18.9 \times 10^{-6} \text{ (heat)}$$
$$D(0) = 21.2 \times 10^{-6} \text{ (water vapor)}$$
$$= 12.9 \times 10^{-6} \text{ (carbon dioxide)}$$

Graham's Law states that the diffusion coefficients of gases are inversely proportional to the square roots of their densities; (i.e., $D \propto (M)^{-0.5}$) since density is proportional to molecular weight. Consequently, the diffusion coefficient D_x for an unknown gas of molecular weight M_x can be estimated from values for a known gas (D_y, M_y) using the relation $D_x = D_y(M_y/M_x)^{0.5}$.

RESISTANCES TO TRANSFER

Equations 3.6, 3.8, and 3.9 have the same form

$$\text{flux} = \text{diffusion coefficient} \times \text{gradient}$$

which is a general statement of Fick's Law of Diffusion. This law can be applied to problems in which diffusion is a one-, two-, or three-dimensional process but only one-dimensional cases will be considered here. Because the gradient of a quantity at a point is often difficult to estimate accurately, Fick's law is generally applied in an integrated form. The integration is straightforward in cases where the (one-dimensional) flux can be treated as constant in the direction specified by the coordinate z (e.g., at right angles to a surface). Then the integration of (3.9) for example gives

$$\mathbf{E} = -\frac{\int_{z_1}^{z_2} d(\rho q)}{\int_{z_1}^{z_2} dz/D} = \frac{\rho q\,(z_1) - \rho q\,(z_2)}{\int_{z_1}^{z_2} dz/D} \qquad (3.11)$$

where $\rho q\,(z_1)$ and $\rho q\,(z_2)$ are concentrations of water vapor at distances z_1 and z_2 from a surface absorbing or releasing water vapor at a rate $\mathbf{E}$. Usually, $\rho q\,(z_1)$ is taken as the concentration at the surface so that $z_1 = 0$.

Equation 3.11 and similar equations derived by integrating Equations 3.6 and 3.8 are analogous to Ohm's Law in electrical circuits; i.e.,

$$\text{current through resistance} = \frac{\text{potential difference across resistance}}{\text{resistance}}$$

Equivalent expressions for diffusion can be written as:

$$\text{rate of transfer of entity} = \frac{\text{potential difference}}{\text{resistance}}$$

i.e.,

$$\text{rate of momentum transfer } \tau = \rho u \Big/ \int dz/v \qquad (3.12a)$$

$$\text{rate of heat transfer } \mathbf{C} = \rho c_p T \Big/ \int dz/\kappa \qquad (3.12b)$$

$$\text{rate of mass transfer } \mathbf{E} = \rho q \Big/ \int dz/D \qquad (3.12c)$$

The definition of *resistance r* to mass transfer is therefore

$$r = \int dz/D \qquad (3.13)$$

and similar equations define resistances to heat and momentum transfer.

Diffusion coefficients have dimensions of $(\text{length})^2 \times (\text{time})$ so the corresponding resistances have dimensions of $(\text{time})/(\text{length})$ or $1/(\text{velocity})$. In a system where rates of diffusion are governed purely by molecular processes, the coefficients can usually be assumed independent of z so that $\int_{z_1}^{z_2} dz/D$, for example, becomes simply $(z_2 - z_1)/D$ or (diffusion pathlength)/(diffusion coefficient).

EXAMPLE 3.1 What is the resistance to water vapor diffusion for a pathlength of 1 mm of air at $20°C$ and 101.3 kPa?

SOLUTION ► The molecular diffusion coefficient of water vapor in air for the specified temperature and pressure is given in Appendix A.3, and is 24.9×10^{-6} m^2 s^{-1}. Consequently, the resistance for a pathlength of 1 mm (1×10^{-3} m) is $1 \times 10^{-3}/24.9 \times 10^{-6} = 40$ s m^{-1}. ◄

It is often convenient to treat the process of diffusion in laminar boundary layers in terms of resistances, and in the remainder of this book the following symbols are used:

r_M resistance for momentum transfer at the surface of a body

r_H resistance for convective heat transfer

r_V resistance for water vapor transfer

r_C resistance for CO_2 transfer

The concept of resistance is not limited to molecular diffusion but is applicable to any system in which fluxes are uniquely related to gradients. In the atmosphere where turbulence is the dominant mechanism of diffusion, diffusion coefficients are several orders of magnitude larger than the corresponding molecular value and increase with height above the ground (Chapter 16). Diffusion resistances for momentum, heat, water vapor, and carbon dioxide in the atmosphere will be distinguished by the symbols r_{aM}, r_{aH}, r_{aV} and r_{aC}; the measurement of these resistances is discussed in Chapters 9 to 11. The concept of resistances applied to atmospheric transfer was initially developed in a report on data collected at the University of California, Davis [Monteith 1963].

In studies of the deposition of radioactive material and pollutant gases from the atmosphere to the surface, the rate of transfer is sometimes expressed as a *deposition velocity*, which is the reciprocal of a diffusion resistance. In this case, the surface concentration is often assumed zero and the deposition velocity is found by dividing the rate of deposition of the material by its concentration at an arbitrary height.

Plant physiologists also frequently use the reciprocal of resistance (in this context termed *conductance*) to describe transfer between leaves and the atmosphere, arguing that the direct proportionality between *flux* and *conductance* is a more intuitive concept than the inverse relationship between flux and resistance. In this book, we generally prefer the resistance formulation because of its familiarity to physicists, particularly when combinations of resistances in parallel and series must be calculated.

Alternative Units for Resistance and Conductance

Units of s m^{-1} for resistance and m s^{-1} for conductance are the result of expressing mass fluxes as mass flux density (e.g., kg m^{-2} s^{-1}) and driving potentials as concentrations (e.g., kg m^{-3}). The forms of the equations for momentum, heat, and mass transfer (3.12a) ensure that resistance units for these variables are also s m^{-1}. A criticism of this convention is that, when resistance is defined as (diffusion pathlength)/(diffusion coefficient), r is proportional to pressure P and inversely approximately proportional to T^2 (see Equation 3.10). Thus, for example, analyzing the effects of altitude on fluxes can be confusing. Alternative definitions of resistance units, less sensitive to temperature and pressure, are sometimes used, particularly by plant physiologists, as follows.

Since biochemical reactions concern numbers of molecules reacting, rather than the mass of substances, it is convenient to express fluxes in *mole flux density*, J (mol m^{-2} s^{-1}). Similarly, amount of the substance can be expressed as the *mole fraction*; i.e., the number of moles of the substance as a fraction of the total number of moles x in the mixture (mol mol^{-1}). Using the Gas Laws, it is readily shown that mass concentration χ (kg m^{-3}), or its equivalent ρq, is related to x by

$$\chi = \rho q = x P / RT$$

so Equation 3.9 may be written

$$J = \frac{x(z_1) - x(z_2)}{\frac{RT}{P} \int dz / D}$$

or, since $x = p/P$, where p is the partial pressure,

$$J = \frac{p(z_1) - p(z_2)}{RT \int dz / D}$$

The molar resistance r_m (m^2 s mol^{-1}) is defined as

$$r_m = \frac{RT}{P} \int \frac{dz}{D} = \frac{RT}{P} r \tag{3.14}$$

demonstrating that r_m is independent of pressure and less dependent on temperature than r. Cowan [1977] pointed out that it is important to use partial pressure or mole fraction to describe potentials that drive diffusion when systems are not isothermal.

At 20° C and 101.3 kPa, the approximate conversion between resistance units is

$$r_m \; (\mathrm{m^2 s \; mol^{-1}}) = 0.024r \; (\mathrm{s \; m^{-1}}) \tag{3.15}$$

DIFFUSION OF PARTICLES (BROWNIAN MOTION)

The random motion of particles suspended in a fluid or gas was first described by the English botanist Brown in 1827, but it was nearly 80 years before Einstein used the kinetic theory of gases to show that the motion was the result of multiple collisions with the surrounding molecules. He found that the mean square displacement $\overline{x^2}$ of a particle in time t is given by

$$\overline{x^2} = 2Dt \tag{3.16}$$

where D is a diffusion coefficient (dimensions $\mathrm{L^2 \; T^{-1}}$) for the particle, analogous to the coefficient for gas molecules. The quantity D depends on the intensity of molecular bombardment (a function of absolute temperature), and on the viscosity of the fluid, as follows.

Suppose that particles, each with mass m (kg), are dispersed in a container where they neither stick to the walls nor coagulate. The Boltzmann statistical description of concentration, derived from kinetic theory, requires that, as a consequence of the Earth's gravitational field, the particle concentration n should decrease exponentially with height z(m) according to the relation

$$n = n(0) \exp(-mgz/kT) \tag{3.17}$$

where

$$n = n(0) \text{ at } z = 0$$

$$g = \text{gravitational acceleration } \left(\mathrm{m \; s^{-2}}\right)$$

$$T = \text{absolute temperature (K)}$$

$$k = \text{Boltzmann's constant (J K}^{-1})$$

Across a horizontal area at height z within the container, the flux of particles upward by diffusion (cf. diffusion of gases) is:

$$\mathbf{F}_1 = -D\frac{dn}{dz} = n\frac{Dmg}{kT} \tag{3.18}$$

from Equation 3.17. Because all particles tend to move downward in response to gravity, there must be a downward flux through the area of

$$F_2 = nV_s$$

where V_s is the *sedimentation velocity* (see Chapter 12). Since the system is in equilibrium, $F_1 = F_2$ and so

$$D = kTV_s/mg \qquad (3.19)$$

For spherical particles, radius r, obeying Stokes' Law, the downward force mg due to gravity is balanced by a drag force $6\pi \mu r V_s$, and so

$$D = kT/6\pi \mu r \qquad (3.20)$$

Thus D depends inversely on particle radius (Appendix A.6); its dependence on temperature is dominated by the rapid decrease in dynamic viscosity μ with increasing temperature.

Equations 3.16 and 3.20 show that the root mean square displacement of a particle by Brownian motion is proportional to $T^{0.5}$ and to $r^{-0.5}$. Surprisingly, $\overline{x^2}$ does not depend on the mass of the particle, an inference confirmed by experiment.

3.4 Problems

1. Calculate the resistance (in s m^{-1}) for carbon dioxide diffusion in air over a pathlength of 1 mm, assuming air temperature is 20°C and atmospheric pressure is 101 kPa. Recalculate the resistance assuming that atmospheric pressure decreased to 70 kPa (keeping temperature constant). Repeat the two calculations using molar units.

2. Use the data in Appendix A.6 to investigate how the diffusion coefficient of particles depends on temperature.

▶ Chapter 4

Transport of Radiant Energy

4.1 The Origin and Nature of Radiation

Electromagnetic radiation is a form of energy derived from oscillating magnetic and electrostatic fields, and is capable of transmission through empty space where its velocity is $c = 3.0 \times 10^8$ m s^{-1}. The frequency of oscillation ν is related to the wavelength λ by the standard wave equation $c = \lambda \nu$, and the wave number $1/\lambda = \nu/c$ is sometimes used as an index of frequency.

The ability to emit and absorb radiation is an intrinsic property of solids, liquids, and gases and is always associated with changes in the energy state of atoms and molecules. Changes in the energy state of atomic electrons are associated with line spectra confined to a specific frequency or set of frequencies. In molecules, the energy of radiation is derived from the vibration and rotation of individual atoms within the molecular structure, and numerous possible energy states allow radiation to be emitted or absorbed over a wide range of frequencies to form band spectra. The principle of energy conservation is fundamental to the material origin of radiation. The amount of radiant energy emitted by an individual atom or molecule is equal to the decrease in the potential energy of its constituents.

ABSORPTION AND EMISSION OF RADIATION

All molecules possess a certain amount of "internal" energy (i.e., not associated with their motion in the atmosphere). Most of the energy is associated with electrons orbiting around the nucleus, but part is related to vibration of atoms in the molecular structure and to rotation of the molecule. Quantum physics predicts that only certain electron orbits, vibration frequencies, and rotation rates are allowed for a particular molecule, and each combination of orbits, vibrations, and rotations corresponds to a particular amount of energy associated with the three features. Molecules may make a transition to a higher or lower energy level by absorbing or emitting electromagnetic radiation, respectively. Quantum theory allows only certain discrete changes in energy levels, which are the same whether the energy is being absorbed or emitted. Since the energy associated with a photon is related to its wavelength by $E = hc/\lambda$, it follows that molecules can interact only with certain wavelengths of radiation. Thus the variation of absorption and emission of molecules with wavelength takes the form of a *line spectrum*, consisting of a finite number of wavelengths where interaction is allowed, interspaced by gaps where there is no interaction.

Most absorption lines associated with orbital changes are in the X-ray, UV, and visible spectrum. Vibrational changes are associated with absorption at infrared wavelengths, and rotational changes correspond to lines at even longer, microwave wavelengths. Molecules such as CO_2, H_2O, and O_3 have structures that allow vibration-rotation transitions simultaneously, and these correspond to clusters of very closely spaced lines in the infrared region. Molecules such as O_2 do not interact this way, and so have only small numbers of absorption lines.

When large numbers of molecules are present in a gas, the width of their absorption and emission lines is greatly enhanced by broadening associated with random molecular motions (*Doppler broadening*—depending on the square root of absolute temperature) and with interactions during collisions (*collision broadening*—depending on the frequency of collisions, which is proportional to gas pressure). Collision broadening is most important for atmospheric molecules below about 30 km, and results in overlapping of lines in the clusters associated with vibration-rotational transitions in CO_2 and H_2O, creating absorption bands in the infrared for these gases. Since pressure decreases with increasing height, the absorptivity and emissivity of a gas distributed in the lower atmosphere with a constant mixing ratio changes with height, making calculations of radiative transfer complex.

FULL OR BLACK BODY RADIATION

Relations between radiation absorbed and emitted by matter were examined by Kirchhoff. He defined the absorptivity of a surface $\alpha(\lambda)$ as the fraction of incident radiation absorbed at a specific wavelength λ. The emissivity $\varepsilon(\lambda)$ was defined as the ratio of the actual radiation emitted at the wavelength λ to a hypothetical amount of radiant flux $\mathbf{B}(\lambda)$. By considering the thermal equilibrium of an object inside an enclosure at a uniform temperature, he showed that $\alpha(\lambda)$ is always equal to $\varepsilon(\lambda)$. For an object completely absorbing radiation at wavelength λ, $\alpha(\lambda) = 1$, $\varepsilon(\lambda) = 1$ and the emitted radiation is $\mathbf{B}(\lambda)$. In the special case of an object with $\varepsilon = 1$ at *all* wavelengths, the spectrum of emitted radiation is known as the "full" or *black body spectrum*. Within the range of temperatures prevailing at the Earth's surface, nearly all the radiation emitted by full radiators is confined to the waveband 3 to 100 μm, and most natural objects—soil, vegetation, water—behave radiatively almost like full radiators in this restricted region of the spectrum (but not in the visible spectrum). Even fresh snow, one of the whitest surfaces in nature, emits radiation like a black body between 3 and 100 μm. The statement "snow behaves like a black body" refers therefore to the radiation *emitted* by a snow surface and not to solar radiation *reflected* by snow. The semantic confusion inherent in the term *black body* can be avoided by referring to "full radiation" and to a "full radiator".

After Kirchhoff's work was published in 1859, the emission of radiation by matter was investigated by a number of experimental and theoretical physicists. By combining a spectrometer with a sensitive thermopile, it was established that the spectral distribution of radiation from a full radiator resembles the curve in Figure 4.1 in which the chosen temperatures of 6000 and 300°K correspond approximately to the mean black body temperatures of the sun and the earth's surface. A theoretical explanation of the distribution eluded physicists until Plank's quantum hypothesis emerged (page 42).

WIEN'S LAW

Wien deduced from thermodynamic principles that the energy emitted per unit wavelength $\mathbf{E}(\lambda)$ should be a function of absolute temperature T and of λ such that

$$\mathbf{E}(\lambda) = f(\lambda T)/\lambda^5 \qquad (4.1)$$

Deductions from this relation are:

1. that when spectra from full radiators at different temperatures are compared, the wavelength λ_m at which $\mathbf{E}(\lambda)$ reaches a maximum should be inversely proportional to T so that $\lambda_m T$ is the same for all values of T. The value of this constant is 2897 μm K so, in Figure 4.1, $\lambda_m = 0.48$

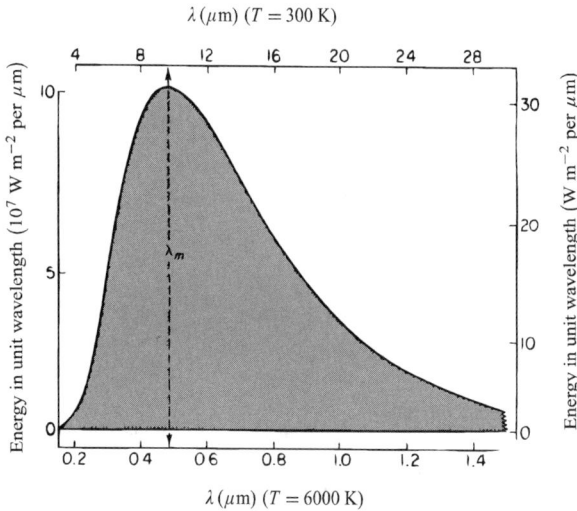

FIGURE 4.1 Spectral distribution of radiant energy from a full radiator at a temperature of (a) 6000 K, left-hand vertical and lower horizontal axis, and (b) 300 K, right-hand vertical and upper horizontal axis. About 10% of the energy is emitted at wavelengths longer than those shown in the diagram. If this tail were included, the total area under the curve would be proportional to σT^4 (W m^{-2}). λ_m is the wavelength at which the energy per unit wavelength is maximal.

μm for a solar spectrum ($T = 6000$ K) and 9.7 μm for a terrestrial spectrum ($T = 300$ K).

2. that the value of $E(\lambda)$ at λ_m is proportional to λ_m^{-5} and therefore to T^5 because $\lambda_m \propto 1/T$. The ratio of $E(\lambda_m)$ for solar and terrestrial radiation is therefore $(6000/300)^5 = 3.2 \times 10^6$.

STEFAN'S LAW

Stefan and Boltzmann showed that the rate of energy emission from a full radiator, integrated over all wavelengths, was proportional to the fourth power of its absolute temperature: in symbols,

$$\mathbf{B} = \sigma T^4 \qquad (4.2)$$

where $\mathbf{B}$ (W m^{-2}) is the radiant flux emitted by unit area of a plane surface into an imaginary hemisphere surrounding it. The Stefan-Boltzmann constant σ is a fundamental constant that can be deduced from quantum theory, and has the value 56.7×10^{-9} W m^{-2} K^{-4}. The spatial distribution of this flux is considered on p. 44.

As a generalization from Equation 4.2, the radiation emitted from unit area of any plane surface with an emissivity of $\varepsilon(< 1)$ can be written in the form

$$\Phi = \varepsilon \sigma T^n$$

where n is a numerical index. For a "gray" surface whose emissivity is almost independent of wavelength, $n = 4$. When radiation is emitted predominantly at wavelengths less than λ_m, n exceeds 4, and conversely, when the bulk of emitted radiation appears in a waveband above λ_m, n is less than 4.

PLANCK'S LAW

Prolonged attempts to establish the shape of the black-body spectrum (i.e., the function $f(\lambda T)$ in Equation 4.1) culminated in a theory developed by Max Planck who laid the foundation for much of modern physics by introducing the quantum hypothesis. Planck found that the spectrum could not be predicted from classical mechanics because classical principles imposed no restriction on the amount of radiant energy that a molecule could emit. He postulated that energy was emitted in discrete packets which he called "quanta" and that the energy of a single quantum E_q was proportional to the frequency of the radiation; i.e.,

$$E_q = h\upsilon = hc/\lambda \tag{4.3}$$

where $h = 6.63 \times 10^{-34}$ J s is *Planck's constant*.

The formula derived by Planck on the basis of this hypothesis was

$$E(\lambda) = \frac{2\pi hc^2 \lambda^{-5}}{\exp\{hc/(k\lambda T)\} - 1} \tag{4.4}$$

where k is the Boltzmann constant (see p. 6) and T is absolute temperature. Equation 4.4 was used to calculate the data for the curves in Figure 4.1. The units of E in Equation 4.4 are energy per second per unit area per unit wavelength; i.e., W m^{-2} nm^{-1} when the constants and variables are expressed appropriately. Differentiation of Equation 4.4 with respect to λ gives the constant of Wien's Law, and integration reveals that the Stefan-Boltzmann constant is given by $(2\pi^5/15)k^4c^{-2}h^{-3}$.

QUANTUM UNIT

Equation 4.3 shows that the energy in a quantum of red light ($\lambda = 660$ nm) is about 3×10^{-19} J, and that of a quantum of blue light (400 nm) is about 5×10^{-19} J. (Quanta of light are also termed *photons*). Because the amount of energy in a single quantum is inconveniently small, photochemists often refer to the energy in a mole of quanta (i.e., Avogadro's number, $N = 6.02 \times 10^{23}$). This number of quanta was originally called an *Einstein* in recognition of Einstein's contribution to the foundations of photochemical theory. It is now usually called a *mole*. (Although the mole is formally defined as "an amount of substance," its application in this context suggests that it should be redefined simply as "a number, namely, the number of molecules in 0.012 kg of carbon 12"). A photochemical reaction requiring one quantum per molecule of a compound therefore requires one mole of quanta per mole of compound.

RADIATIVE EXCHANGE: SMALL TEMPERATURE DIFFERENCES

Although the total radiation emitted by a full radiator is proportional to T^4, the exchange of energy between radiators at temperatures T_1 and T_2 becomes nearly proportional to $(T_2 - T_1)$ when this difference is small, a common case in natural environments.

Writing $(T_2 - T_1) = \delta T$, the difference between full radiation at the two temperatures is

$$
\begin{aligned}
\mathbf{R} &= \sigma \left\{ (T_1 + \delta T)^4 - T_1^4 \right\} \\
&= \sigma \left\{ 4T_1^3 \delta T + 6T_1^2 \delta T^2 + \ldots \right\} \\
&\approx 4\sigma T_1^3 \delta T \left\{ 1 + 6\delta T / 4T_1 \right\}
\end{aligned}
\tag{4.5}
$$

where negligible terms in δT^3 and δT^4 are omitted. The value of $\mathbf{R}$ may therefore be taken as $4\sigma T_1^3 \delta T$ with an error given by $-1.5 \, \delta T / T_1$. For $T_1 = 298$ K ($25°$C), the error is only 0.005 or 0.5% per degree temperature difference, and the value of $4\sigma T_1^3$ is 6.0 W m^{-2} K^{-1}, a useful approximation for estimating radiative exchange between full radiators with similar temperatures.

By analogy with the equation already derived for heat transfer in gases, a notional resistance to radiative transfer r_R may be derived by writing

$$
\mathbf{R} = \rho c_p \delta T / r_R
\tag{4.6}
$$

where the introduction of a volumetric specific heat allows r_R to have the same dimensions as resistances for momentum, heat, and mass transfer. From Equations 4.5 and 4.6

$$
r_R \approx \rho c_p / \left(4\sigma T_1^3 \right)
\tag{4.7}
$$

which has a value of almost exactly 300 s m^{-1} at 298 K.

4.2 Spatial Relations

An amount of radiant energy emitted, transmitted, or received per unit time is known as a *radiant flux* and in most problems of environmental physics, the watt is a convenient unit of flux. The term *radiant flux density* means flux per unit area, usually quoted in watts per m^2. *Irradiance* is the radiant flux density incident on a surface, and *emittance* (or radiant excitance) is the radiant flux density emitted by a surface.

For a beam of parallel radiation, flux density is defined in terms of a plane at right angles to the beam, but several additional terms are needed to describe the spatial relationships of radiation dispersing in all directions from a point source or from a radiating surface. Figure 4.2(a) represents the flux $d\mathbf{F}$ emitted from a point source into a solid angle $d\omega$ where $d\mathbf{F}$ and $d\omega$ are both very small quantities. The *intensity* of radiation $\mathbf{I}$ is defined as the flux per unit solid angle or $\mathbf{I} = d\mathbf{F}/d\omega$). This quantity may be expressed in watts per steradian.

Figure 4.2(b) illustrates the definition of a closely related quantity— *radiance*. An element of surface with an area dS emits a flux $d\mathbf{F}$ in a direction specified by an angle ψ with respect to the normal. When the element is projected at right angles to the direction of the flux its projected area is $dS \cos \psi$, which is the apparent or effective area of the surface viewed from an angle ψ. The radiance of the element in this direction is the flux emitted in the direction per unit solid angle, or $d\mathbf{F}/\omega$, divided by the projected area $dS \cos \psi$. In other words, radiance is equivalent to the intensity of radiant flux observed in a particular direction divided by the apparent area of the source in the same direction. This quantity may be expressed in watts per m^2 per steradian.

The term "intensity" is often used colloquially as a synonym for flux density and is also confused with radiance.

COSINE LAW FOR EMISSION AND ABSORPTION

The concept of radiance is linked to an important law describing the spatial distribution of radiation emitted by a full radiator that has a uniform surface temperature T. This temperature determines the total flux of energy emitted by the surface (σT^4) and can be estimated by measuring the radiance of the surface with a radiometer.

(a)

Point source

dA

$d\omega$

$d\mathbf{F}$

r

Solid angle $d\omega = dA/r^2$

Intensity $\mathbf{I} = d\mathbf{F}/d\omega$

(b)

ω

$d\mathbf{F}$

ψ

Plane source $d\mathbf{S}$

$d\mathbf{S}\cos\psi$

Intensity $d\mathbf{I} = d\mathbf{F}/\omega$

Radiance $= (d\mathbf{F}/\omega) \div d\mathbf{S}\cos\psi$

$= d\mathbf{I}/(d\mathbf{S}\cos\psi)$

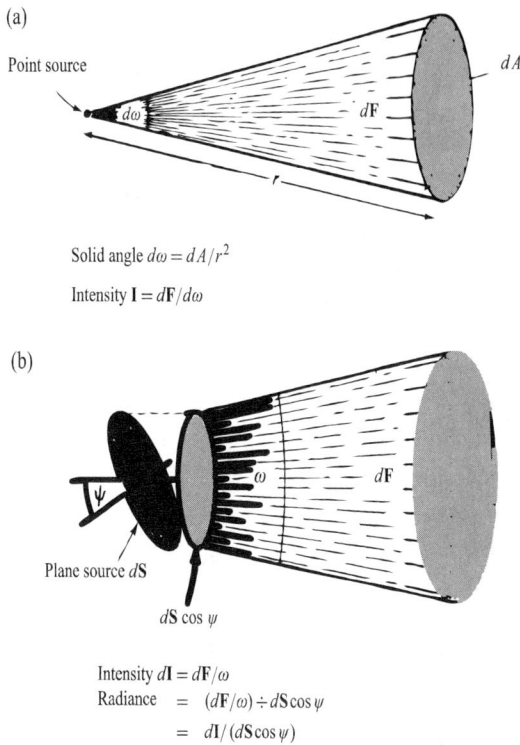

FIGURE 4.2 (a) Geometry of radiation emitted by a point source. (b) Geometry of radiation emitted by a surface element. In both diagrams a portion of a spherical surface receives radiation at normal incidence, but when the distance between the source and the receiving surface is large, it can be treated as a plane.

As the surface of a full radiator must appear to have the same temperature whatever angle ψ it is viewed from, the intensity of radiation emitted from a point on the surface and the radiance of an element of surface must both be independent of ψ. On the other hand, the flux per unit solid angle divided by the *true* area of the surface must be proportional to $\cos\psi$.

Figure 4.3 makes this point diagrammatically. A radiometer R mounted vertically above an extended horizontal surface XY "sees" an area dA and measures a flux that is proportional to dA. When the surface is tilted through an angle ψ, the radiometer now sees a larger surface $dA/\cos\psi$,

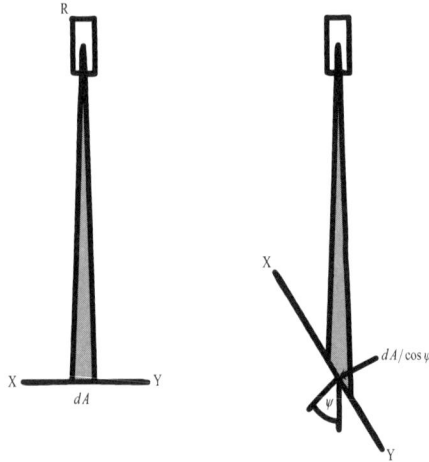

FIGURE 4.3   The amount of radiation received by a radiometer from the surface XY is independent of the angle of emission, but the flux emitted per unit area is proportional to cos ψ.

but provided the temperature of the surface stays the same, its radiance will be constant and the flux recorded by the radiometer will also be constant. It follows that the flux emitted per unit area (the emittance of the surface) at an angle ψ must be proportional to cos ψ so that the product of emittance ($\propto$ cos ψ) and the area emitting to the instrument ($\propto 1/\cos\psi$) stays the same for all values of ψ.

This argument is the basis of *Lambert's Cosine Law*, which states that when radiation is emitted by a full radiator at an angle ψ to the normal, the flux per unit solid angle emitted by unit surface area is proportional to cos ψ. As a corollary to Lambert's Law, it can be shown by simple geometry that when a full radiator is exposed to a beam of radiant energy at an angle ψ to the normal, the flux density of the absorbed radiation is proportional to cos ψ. In remote sensing it is often necessary to specify the directions of both incident and reflected radiation, and reflectivity is then described as "bi-directional."

REFLECTION

The *reflectivity* of a surface $\rho(\lambda)$ is defined as the ratio of the incident flux to the reflected flux at the same wavelength. Two extreme types of behaviour can be distinguished. For surfaces exhibiting specular or mirror-like reflection, a beam of radiation incident at an angle ψ to the normal is reflected at the complementary angle $(-\psi)$. On the other hand, the radiation scattered by a perfectly diffuse reflector (also called a *Lambertian surface*) is distributed in all directions according to the Cosine Law; i.e., the intensity of the scattered radiation is independent of the angle of reflection, but the flux scattered from a specific area is proportional to $\cos \psi$.

The nature of reflection from the surface of an object depends in a complex way on its electrical properties and on the structure of the surface. In general, specular reflection assumes increasing importance as the angle of incidence increases, and surfaces acting as specular reflectors absorb less radiation than diffuse reflectors made of the same material.

Most natural surfaces act as diffuse reflectors when ψ is less than 60 or 70°, but as ψ approaches 90°, a condition known as *grazing incidence*, the reflection from open water, waxy leaves, and other smooth surfaces becomes dominantly specular and there is a corresponding increase in reflectivity. The effect is often visible at sunrise and sunset over an extensive water surface, or a lawn, or a field of barley in ear.

When surfaces are observed by techniques of remote sensing, the direction of the radiation received by the radiometer is significant, and several additional definitions are necessary:

Bi-directional reflectance (sr^{-1}) is the ratio of the radiation *reflected* in a specific direction of view to the radiation *incident* in that direction.

The *Bi-directional reflectance factor* of a surface (BRF) is the ratio of the reflected radiance from a specific view direction to the radiance that would be observed from a perfectly diffuse surface at the same location. Most examples in this book deal with a plane surface below a uniform hemispherical source of radiation (e.g., an overcast sky). The fraction of incident radiation reflected from such a surface is sometimes referred to as the *bi-hemispherical reflectance*, or simply the *reflection coefficient*.

RADIANCE AND IRRADIANCE

When a plane surface is surrounded by a uniform source of radiant energy, a simple relation exists between the irradiance of the surface (the flux incident per unit area) and the radiance of the source. Figure 4.4 displays a surface of unit area surrounded by a radiating hemispherical shell so large that the surface can be treated a point at the center of the hemisphere. The shaded area dS is a small element of radiating surface, and the radiation reaching the center of the hemisphere from dS makes an angle β with the normal to the plane. As the projection of unit area in the direc-

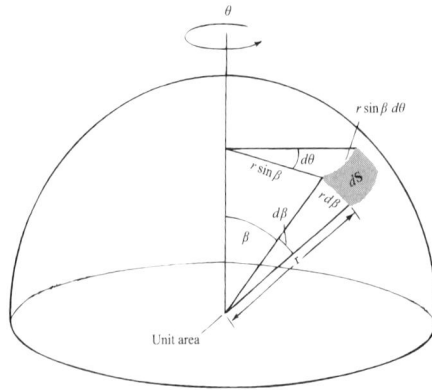

FIGURE 4.4 Method for calculating irradiance at the centre of an equatorial plane from a surface element dS at angle β to vertical axis (see text).

tion of the radiation is $1 \times cos\beta$, the solid angle the area subtends at dS is $\omega = \cos\beta/r^2$. If the element dS has a radiance $\mathbf{N}$, the flux emitted by dS in the direction of the disc must be $\mathbf{N} \times dS \times \omega = \mathbf{N}dS\cos\beta/r^2$. To find the total irradiance of the disc, this quantity must be integrated over the whole hemisphere, but if the radiance is uniform, conventional calculus can be avoided by noting that $dS\cos\beta$ is the area dS projected on the equatorial plane. It follows that $\int dS\cos\beta$ is the area of the whole plane or πr^2, so that the total irradiance at the center of the plane becomes

$$\left(\mathbf{N}/r^2\right) \int \cos\beta dS = \pi \mathbf{N} \tag{4.8}$$

The irradiance expressed in watts m^{-2} is therefore found by multiplying the radiance in watts m^{-2} steradian^{-1} by the solid angle π.

A more rigorous treatment is needed if the radiance depends on the position of dS with respect to the surface receiving radiation. It is necessary to treat dS as a rectangle whose sides are $rd\beta$ and $r\sin\beta d\theta$ where θ is an azimuth angle with respect to the axis of the hemisphere, radius r. Given that $dS = r^2\sin\beta d\beta d\theta$, the integral becomes

$$\int_{\theta=0}^{2\pi} \int_{\beta=0}^{\pi/2} N(\beta, \theta) \left(\frac{\cos\beta}{r^2}\right) r^2 \sin\beta d\beta d\theta \tag{4.9}$$

If N is independent of azimuth (i.e., only a function of β), Equation 4.9 simplifies to

$$= 2\pi \int_{\beta=0}^{\pi/2} N(\beta) \sin\beta \cos\beta d\beta$$

$$= \pi \int_{\beta=0}^{\pi/2} N(\beta) \sin 2\beta d\beta \qquad (4.10)$$

ATTENUATION OF A PARALLEL BEAM

When a beam of radiation consisting of parallel rays of radiation passes through a gas or liquid, quanta encounter molecules of the medium or particles in suspension. After interacting with the molecule or particle, a quantum may suffer one of two fates: it may be absorbed, thereby increasing the energy of the absorbing molecule or particle; or it may be scattered, i.e., diverted from its previous course either forward (within $90°$ of the beam) or backward in a process akin to reflection from a solid. After transmission through the medium, the beam is said to be "attenuated" by losses caused by absorption and scattering.

Beer's Law, frequently invoked in environmental physics, describes attenuation in a very simple system where radiation of a single wavelength is absorbed but not scattered when it passes through a homogeneous medium. Suppose that at some distance x into the medium the flux density of radiation is $\Phi(x)$ (Figure 4.5).

Absorption in a thin layer dx, assumed proportional to dx and to $\Phi(x)$, may be written

$$d\Phi = -k\Phi(x)\,dx \qquad (4.11)$$

where the constant of proportionality k, described as an "*attenuation coefficient*," is the probability of a ray being intercepted within the small distance dx. Integration gives

$$\Phi(x) = \Phi(0)\exp(-kx) \qquad (4.12)$$

where $\Phi(0)$ is the flux density at $x = 0$.

Beer's Law can also be applied to radiation in a waveband over which k is constant; or to a system in which the concentration of scattering centers is so small that a quantum is unlikely to interact more than once (*single scattering*).

The treatment of *multiple scattering* is much more complex for two main reasons: (a) radiation in a beam scattered backward must be considered as well as the forward beam; (b) if k depends on beam direction,

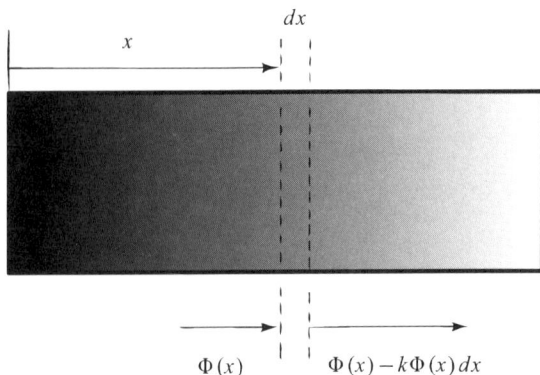

FIGURE 4.5   The absorption of a parallel beam of monochromatic radiation in a homogeneous medium with an absorption coefficient k. $\Phi(0)$ is the incident flux, $\Phi(x)$ is the flux at depth x and the flux absorbed in a thin layer dx is $k\Phi(x)\,dx$.

the angular distribution of scattering must be taken into account. For the simplest case where k is independent of beam direction, equations of a type developed by Kubelka and Munk are valid. They allow the attenuation coefficient to be expressed as a function of a reflection coefficient ρ, which is the probability of an interacting quantum being reflected backward and τ, which is the probability of forward scattering, implying that the probability of absorption is $\alpha = 1 - \rho - \tau$.

In a system with multiple scattering, it is necessary to distinguish two streams of radiation. One moves *into* the medium, and at a distance x from the boundary has a flux density $\Phi_+(x)$. The other, generated by scattering, moves *out* of the medium with a flux density $\Phi_-(x)$. The inward flux is depleted by absorption and reflection but is augmented by reflection of a fraction of the outward stream. The *net* loss of inward flux at a depth x and in a distance dx is therefore

$$d\Phi_+(x) = (-(\alpha + \rho)\Phi_+(x) + \rho\Phi_-(x))dx \qquad (4.13)$$

where $(\alpha + \rho)$ is a probability of interception.

The outward stream is also weakened by absorption and reflection but is augmented by reflection of the inward stream to give a net outward flux of

$$d\Phi_-(x) = -((\alpha + \rho)\Phi_-(x) - \rho\Phi_+(x))dx \qquad (4.14)$$

where the minus sign in front of the brackets is a reminder that the outward flux is moving in a negative direction with respect to the x axis.

For a system in which the strength of the forward-scattered beam is reduced virtually to zero, and where there is uniform scattering in all directions (isotropic scattering), $\rho = \tau = \alpha/2$. It may then be shown that the bulk reflection coefficient ρ' is given by

$$\rho' = \left(1 - \alpha^{0.5}\right) / \left(1 + \alpha^{0.5}\right) \qquad (4.15)$$

and the bulk attenuation coefficient k' is given by

$$k' = \alpha^{0.5} \qquad (4.16)$$

(These equations are not relevant to the limiting case $\alpha = 1$ when Beer's Law applies.)

When the forward beam strikes a boundary before it is completely attenuated, a fraction ρ_b may be reflected. The fluxes of radiation in the medium, both forward and backward, can then be expressed as functions of ρ_b, ρ, τ and of the concentration of the medium.

Complex numerical methods must be deployed to obtain fluxes when the attenuation coefficient is a function of the direction of scattering [Chandrasekhar, 1960]. Chapter 5 contains examples of the application of Beer's Law to the atmosphere (where the assumption of single scattering is usually valid) and to crop canopies where single scattering is a useful working approximation. In Chapter 6, application of the Kubelka–Munk equations is discussed with reference to canopies and animal coats.

4.3 Problems

1. Ultra violet radiation in the waveband 280–320 nm incident at the top of the earth's atmosphere supplies about 20 W m^{-2} of radiant energy at normal incidence. Assuming a mean wavelength of 300 nm, calculate the photon flux at normal incidence.

2. At what wavelength does the peak emission from a tungsten light bulb burning at 3000 K occur?

3. Differentiate Planck's Law to demonstrate that the constant in Wien's Law is 2897 μm K.

4. Show by integration of Planck's Law that the Stefan Boltzmann constant is 56.7 W m^{-2} K^{-4}.

Chapter 5

Radiation Environment

All the energy for physical and biological processes at the earth's surface comes from the sun, and much of environmental physics is concerned with ways in which this energy is dispersed or stored in thermal, mechanical, or chemical form. This chapter considers the quantity and quality of solar (short-wave) radiation received at the ground and the exchange of terrestrial (long-wave) radiation between the ground and the atmosphere.

5.1 Solar Radiation

SOLAR CONSTANT

At the mean distance of the earth from the sun R, which is 1.50×10^8 km, the irradiance of a surface held perpendicular to the solar beam and just outside the earth's atmosphere is known as the *Solar Constant*. The name is somewhat misleading because this quantity is known to change by small amounts over periods of weeks to years in response to changes within the sun, and the preferred term to describe the irradiance at mean earth-sun distance is now the *Total Solar Irradiance*, TSI.

Increasingly precise determinations of the TSI have been made from mountaintops, balloons, rocket aircraft flying above the stratosphere, and since the late 1970s, from satellites. The satellite observations have clearly demonstrated that the annual mean TSI varies by about 1.1 W m^{-2} between the minimum and maximum of the 11-year cycle of solar activity (Figure 5.1, see color plates) [Frohlich and Lean 1995, IPCC 2001]. Although satellite radiometers are capable of great precision, their absolute accuracy is much poorer: the absolute value of the TSI is believed to be about 1366 W m^{-2} ±2 W m^{-2} [Crommelynck et al, 1995]. It has been estimated from indirect proxies that vary with TSI (e.g., sunspot number) that the Total Solar Irradiance (Solar Constant) may have increased by about 0.3 W m^{-2} since 1750, contributing to global warming, but reliability of this estimate is poor [IPCC 2001].

EXAMPLE 5.1 Find the rate at which energy is emitted from the sun, and estimate the mean radiative temperature of the sun's surface.

SOLUTION ▶ The total energy emitted by the sun can be calculated by multiplying the TSI by the area of a sphere at the earth's mean distance R (1.50 × 10^{11} m), i.e., it is

$$E = 4\pi R^2 \times 1366 = 3.86 \times 10^{26} \text{ W}$$

The radius of the sun is $r = 6.96 \times 10^8$ m so, assuming the surface behaves like a full radiator (a close approximation), its effective temperature is given by

$$\sigma T^4 = 3.86 \times 10^{26} / \left(4\pi r^2\right)$$

from which the (rounded off) value of T is 5780 K.　　◀

SUN-EARTH GEOMETRY

Major features of radiation at the surface of the earth are determined by the earth's rotation about its own axis and by its elliptical orbit around the sun. The polar axis about which the earth rotates is fixed in space (pointing at the Pole Star) at a mean angle of 66.5° to the plane of its orbit (termed the *obliquity)* but with a small top-like wobble (the *precession* of the axis). The angle between the orbital plane and the earth's equatorial plane therefore oscillates between a maximum of 90 − 66.5 = 23.5° in midsummer, and a minimum of −23.5° in midwinter with small deviations attributable to the wobble. This angle is known as the *solar declination* (δ) and its value for any date and year can be found from astronomical tables.

The shape of the earth's orbit (the *eccentricity*), obliquity, and precession each vary over millenia, with cycles ranging from about 100,000 years to 23,000 years. The Russian astronomer Milutin Milankovitch proposed in the 1930s that these orbital variations were responsible for long-term cyclical changes in the earth's climate. Modern evidence from the analysis of layers of deep ocean sediments has confirmed that the *Milankovitch theory* explains part, but not all, of the variations in climate over the past few hundred thousand years.

Currently, the eccentricity of the earth's orbit is relatively small. The earth is about 3% closer to the sun in January than in July, so the irradiance at the top of the atmosphere is almost 7% larger in January (irradiance is proportional to the inverse square of the sun-earth radius). At any point on the earth's surface, the angle between the direction of the sun and a vertical axis depends on the latitude of the site, and on time t(h), most conveniently referred to the time when the sun reaches its zenith. The *hour angle* θ (radians) of the sun is the fraction of 2π through which the earth has turned after local solar noon; i.e., $\theta = 2\pi t/24$. Since $2\pi \equiv 360°$, each hour corresponds to $15°$ rotation.

Three-dimensional geometry is needed to show that the *zenith angle* of the sun ψ at latitude φ is given by

$$\cos \psi = \sin \varphi \sin \delta + \cos \varphi \cos \delta \cos \theta \qquad (5.1)$$

and that the *azimuth angle* of the sun A with respect to south is given by

$$\sin A = -\cos \delta \sin \theta / \sin \psi \qquad (5.2)$$

The angle of the sun at its zenith (local solar noon) is found by putting $\theta = 0$ in Equation 5.1 to give

$$\cos \psi = \sin \varphi \sin \delta + \cos \varphi \cos \delta$$
$$= \cos (\varphi - \delta)$$

so

$$\psi = \varphi - \delta \qquad (5.3)$$

The time from solar noon to sunset (i.e., the half-day length) is found by putting $\psi = \pi/2$ radians ($90°$) in Equation 5.1, and rearranging to give

$$\cos \theta = -\frac{\sin \varphi}{\cos \varphi} \frac{\sin \delta}{\cos \delta} = -\tan \varphi \tan \delta \qquad (5.4)$$

from which the daylength, defined as the period for which the sun is above the horizon, is

$$2t = 24\theta/\pi = (24/\pi) \cos^{-1} (-\tan \varphi \tan \delta) \qquad (5.5)$$

Some developmental processes in plants and activity in animals occur at the very weak levels of radiation received during twilight before sunrise and after sunset. The biological length of a day may therefore exceed the daylength given by Equation 5.5 to an extent that can be estimated from the time of *civil twilight* (sun dropping to $6°$ below horizon with a correction for refraction; i.e., $\psi = 96°$ or 1.68 radians) or *astronomical twilight* (down to $18°$ below horizon, $\psi = 108°$ or 1.89 radians). The beginning of civil and astronomical twilight can be calculated from Equation 5.1 by substituting these values of ψ. At the equator, the interval between civil twilight and sunrise is close to 22 min throughout the year, but at latitude $50°$ it ranges from 45 min at midsummer to 32 min at the equinoxes.

For the application of Equations 5.1 to 5.5, values of δ may be found in astronomical tables (e.g., List 1966) or δ may be calculated from empirical expressions such as

$$\sin \delta = a \sin[b + cJ + d \sin(e + cJ)] \qquad (5.6)$$

where J is the calendar day with $J = 1$ on January 1, and the constants are $a = 0.39785$; $b = 278.97$; $c = 0.9856$; $d = 1.9165$; $e = 356.6$ [Campbell and Norman 1998].

EXAMPLE 5.2 Find the solar zenith angle at local solar noon, and the daylength at Edinburgh, Scotland ($55.0°N$) on June 21 (calendar day 172).

SOLUTION ▶ Using astronomical tables or Equation 5.6, the solar declination δ on June 21 is $23.5°$. Equation 5.1 could be used to calculate the solar zenith angle ψ, but since the time is solar noon, when $\theta = 0$, the much simpler Equation 5.3 applies, and

$$\psi = \phi - \delta = 55.0 - 23.5 = 31.5°$$

The daylength is given by Equation 5.5,

$$2t = (24/\pi) \cos^{-1} (-\tan \varphi \tan \delta)$$
$$= (24/\pi) \cos^{-1}(-\tan 55.0 \tan 23.5)$$
$$= (24/\pi) \cos^{-1}(-0.621) = 24 \times 2.24/\pi = 17.1 \text{ hours}$$

(note that $\cos^{-1}(-0.621)$ is 2.24 radians). ◀

SPECTRAL QUALITY

For biological work, the spectrum of solar radiation outside the earth's atmosphere can be divided into several major wavebands, shown in Table 5.1 with the corresponding fraction of the TSI. Most measurements of solar energy at the ground are confined to the visible and near-infrared wavebands that contain energy in almost equal proportions.

Ultra-violet radiation contains sufficient energy per quantum to damage living cells. The ultra-violet spectrum is divided into UVA (320–400 nm) responsible for tanning the skin; UVB (280–320 nm) responsible for skin cancer and Vitamin D synthesis; and UVC (200–280 nm), potentially the most harmful waveband but absorbed almost completely by molecular

TABLE 5.1 Distribution of energy in the
spectrum of radiation emitted by the sun.

Waveband (nm)	Energy (%)
0 − 200	0.7
200 − 280 (UV-C)	0.5
280 − 320 (UV-B)	1.5
320 − 400 (UV-A)	6.3
400 − 700 (visible/PAR)	39.8
700 − 1500 (near infra-red)	38.8
1500 − ∞	12.4

oxygen in the stratosphere. Human skin is about 1000 times more sensitive
to the UVB range than to the UVA.

The waveband to which the eyes of humans and most terrestrial animals are sensitive ranges from blue (400 nm) through green (550 nm)
to red (700 nm), with maximum sensitivity at around 500 nm. Eyes of
aquatic mammals have peak sensitivity at slightly shorter wavelengths,
around 488 nm, perhaps a consequence of the "blueness" of the ocean
habitat [McFarland and Munz, 1975]. Eyes of fish have similar peak sensitivity.

Photosynthesis is stimulated by radiation in the same waveband as human vision, and this is referred to as *Photosynthetically Active Radiation*
(PAR), a misnomer because it is green cells that are active, not radiation.
Initially, the term PAR was applied to radiation measured in units of energy flux density (W m^{-2}), but for two reasons it is more appropriately
expressed as quantum flux density (mol m^{-2} s^{-1}): (i) when photosynthesis rates are compared for light of different quality (e.g., from the sun and
from lamps), they are more closely related to the quantum content than to
the energy content of the radiation [Jones 1992]; and (ii) because the number of moles of carbon dioxide fixed in photosynthesis is closely proportional to the number of moles of photons absorbed in the PAR waveband.
The fraction of PAR to total energy in the extraterrestrial solar spectrum
is about 0.40 (Table 5.1), but it is closer to 0.50 for solar radiation at the
earth's surface (p. 71).

Many developmental processes in green plants have been found to depend on the state of the pigment phytochrome, which exists in two photo-
interconvertible forms that absorb radiation in wavebands centred at 660
nm (red light—the Pr form) and 730 nm (far-red light—the Pfr form).
The ratio of the two forms present in plant tissue depends on the ratio of
spectral irradiance at these wavelengths, known as the red:far-red ratio, so

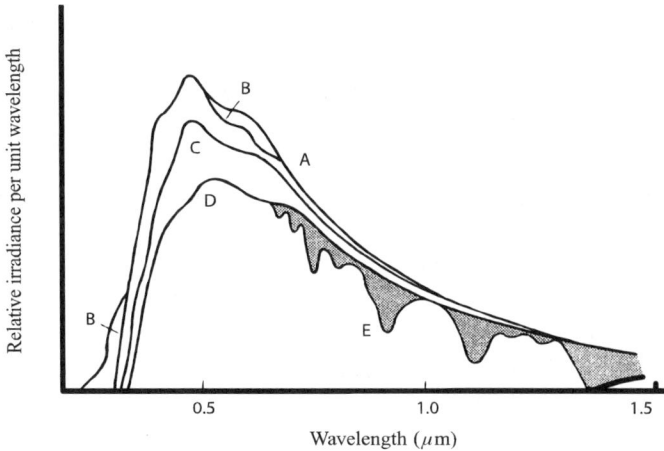

FIGURE 5.2   Successive processes attenuating the solar beam as it penetrates the atmosphere. A—extraterrestrial radiation, B—after ozone absorption, C—after molecular scattering, D—after aerosol scattering, E—after water vapour and oxygen absorption (from Henderson, 1977).

phytochrome is an effective detector of the quality of radiation (for example the presence of shade from other plants). Physiologists and ecologists now regard the red:far-red ratio as an environmental signal of major significance for germination, competitive survival, and reproduction [Smith and Morgan, 1981].

5.2   Attenuation of Solar Radiation in the Atmosphere

As the solar beam passes through the earth's atmosphere, it is modified in quantity, quality, and direction by processes of scattering and absorption (Figure 5.2).

Scattering has two main forms. First, individual quanta striking molecules of any gas in the atmosphere are diverted in all directions, a process known as *Rayleigh scattering* after the physicist who showed theoretically that the effectiveness of molecular scattering was proportional to the inverse fourth power of the wavelength.

The scattering of blue light ($\lambda = 400$ nm) therefore exceeds the scattering of red (700 nm) by a factor of $(7/4)^4$, or about 9. This is the physical basis of the blue color of the sky as seen from the ground and the blue haze surrounding the earth when viewed from space. The apparent redness of the sun's disc near sunrise or sunset is further evidence that blue light has been removed from the beam preferentially. Rayleigh also showed that the spatial distribution of scattered radiation was proportional to $(1 + \cos^2 \theta)$, where θ is the angle between the initial and final directions of the radiation. Thus, the probability of forward- and back-scattering is twice that at $90°$.

Rayleigh scattering is confined to systems in which the diameter (d) of the scatterer is much smaller than the wavelength of the radiation λ. This condition is not met for particles of dust, smoke, pollen, etc. in the atmosphere, referred to as "aerosol," which often fall in the range $0.1\lambda < d < 25\lambda$. Theory developed by Mie predicts that the wavelength dependence of scattering for aerosol particles should be a function of d/λ, and that, for some values of the ratio, longer wavelengths should be scattered more efficiently than short—the reverse of Rayleigh scattering. This happens rarely, but smoke with the appropriate narrow range of particle sizes (e.g., from forest fires) can occasionally cause the sun and moon viewed from the earth to appear blue!

Usually, aerosol contains such a wide range of particle sizes that scattering is not strongly dependent on λ. Angstrom [1929] proposed that the dependence could often be described by a power law; i.e., $\propto \lambda^{-\alpha}$ where α had an average value of 1.3, and many investigators have confirmed the power law dependence. For example, a set of measurements in the English Midlands corresponded to α between 1.3 and 2.0 [McCartney and Unsworth, 1978]. When particles are large and sufficiently dense for multiple scattering, there is essentially no wavelength dependence and the scattered light appears white; for example, with very hazy skies in summer. Aerosol scattering is usually predominantly "forwards"; i.e., in a narrow cone around the direction of the incident radiation.

The second process of attenuation is *absorption* by atmospheric gases and aerosols. The most important gases absorbing solar radiation are ozone, and oxygen (particularly in the ultra-violet spectrum), and water vapor and carbon dioxide (with strong bands in the infra-red). Absorption by aerosols is very variable, depending on the source of the material (see textbox).

In contrast to scattering, which simply changes the direction of radiation, absorption removes energy from the beam so that the aerosol, and atmosphere containing it, is heated. In the ultraviolet, absorption by oxygen and ozone in the stratosphere removes all UVC and most UVB radiation before it reaches the ground, resulting in stratospheric heating. The

Sources and Radiative Properties of Aerosols

Aerosols are solid or liquid particles small enough to remain suspended in the atmosphere for long periods. They have an important *direct* influence on radiation reaching the ground because they scatter and absorb solar and long-wave radiation. Aerosols also influence cloud formation and duration by increasing the number of droplets in clouds and altering the efficiency of precipitation production, thus also having an *indirect* effect on radiation [IPCC 2001]. These indirect effects are hard to isolate in the free atmosphere, but can be seen in "ship tracks" observed from space (Figure 5.3, see color plates). When there is a thin layer of marine stratus cloud, aerosols emitted from ships' smokestacks provide additional condensation nuclei around which droplets condense. The polluted clouds have more droplets than adjacent clean clouds, and their droplets are smaller. The smaller droplets increase the reflectivity of the polluted clouds at visible and near infrared wavelengths, so that ship tracks can readily be discerned from satellite images of scattered radiation at 2.1 μm [Coakley et al 1987].

Aerosols in the lower atmosphere have relatively short lifetimes before they are removed by precipitation and turbulent transfer (see Chapter 12). Aerosols in the stratosphere, for example as a result of explosive volcanic eruptions, have much longer lifetimes and may be dispersed around the globe, causing observable effects on the earth's climate [Hansen 1996].

The size distribution of aerosols is critical for their radiative effects. Particles in the "accumulation size range" (Chapter 12) (i.e., with diameters between about 0.1 μm and 1.0 μm), scatter more light per unit mass than larger particles, and have longer lifetimes in the atmosphere, so are particularly important in influencing the irradiance at the ground. Some aerosols are *hygroscopic*; i.e., they absorb water depending on atmospheric humidity, thus changing the aerosol size distribution. Examples are sea salt particles and ammonium sulphate.

Primary aerosols are generated at the surface by natural processes and human activity. Desert dust storms generate aerosols that can be transported across the Atlantic and Pacific oceans [Kaufman et al., 2002]. Inefficient combustion of wood or fossil fuels releases organic and black carbon aerosols. *Secondary aerosols* are created in the atmosphere by chemical reactions. A ubiquitous example is ammonium sulphate aerosol, formed by reactions that include the gases ammonia, sulphur dioxide and dimethyl sulphide, which have natural sources (volcanoes, ocean plankton) and human sources (fossil fuel, animal production). Other secondary aerosols are created during photochemical smog episodes.

Absorption of radiation by aerosols is very variable. For example, desert dust may absorb little, but black carbon aerosols from wildfires and human activity are much stronger absorbers [Hansen et al., 2004]. Absorption by secondary aerosols can be greatly increased when black carbon particles are incorporated into the aerosol. For several reasons, radiative effects of aerosols are generally much more difficult to assess than those of atmospheric gases: aerosols are not uniformly distributed, and may be formed and transformed in the atmosphere; some aerosol types (e.g., dust and sea salt) consist of particles whose physical properties that influence scattering and absorption have wide ranges; and aerosol species may combine to form mixed particles with optical properties different from their precursors. IPCC [2001] included a good review of progress in understanding aerosol effects on the earth's energy balance.

remaining UVB and UVA radiation is scattered very effectively, however, so that it is possible to suffer sunburn beneath a cloudless sky even if not exposed to the direct solar beam.

In the visible region of the spectrum, absorption by atmospheric gases is much *less* important than scattering in determining the spectral distribution of solar energy at the ground. In the infra-red spectrum, however, absorption is much *more* important than scattering, because several atmospheric constituents absorb strongly, notably water vapor with absorption bands between 0.9 and 3 μm. The presence of water vapor in the atmosphere thus increases the amount of visible radiation relative to infra-red radiation.

The scale of absorption and scattering in the atmosphere depends partly on the pathlength of the solar beam and partly on the amount of the attenuating constituent present in the path. The pathlength is usually specified in terms of an *air mass number m*, which is the length of the path relative to the depth of the atmosphere. Air mass number therefore depends on altitude (represented by the pressure exerted by the atmospheric column above the site), and on zenith angle ψ. For values of zenith angle ψ less than 80°, the air mass number at a location where atmospheric pressure is P is simply $m = (P/P_0)\sec\psi$, where P_0 is standard atmospheric pressure at sea level, but for values between 80° and 90°, m is smaller than $\sec\psi$ because of the earth's curvature. Values of m, corrected for refraction, can be obtained from tables (e.g., List, 1966).

The most variable absorbing gas in the atmosphere is water vapor, the amount of which can be specified by a depth of *precipitable water u*, defined as the depth of water that would be formed if all the vapor were condensed (u is typically between 5 and 50 mm at most locations). If the precipitable water is u, the pathlength for water vapor is um. Similarly, the total amount of ozone in the atmosphere (the *ozone column*) is specified by an equivalent depth of the pure gas at a standard pressure of one atmosphere (101.3 kPa). At mid-latitudes, the ozone column is typically about 3 mm, and varies only slightly with season. However, over some parts of the Antarctic, up to 60% of the ozone column is lost during the Antarctic Spring (September–October). When this phenomenon was first reported by Joe Farman of the British Antarctic Survey, who analyzed surface observations of irradiance in the UVB waveband [Farman 1985], it could not be explained by atmospheric chemists and had not been detected by satellite monitoring. In retrospect, it turned out that a computer program had caused the anomalous Antarctic Spring data from the satellites to be ignored, and chemists traced the cause of the alarming decrease in ozone to the presence in the Antarctic stratosphere of frozen particles of nitric acid that served as catalytic sites for reactions destroying ozone. Figure 5.4 (see color plates) shows the decline of ozone column thickness in the Antarctic Spring from the mid 1950s. The stratospheric conditions necessary for ozone destruction by this mechanism are less common over the Arctic, and fortunately do not exist over the large parts of the earth where plant and animal populations would be vulnerable to the extra UVB radiation that would reach the surface if ozone were depleted.

In contrast with the mainly regional effects of ozone depletion, the consequences of radiative absorption by the steadily increasing amount of carbon dioxide in the earth's atmosphere are apparent on a global scale (Figure 2.3), and the resulting disturbances to the earth's climate by this human-caused disturbance are the subject of intensive investigation [IPCC 2001].

Clouds, consisting of water droplets or ice crystals, scatter radiation both forward and backward, but when the depth of cloud is substantial, back-scattering predominates and thick stratus can reflect up to 70% of incident radiation, appearing snow-white from an aircraft flying above it. About 20% of the radiation may be absorbed, leaving only 10% for transmission, so that the base of such a cloud seems gray. At the edge of a cumulus cloud, however, where the concentration of droplets is small, forward scattering is strong—the silver lining effect—and under a thin sheet of cirrus, the reduction of irradiance can be less than 30% (see Figure 5.10).

5.3   Solar Radiation at the Ground

As a consequence of attenuation, radiation has two distinct directional properties when it reaches the ground. *Direct* radiation arrives from the direction of the solar disc and includes a small component scattered directly forward. The term *diffuse* describes all other scattered radiation received from the blue sky (including the very bright aureole surrounding the sun) and from clouds, either by reflection or by transmission. The sum of the energy flux densities for direct and diffuse radiation is known as *total* or *global* radiation, and for climatological purposes is measured on a horizontal surface. The symbols S_b, S_d, and S_t describe direct, diffuse and total irradiance respectively on a horizontal surface, and S_p signifies direct irradiance at right angles to the solar beam.

DIRECT RADIATION

At sea level, the direct irradiance S_p rarely exceeds 75% of the Solar Constant; i.e., about 1000 Wm^{-2}. The minimal loss of 25% is attributable to molecular scattering and absorption in almost equal proportions with a negligible contribution from aerosol when the air mass is clean. Aerosol increases the ratio of diffuse to global radiation by forward scattering, and also changes the spectral composition. Several expressions are available to describe atmospheric transmissivity as affected by molecular and aerosol components. A simple practical relation is

$$S_p = S^* T^m \tag{5.7}$$

where S^* is the Solar Constant, T is the *atmospheric transmissivity*, and m is the air mass number. Liu and Jordan [1960] found that T ranged from about 0.75 to 0.45 on cloudless days, implying, as previously, that aerosol attenuation was insignificant on the clearest days.

To illustrate more directly the combined impact of attenuation by aerosols and molecules, Beer's Law may be applied to give

$$S_p(\tau) = S^* \exp(-\tau m) \tag{5.8}$$

where τ is an *extinction coefficient* or *optical thickness*, and m is air mass number. When the value of τ is expressed as the sum of molecular extinction (τ_m) and aerosol extinction (τ_a), Equation 5.8 may be written in the form

$$S_p(\tau) = S^* \exp(-\tau_m m) \exp(-\tau_a m) = S_p(0) \exp(-\tau_a m) \tag{5.9}$$

where $S_p(0)$ is the irradiance of the direct beam below an atmosphere free of aerosol.

Comparison of Equations 5.7, 5.8, and 5.9 yields

$$T = \exp(-\tau) = \exp(-(\tau_m + \tau_a)) \qquad (5.10)$$

implying that τ_m was about 0.3 in Liu and Jordan's measurements (assuming $\tau_a = 0$ when $T = 0.75$), and τ_a was about 0.5 on the most turbid days.

The value of τ_a at a site can be determined from Equation 5.9 by measuring $S_p(\tau)$ as a function of m (= sec ψ) and by calculating $S_p(0)$ (also a function of m) from the properties of a clean atmosphere containing appropriate amounts of gases and water vapor. For example, a series of measurements in Britain gave values of τ_a ranging from 0.05 for very clear air of Arctic origin to 0.6 for very polluted air in the English Midlands during a stagnant anticylone [Unsworth and Monteith, 1972]. Corresponding values of $\exp(-\tau_a m)$ for $\psi = 30^\circ$ are 0.92 and 0.50, indicating radiant energy losses of up to 50% from the direct solar beam due to aerosol scattering and absorption.

The spectrum of direct radiation depends strongly on the pathlength of the beam and therefore on solar zenith angle. A spreadsheet model for calculating direct and diffuse spectral irradiance (based on work by Bird and Riordan 1986) is available from the Solar Energy Research Institute, Golden, Colorado (http://rredc.nrel.gov/solar/). Figure 5.5 shows results at the earth's surface calculated from the model, indicating that the spectral irradiance calculated for the solar beam is almost constant between 500 and 700 nm, whereas in the corresponding extraterrestrial spectrum, irradiance decreases markedly as wavelength increases beyond $\lambda_m \approx 500$ nm (Figure 5.2). The difference is mainly a consequence of energy removed from the beam by Rayleigh scattering, which increases as wavelength decreases; ozone absorption is implicated, too. As solar zenith angle increases, attenuation by scattering becomes very pronounced, and the wavelength for maximum direct solar irradiance moves into the infrared waveband when the sun is more than 70° from the zenith.

The measurements by Unsworth and Monteith [1972] also showed that, for zenith angles between 40° and 60°, the ratio of visible to all-wavelength radiation in the direct solar beam decreased from a maximum of about 0.5 in clean air to about 0.4 in very turbid air. The maximum ratio exceeds the figure of 0.40 for extraterrestrial radiation (Table 5.1) because losses of visible radiation by scattering are more than offset by losses of infra-red radiation absorbed by water vapor and oxygen (Figure 5.2). The

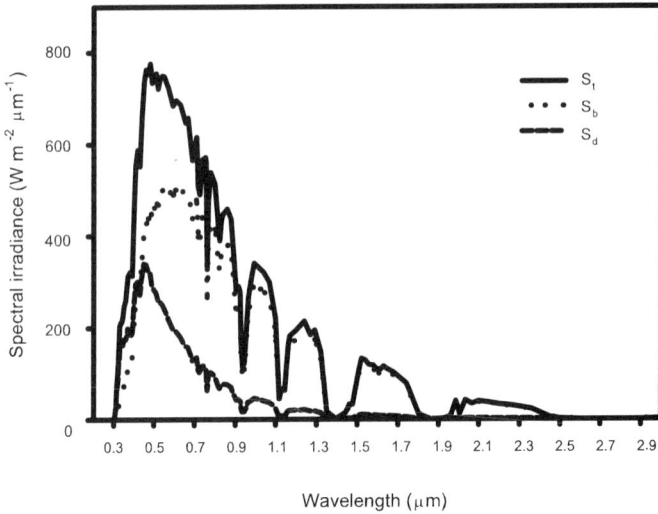

FIGURE 5.5   Spectral distribution of direct, diffuse, and total solar radiation calculated using a simple model of a cloudless atmosphere [Bird and Riordan, 1984]. Solar zenith angle is 60° ($m = 2$), precipitable water is 20 mm, ozone thickness is 3 mm, and aerosol optical depth is 0.2. Note that the diffuse flux has maximum energy per unit wavelength at about 0.46 μm.

quantum content of *direct radiation* increased with turbidity from a minimum of about 2.7 μmol J^{-1} total radiation in clean air to about 2.8 μmol J^{-1} in turbid air. Theoretical values can be calculated using the spectral distribution model (p. 63). (Confusion has arisen in the literature where quantum content values for *direct* radiation are compared with values referring to *global* radiation, as given later (p. 71.)

DIFFUSE RADIATION

Beneath a clean, cloudless atmosphere, the diffuse irradiance S_d reaches a broad maximum somewhat less than 200 W m^{-2} when ψ is less than about 50°, and the ratio of diffuse to global irradiance falls between 0.1 and 0.15. As turbidity increases, so does S_d/S_t and for $\psi < 60°$, observations in the English Midlands fit the relation

$$S_d/S_t = 0.68\tau + 0.10 \tag{5.11}$$

For $\psi > 60°$, S_d/S_t is a function of ψ, and is larger than Equation 5.11 predicts.

With increasing cloud amount also, S_d/S_t increases and reaches unity when the sun is obscured by dense cloud: but the absolute level of S_d is maximal when cloud cover is about 50%. The spectral composition of diffuse radiation is also strongly influenced by cloudiness. Beneath a cloudless sky, diffuse radiation is predominantly within the visible spectrum (Figure 5.5), but as cloud increases, the ratio of visible/all-wavelength radiation decreases toward the value of about 0.5 characteristic of global radiation.

ANGULAR DISTRIBUTION OF DIFFUSE RADIATION

Under an overcast sky, the flux of solar radiation received at the ground is almost completely diffuse. If it were perfectly diffuse, the radiance of the cloud base observed from the ground would be uniform and would therefore be equal to S_d/π from Equation 4.8. The source providing this distribution is known as an *Uniform Overcast Sky* (UOS).

In practice, the average radiance of a heavily overcast sky is between two and three times greater at the zenith than at the horizon (because multiple-scattered radiation is attenuated by an air mass depending on its perceived direction, so regions near the horizon appear relatively depleted). To allow for this variation, ambitious architects and pedantic professors describe the radiance distribution of overcast skies as a function of zenith angle given by

$$\mathbf{N}(\psi) = \mathbf{N}(0)(1 + b\cos\psi)/(1 + b) \qquad (5.12)$$

This distribution defines a *Standard Overcast Sky*. Measurements indicate that the number $(1 + b)$, which is the ratio of radiance at the zenith to that at the horizon, is typically in the range 2.1–2.4 [Steven and Unsworth, 1979], values supported by theoretical analysis [Goudriaan, 1977], which also shows the dependence on surface reflectivity. The value $(1 + b) = 3$, in common use, is based on photometric studies and significantly overestimates the diffuse irradiance of surfaces.

Under a cloudless sky, the angular distribution of skylight depends on the position of the sun and cannot be described by any simple relation. In general, the sky around the sun is much brighter than elsewhere because there is a preponderance of scattering in a forward direction, but there is a sector of sky about 90° from the sun where the intensity of skylight is below the average for the hemisphere (Figure 5.6). On average, the diffuse radiation from a blue sky tends to be stronger nearer the horizon than at the zenith. As the atmosphere becomes dustier, the general effect is to reduce the radiance of the circumsolar region and to increase the relative

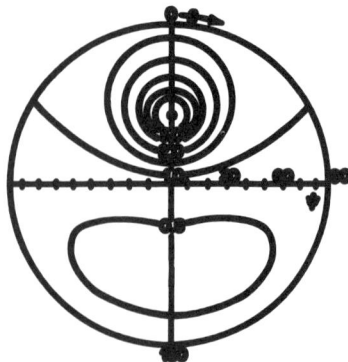

FIGURE 5.6   Standard distribution of normalized sky radiance $\pi\,\mathbf{N}/\mathbf{S}_d$ for solar zenith angle $35°$, where $\mathbf{N}$ is the value of sky radiance at a point and $\pi\,\mathbf{N}$ is the diffuse flux which the surface would receive if the whole sky were uniformly bright (see p. 66) [from Steven, 1977].

radiance of the upper part of the sky at the expense of regions near the horizon. Consequently, the angular distribution of radiance becomes more uniform as turbidity increases.

TOTAL (GLOBAL) RADIATION

The total radiation on a horizontal surface is given formally by

$$S_t = S_p \cos \psi + S_d$$
$$= S_b + S_d \qquad (5.13)$$

where $S_b = S_p \cos \psi$ is the contribution from the direct beam.

Figure 5.7 contains an example of measured values of S_t, S_d, and S_b as a function of solar zenith angle at Sutton Bonington, England, and Figure 5.8 shows similar data (with the addition of S_p) from a more southerly site at Eugene, Oregon.

Estimating radiation under cloudless skies   A simple spreadsheet model that enables calculations of S_p, S_d, and S_t under cloudless skies at any location and date for specified values of water vapor, ozone, and aerosol optical thickness, is based on the work of Bird and Hulstrom [1981] and is available from the Solar Energy Research Institute, Golden, Colorado (http://rredc.nrel.gov/solar/). The following simpler approach for direct and diffuse (and hence total) radiation often gives estimates of ap-

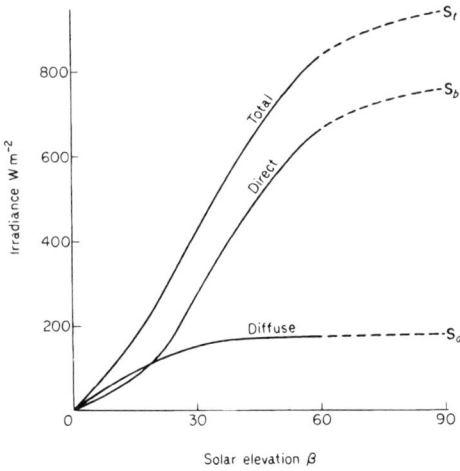

FIGURE 5.7 Solar irradiance on a cloudless day (16 July, 1969) at Sutton Bonington (53°N, 1°W): S_t total flux; S_d diffuse flux, S_b direct flux on a horizontal surface.

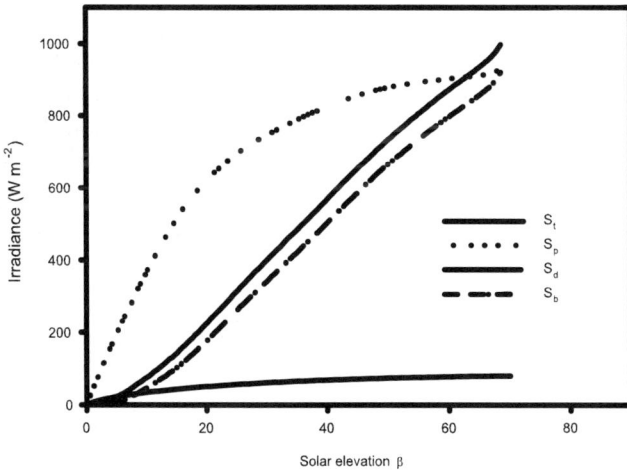

FIGURE 5.8 Solar irradiance on a cloudless day (12 June, 2002) at Eugene, Oregon (44°N, 123°W): S_t total flux; S_p direct flux at normal incidence; S_d diffuse flux; S_b direct flux on a horizontal surface (data courtesy of Frank Vignola, University of Oregon).

propriate accuracy for biological calculations. Using Equation 5.9, it can be shown that

$$S_p = S^* \exp(-\tau_m m) \exp(-\tau_a m) \tag{5.14}$$

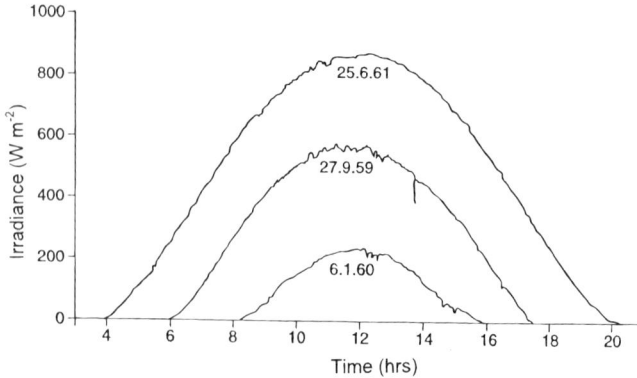

FIGURE 5.9   Solar radiation on three cloudless days at Rothamsted, England (52° W). During the middle of the day the record tends to fluctuate more than in the morning and evening, suggesting a diurnal change in the amount of dust in the lower atmosphere, at least in summer and autumn.

The Solar Constant S^* is 1366 W m^{-2} (p. 53), the optical thickness for molecular attenuation τ_m is typically about 0.3 (but changes with the amount af water vapor and other absorbing gases in the atmosphere), and the aerosol optical thickness τ_a is typically in the range 0.05 to 0.50. Given appropriate values of these parameters, S_p can be calculated for a specific air mass number m (where the dependence of m on solar zenith angle ψ is given by $m = (P/P_0)\sec\psi$ (p. 60).

Approximate values of S_d as a function of m on cloudless days can be estimated from an empirical equation based on measurements by Liu and Jordan [1960]; i.e.,

$$S_d = 0.3S^*[1 - \exp(-(\tau_m + \tau_a)m]\cos\psi \qquad (5.15)$$

The total irradiance is then given by

$$S_t = S_p \cos\psi + S_d$$

On cloudless days, illustrated by Figure 5.9, the change of S_t with time is approximately sinusoidal. This form is distorted by cloud, but in many climates the degree of cloud cover, averaged over a period of a month, is almost constant throughout the day, so the monthly average variation of irradiance over a day is again sinusoidal. In both cases, the irradiance at t

hours after sunrise can be expressed as

$$S_t = S_{tm} \sin(\pi t / n) \qquad (5.16)$$

where S_{tm} is the maximum irradiance at solar noon and n is the daylength in hours. This equation can be integrated to give an approximate relation between maximum irradiance and the daily integral of irradiation (the *insolation*) by writing

$$\int_0^n S_t dt \approx 2S_{tm} \int_0^{n/2} \sin(\pi t / n) \, dt = (2n/\pi) S_{tm} \qquad (5.17)$$

For example, over southern England in summer, S_{tm} may reach 900 W m^{-2} on a clear cloudless day and with $n = 16$ h $= 58 \times 10^3$ s, the insolation calculated from Equation 5.17 is 33 MJ m^{-2} compared with a measured maximum of about 30 MJ m^{-2}. In Israel, where S_{tm} reaches 1050 W m^{-2} in summer for a daylength of 14 h, the equation gives an insolation of 34 MJ m^{-2} compared with 32 MJ m^{-2} by measurement.

At higher latitudes in summer when dawn and dusk are prolonged, a full sine-wave may be more appropriate than Equation 5.17. If S_t is given by

$$S_t \approx S_{tm}(1 - \cos 2\pi t / n) = S_{tm} \sin^2(\pi t / n) \qquad (5.18)$$

integration yields

$$\int_0^n S_t dt = S_{tm} n / 2 \qquad (5.19)$$

Gloyne [1972] showed that the radiation regime at Aberdeen (57°N) was described best by the average of values given by Equations 5.17 and 5.19. In most climates, the daily receipt of total solar radiation is greatly reduced by cloud for at least part of the year. Figure 5.10 shows the extent to which the total irradiance beneath continuous cloud depends on cloud type and solar elevation β ($= \pi/2 - \psi$). The fraction of extraterrestrial radiation can be read from the full lines, and the corresponding irradiance by interpolation between the dashed lines.

The formation of a small amount of cloud in an otherwise clear sky always increases the diffuse flux, but the direct component remains unchanged provided neither the sun's disc nor its aureole are obscured. With a few isolated cumulus, the total irradiance can therefore exceed the flux beneath a cloudless sky by 5 to 10%. On a day of broken cloud (Figure 5.11), the temporal distribution of radiation is strongly bimodal: the irradiance is very weak when the sun is completely occluded and strong when it is exposed. For a few minutes before and after occlusion, the irradiance

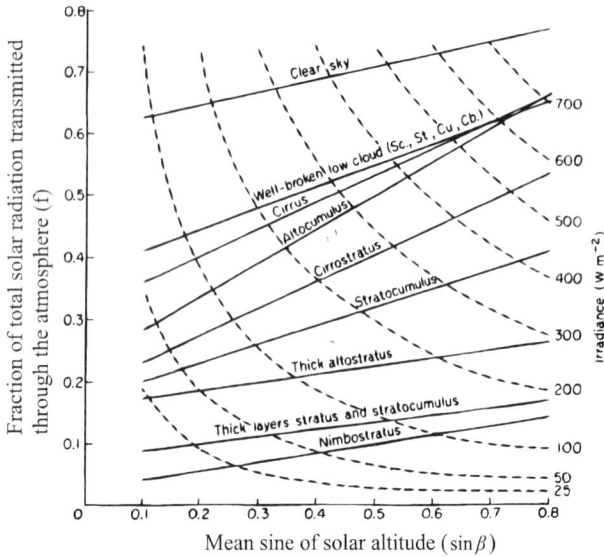

FIGURE 5.10   Empirical relations between solar radiation and solar angle for different cloud types from measurements in the N. Atlantic (52°N, 20°W). The curves are isopleths of irradiance [from Lumb, 1964]. Sc, stratocumulus; St, stratus; Cu, cumulus; Cb, cumulonimbus.

commonly reaches 1000 W m^{-2} in temperate latitudes and even exceeds the Solar Constant in the tropics. This effect is a consequence of strong forward scattering by the small concentration of water droplets present at the edge of a cloud.

As a consequence of cloud, the average insolation over most of Europe in summer is restricted to between 15 and 25 MJ m^{-2}, about 50 to 80% of the insolation on cloudless days. Comparable figures in the United States range from 23 MJ m^{-2} around the Great Lakes to 31 MJ m^{-2} under the almost cloudless skies of the Sacramento and San Joaquim valleys. Winter values range from 1 to 5 MJ m^{-2} over most of Europe and from 6 MJ m^{-2} in the northern United States to 12 MJ m^{-2} in the south. Australian stations record a range of values similar to those of the United States.

Although the difference of irradiance with and without cloud is roughly an order of magnitude, the radiation to which plants are exposed covers a much wider range. In units of micromoles of photons m^{-2} in the PAR waveband, full summer sunshine is approximately 2200, shade on the for-

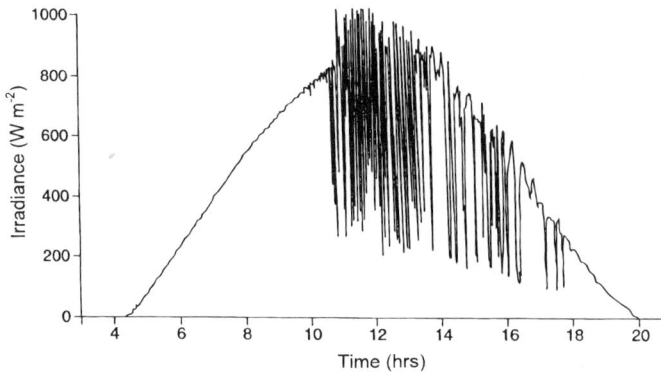

FIGURE 5.11   Solar radiation on a day of broken cloud (11 June, 1969) at Rothamsted, England (52°N, 0°W). Note very high values of irradiance immediately before and after occlusion of the sun by cloud and the regular succession of minimum values when the sun is completely obscured.

est floor 20, twilight 1, moonlight 3×10^{-4}, and radiation from an overcast sky on a moonless night about 10^{-7} [Smith and Morgan, 1981].

At any location, annual changes of insolation depend in a complex way on seasonal changes in the water vapor and aerosol content of the atmosphere and on the seasonal distribution of cloud. Table 5.2 shows the main components of attenuation for four "seasons" at Kew Observatory, a suburban site 10 miles (16 km) west of the center of London. The data were collected in the 1950s when London air was more heavily polluted with smoke particles from inefficiently burned coal than it is now, so the "dust and smoke" losses for winter are relatively large. For the annual average, roughly a third of the radiation received outside the atmosphere is scattered back to space, a third is absorbed, and a third is transmitted to the surface. The flux at the surface is 20 to 25% less than it would be in a perfectly clear atmosphere. Because the climate at Kew is relatively cloudy, the diffuse component is larger than the direct component throughout the year.

SPECTRUM OF TOTAL SOLAR RADIATION

The spectrum of total (global) solar radiation depends, in principle, on solar zenith angle, cloudiness, and turbidity, and the interaction of these three factors limits the usefulness of generalizations. As zenith angle increases

TABLE 5.2   Short-wave radiation balance of atmosphere and surface at Kew Observatory (51.5°N) for 1956–1960 expressed as a percentage of extraterrestrial flux.

	Winter (Nov–Jan)	Spring (Feb–April)	Summer (May–July)	Autumn (Aug–Oct)	Year
Extraterrestrial radiation three month total MJ m^{-2}	800	2050	3720	2340	8910
daily mean MJ m^{-2} day^{-1}	8.7	22.3	40.4	25.4	24.4
Losses in atmosphere (percent) (a) Absorption					
water vapor	15	12	13	15	13
cloud	8	9	9	9	9
dust and smoke	15	10	5	8	8
total	38	31	27	32	30
(b) Scattering (away from surface)	37	35	33	34	34
Radiation at surface direct	8	14	18	14	15
diffuse	17	20	22	20	21
total	25	34	40	34	36
	100	100	100	100	100
Total as MJ m^{-2} day^{-1}	2.2	7.6	16.2	8.7	8.8

beyond $60°$, so does the proportion of scattered radiation and therefore the ratio of visible to all-wavelength radiation. In one record from Cambridge, England, this ratio increased from about 0.49 at $\psi = 60°$ to 0.52 at $\psi = 10°$ [Szeicz, 1974]. Cloud droplets absorb radiation in the infrared spectrum, so with increasing cloud the fraction of visible radiation should increase. Again at Cambridge, the range was between 0.48 in summer and 0.50 in winter. Finally, with increasing turbidity, shorter wavelengths are scattered preferentially, depleting the direct beam but contributing to the diffuse flux, so that the change in global radiation is relatively small. In another set of measurements at a site close to Cambridge, the visible: all-wavelength ratio decreased from 0.53 at $\tau_a = 0.1$ to 0.48 at $\tau_a = 0.6$ [McCartney, 1978]. Elsewhere, a smaller value of the ratio, 0.44, with little seasonal variation was reported for a Californian site by Howell et al [1983] and, in the tropics, Stigter and Musabilha [1982] found that the ratio increased from 0.51 with clear skies to 0.63 with overcast. It is probable that some of the apparent differences between sites reflect differences or errors in instrumentation and technique rather than in the behavior of the atmosphere.

At a given site, the relation between quantum content and energy appears to be conservative. McCartney [1978] reported a value of 4.56±0.05 μmol per J (PAR) in the English Midlands. This is equivalent to about 2.3 μmol per J *total* radiation, but other values reported range from 2.1 μmol J^{-1} in California to 2.9 μmol J^{-1} in Texas [Howell et al, 1983]. The ratio

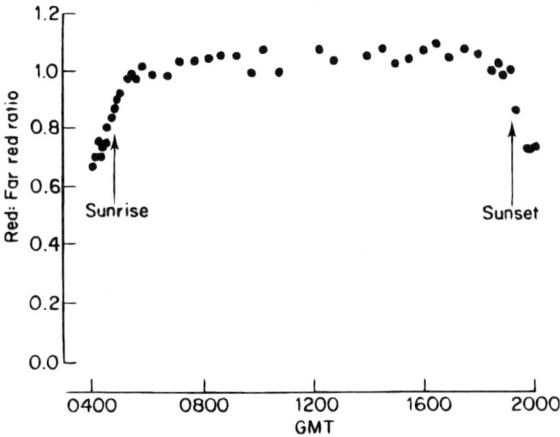

FIGURE 5.12   Ratio of spectral irradiance at 660 nm (red) to irradiance at 730 nm (far red) on an overcast day (25 August 1980) near Leicester in the English Midlands [from Smith and Morgan 1981].

of quantum content to energy in the PAR waveband for diffuse radiation from a cloudless sky is smaller than that for total radiation, about 4.25 μmol J^{-1} PAR, because quanta at shorter wavelengths carry more energy.

The ratio of energy per unit wavelength in red and far-red wavebands is also conservative and has a value of about 1.1 during the day when solar elevation exceeds 10° (Figure 5.12). As the sun approaches the horizon, the ratio decreases because red light is scattered more than far-red and because only a small fraction of the forward-scattered light reaches the ground. There is some evidence that the ratio starts to increase and returns to values greater than 1 when the solar disc falls below the horizon, presumably because skylight alone is relatively rich in shorter wavelengths (Figure 5.5).

5.4 Terrestrial Radiation

At the wavelengths associated with solar radiation, emission from radiatively active gases in the earth's atmosphere and from the earth's surface is negligible, so the previous section considered only absorption and scattering. In contrast, at wavelengths of terrestrial radiation (i.e., long-wave radiation originating in the earth's atmosphere and at its surface), both absorption and emission are important, and will be considered in this section.

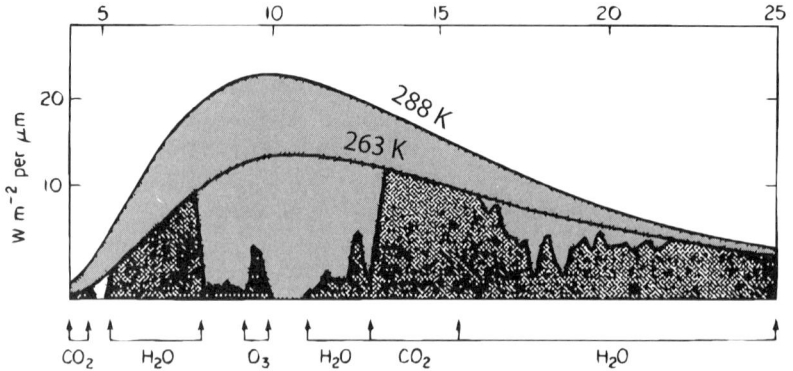

FIGURE 5.13   Spectral distribution of long wave radiation for black bodies at 298 K and 263 K. Dark gray areas show the emission from atmospheric gases at 263 K. The light gray area therefore shows the net loss of radiation from a surface at 288 K to a cloudless atmosphere at a uniform temperature of 263 K [after Gates, 1980].

Most natural surfaces can be treated as "full" radiators that emit long-wave radiation, in contrast to the solar or short-wave radiation emitted by the sun. At a surface temperature of 288 K, the energy per unit wavelength of terrestrial radiation (based on Wien's Law) reaches a maximum at $2897/288$ or 10 μm, and arbitrary limits of 3 and 100 μm are usually set for the long-wave spectrum. Figure 5.13 illustrates the spectrum of radiation that would be emitted to the atmosphere from a surface that was a full radiator at 288 K.

In the absence of cloud, most of the radiation emitted by the earth's surface is absorbed within the atmosphere in specific wavebands by radiatively active atmospheric gases, mainly water vapor and carbon dioxide. A small fraction of radiation from the surface escapes to space, mostly through the *atmospheric window* between 8 and 12 μm. The energy absorbed in atmospheric gases is re-radiated (emitted) in all directions. Atmospheric gases do not emit like full radiators: rather, they have an emission spectrum similar to their absorption spectrum (Kirchhoff's principle, p. 39). Figure 5.13 shows the approximate spectral distribution of the downward flux of atmospheric radiation that would be received at the earth's surface from an atmosphere at 263 K. In reality, much of the atmospheric radiation that reaches the surface arises from gases close to the surface, and consequently close to surface temperature: atmospheric radiation that is lost to space is emitted mainly from gases higher in the troposphere where temperatures are less. Radiation emitted to space is therefore partly surface emission, escaping through the atmospheric window, and partly atmospheric emission from the upper troposphere.

To satisfy the First Law of Thermodynamics for the earth as a planet, assuming that the earth is in equilibrium, the average annual loss of long-wave radiative energy to space must balance the average net gain from solar radiation. If r_E is the radius of the earth, S^* is the Solar Constant, ρ_E is the planetary albedo (the fraction of solar radiation scattered to space from clouds and the surface), and L is the emitted radiation (emittance) to space, this balance may be expressed as

$$(1 - \rho_E)S^* \pi r_E^2 = 4\pi r_E^2 L$$

or

$$L = (1 - \rho_E)S/4$$

Taking $\rho_E = 0.30$ and $S^* = 1366$ W m^{-2} yields $L = 239$ W m^{-2}, and using the Stefan Boltzmann Law (Equation 4.2) this corresponds to an equivalent black body temperature of the earth viewed from space of 255 K ($-18°$C). The low value in comparison to the observed mean temperature at the surface (about 288 K, 15°C) is an indication of the extent to which atmospheric gases and cloud create a favorable climate for life on earth. This phenomenon is commonly called the *greenhouse effect*, though the term is a poor one, as real greenhouses become warm by reducing heat loss by the wind and convection rather than by radiative effects.

Analysis of the exchange and transfer of long-wave radiation throughout the atmosphere is one of the main problems of physical meteorology but micrometeorologists are concerned primarily with the simpler problem of measuring or estimating fluxes at the surface. The upward flux L_u from a surface can be measured with a radiometer or from a knowledge of the surface temperature and emissivity. The downward flux from the atmosphere L_d can also be measured radiometrically, calculated from a knowledge of the temperature and water vapor distribution in the atmosphere, or estimated from empirical formulae.

TERRESTRIAL RADIATION FROM CLOUDLESS SKIES

The radiance of a cloudless sky in the long-wave spectrum (or the effective radiative temperature) is least at the zenith and greatest near the horizon. This variation is a direct consequence of the increase in the pathlength of water vapor and carbon dioxide, the main emitting gases. In general, more than half the radiant flux received at the ground from the atmosphere comes from gases in the lowest 100 m and roughly 90% from the lowest kilometer. The magnitude of the flux is therefore strongly determined by temperature gradients near the ground.

It is convenient to define the apparent emissivity of the atmosphere ε_a as the flux density of downward radiation divided by full radiation at air temperature T_a measured near the ground; i.e.,

$$\mathbf{L_d} = \varepsilon_a \sigma T_a^4 \tag{5.20}$$

Similarly, the apparent emissivity at a zenith angle ψ or $\varepsilon_a(\psi)$ can be taken as the flux density of downward radiation at ψ divided by σT_a^4. Many measurements show that the dependence of $\varepsilon_a(\psi)$ on ψ over short periods can be expressed as

$$\varepsilon_a(\psi) = a + b \ln(u \sec \psi) \tag{5.21}$$

where u is precipitable water (corrected for the pressure dependence of radiative emission) and a and b are empirical constants that change with the vertical gradient of temperature and with the distribution of aerosol [Unsworth and Monteith, 1975]. Integration of this equation over a hemisphere using Equation 4.9 gives the effective (hemispherical) emissivity as

$$\varepsilon_a = a + b(\ln u + 0.5) \tag{5.22}$$

Comparing Equations 5.21 and 5.22 shows that the hemispherical emissivity is identical to the emissivity at a representative angle ψ' such that

$$\ln(u \sec \psi') = \ln u + \ln \sec \psi' = \ln u + 0.5$$

It follows that $\ln \sec \psi' = 0.5$, giving $\psi' = 52.5°$ irrespective of the values of a and b. Hence, a directional radiometer recording the radiance at $52.5°$ could be used to monitor the value of ε_a for cloudless skies, and $\mathbf{L_d}$ would be given by Equation 5.20.

Formulae for estimating ε_a for cloudless skies were reviewed by Prata [1996] and Niemala et al [2001]. The most successful equation was

$$\varepsilon_a = 1 - (1 + aw)\exp{-[(b + cw)^{0.5}]}$$

where w is the precipitable water content of the atmosphere in kg m^{-2}, and the empirical constants are $a = 0.10$ kg^{-1} m^2, $b = 1.2$, $c = 0.30$ kg^{-1}m^2. Values of w were given approximately by

$$w = 4.65 e_a / T_a$$

where e_a is water vapor pressure near the ground (Pa) and T_a is air temperature (K).

An even simpler formulae for estimating L_d was developed by Unsworth and Monteith [1975], viz

$$L_d = c + d\sigma T_a^4 \qquad (5.23)$$

For measurements in the English Midlands, which covered a temperature range from -6 to $26°C$, the empirical constants were $c = -119 \pm 16$ W m^{-2} and $d = 1.06 \pm 0.04$. The uncertainty of a single estimate of L_d was ± 30 W m^{-2}. Measurements in Australia [Swinbank, 1963] gave similar values of c and d but with much less scatter. The lack of an explicit dependence of L_d on humidity in expressions such as Equation 5.23 is probably because there is often a strong correlation between air temperature and humidity in the lower atmospheric layers responsible for most of the radiant emission.

Using a linear approximation to the dependence of full radiation on temperature above a base temperature of 283 K allows Equation 5.23 to be written in the form

$$L_d = 213 + 5.5 T_a' \qquad (5.24)$$

where T_a' is air temperature in °C. Outward long-wave radiation, assumed to be σT_a^4, is given by a similar approximation as

$$L_u = 320 + 5.2 T_a' \qquad (5.25)$$

The net loss of long-wave radiation is therefore

$$L_u - L_d = 107 - 0.3 T_a' \qquad (5.26)$$

implying that 100 W m^{-2} is a good average figure for the net loss to a clear sky (see Figure 5.14).

If the cloudless atmosphere emitted like a full radiator (an incorrect but commonly applied simplification in climatology texts), an expression for the effective radiative temperature of the atmosphere T_b' could be obtained by writing

$$L_d = 320 + 5.2 T_b' = 213 + 5.5 T_a'$$

so that

$$T_b' = \left(5.5 T_a' - 107\right)/5.2 = \left(T_a' - 21\right) + 0.06 T_a'$$

This relation shows that when T_a' is between 0 and 20°C, the mean effective radiative temperature of a cloudless sky is usually about 21 to 19°C below the mean screen temperature.

In general, for climatological work, the more complex formulae for estimating L_d that require humidity or precipitable water data have little additional merit over the previous simpler expressions since the main uncertainty lies in the influence of cloud—the next topic.

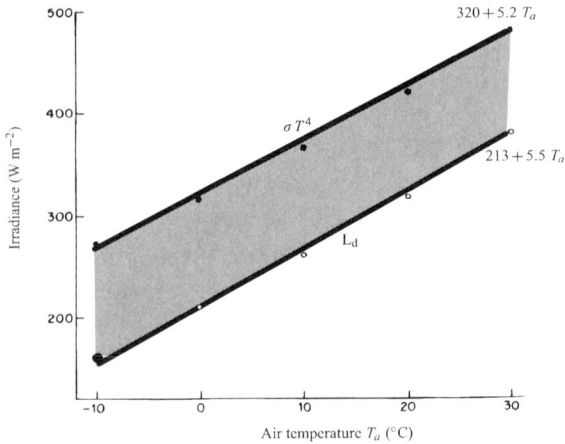

FIGURE 5.14   Black body radiation at T_a (filled circles) and long-wave radiation from clear sky (open circles) from Equation 4.14. Straight lines are approximations from Equations 5.24 and 5.25, respectively.

TERRESTRIAL RADIATION FROM CLOUDY SKIES

Clouds dense enough to cast a shadow on the ground emit like full radiators at the temperature of the water droplets or ice crystals from which they are formed. The presence of cloud increases the flux of atmospheric radiation received at the surface because the radiation from water vapor and carbon dioxide in the lower atmosphere is supplemented by emission from clouds in the waveband which the gaseous emission lacks, particularly from 8 to 13 μm (see Figure 5.12). Because most atmospheric radiation originates below the base of clouds, the gaseous component of the downward flux can be treated as if the sky was cloudless with an apparent emissivity ε_a. From Kirchhoff's Principle, the transmission of radiation through the layer beneath cloud base is $1 - \varepsilon_a$, and if the cloudbase temperature is T_c the downward radiation from a fully overcast sky will be

$$\mathbf{L}_d = \varepsilon_a \sigma T_a^4 + (1 - \varepsilon_a) \sigma T_c^4$$
$$= \sigma T_a^4 \{1 - (1 - \varepsilon_a) 4\delta T / T_a\} \qquad (5.27)$$

using the linear approximation Equation 4.5 with $\delta T = T_a - T_c$.

   The analysis of a series of measurements near Oxford, England [Unsworth and Monteith, 1975] gave an annual mean of $\delta T = 11$ K with a seasonal variation of ± 2 K, figures consistent with a mean cloud base at about 1 km, higher in summer than in winter. Taking 283 K as a mean value of T_a gives $4\delta T / T_a = 0.16$, so that the emissivity of an overcast sky (cloud fraction $c = 1$) at this site is

$$\varepsilon_a(1) = \mathbf{L}_d / \sigma T_a^4 = 1 - 0.16\{1 - \varepsilon_a\} = 0.84 + 0.16\varepsilon_a \qquad (5.28)$$

For a sky covered with a fraction c of cloud, interpolation gives

$$\varepsilon_a(c) = c\varepsilon_a(1) + (1 - c)\varepsilon_a$$
$$= (1 - 0.84c)\varepsilon_a + 0.84c \qquad (5.29)$$

The main limitation to this formula lies in the choice of appropriate values for cloud temperature and for δT, which depend on base height and therefore on cloud type.

It is important to remember that the formulae presented in this section are statistical correlations of radiative fluxes with weather variables at particular sites and do not describe direct functional relationships. For prediction, they are most accurate when the air temperature does not increase or decrease rapidly with height near the surface and when the air is not exceptionally dry or humid. They are therefore appropriate for climatological studies of radiation balance but are often not accurate enough for micrometeorological analyses over periods of a few hours. In particular, the simple equations cannot be used to investigate the diurnal variation of $\mathbf{L}_d$. At most sites, the amplitude of $\mathbf{L}_d$ in cloudless weather is much smaller than the amplitude of $\mathbf{L}_u$, behavior to be expected because changes of atmospheric temperatures are governed by, and follow, changes of surface temperature.

5.5   Net Radiation

All surfaces receive short-wave radiation during daylight and exchange long-wave radiation continuously with the atmosphere. The net amount of radiation received by a surface is called the *net radiation balance*, or *net radiation*. The net radiation received by unit area of a horizontal surface with reflection coefficient ρ is defined by the equation

$$\mathbf{R}_n = (1 - \rho)\,\mathbf{S}_t + \mathbf{L}_d - \mathbf{L}_u \qquad (5.30)$$

*Micro*climatological applications of this equation are considered in Chapter 8. Here, a discussion of the net receipt of radiation by a standard horizontal surface is needed to round off the chapter, although net radiation is

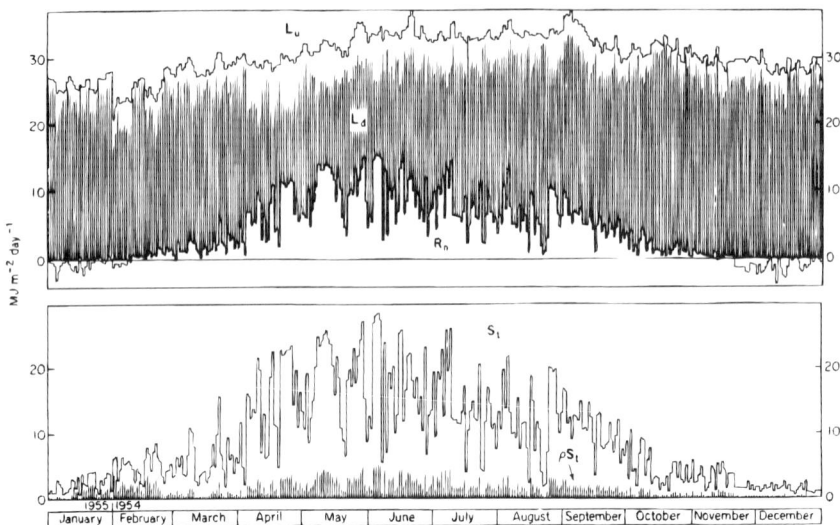

FIGURE 5.15   Annual radiation balance at Hamburg, Germany, 1954/55: S_t, total solar radiation; ρS_t, reflected solar radiation; L_u, upward long wave radiation; L_d, downward long wave radiation; R_n, net radiation (after Fleischer, 1955).

not strictly a *macro*climatological quantity: it depends on the temperature, emissivity, and reflectivity of a surface. Net radiation is measured routinely at very few climatological stations partly because of the problem of providing a standard surface, but also because instruments have been difficult to maintain. A more robust instrument consisting of four independent sensors for the components of R_n has recently become available [Kipp and Zonen, B. V., Delft, The Netherlands] which makes the task of observing and interpreting R_n at climatological sites more feasible. Examples of annual and daily measurements follow.

Figure 5.15 shows the annual change of components in the net radiation balance of a short grass surface at Hamburg, Germany (54°N, 10°E) from February 1954 to January 1955. Each entry in the graph represents the gain or loss of radiation for a period of 24 hours.

The largest term in the balance is L_u, the long-wave emission from the grass surface, ranging between winter and summer from about 23 to about 37 MJ m^{-2} day^{-1} (or a mean emittance of -270 to 430 W m^{-2}). The minimum values of downward long-wave radiation L_d (230 W m^{-2}) were recorded in spring, presumably in cloudless anticyclonic conditions bringing very cold dry air masses; and maximum values (380 W m^{-2}) were recorded during warm humid weather in the autumn. The net loss of long-wave radiation was about 60 W m^{-2} on average (cf. 100 W m^{-2} for cloudless skies on p. 78) and was almost zero on a few, very foggy days in autumn and winter.

In the lower half of the diagram, the income of short-wave radiation forms a Manhattan skyline with much larger day-to-day changes and a much larger seasonal amplitude than the income of long-wave radiation. The maximum value of S_t is about 28 MJ m^{-2} day^{-1} (320 W m^{-2}) and

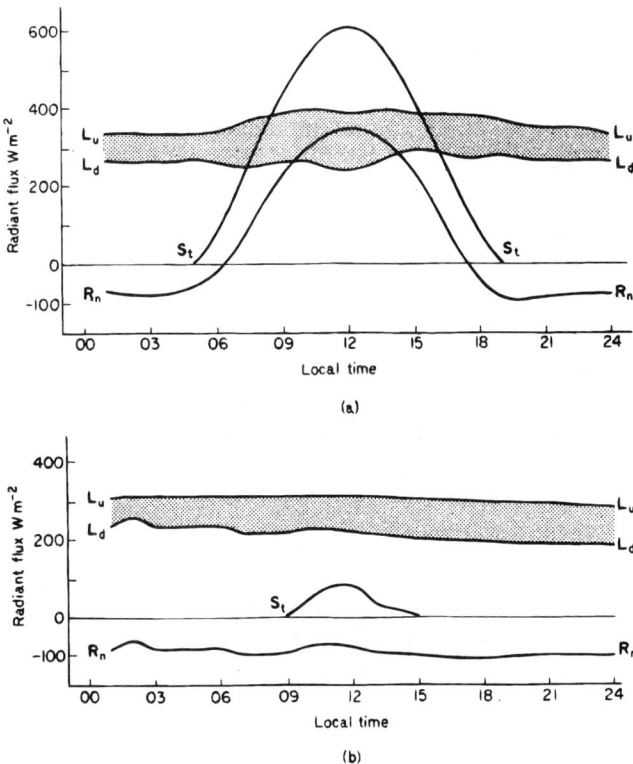

FIGURE 5.16   Radiation balance at Bergen, Norway (60°N, 5°E): (a) on 13 April, 1968, (b) on 11 January, 1968. The grey area shows the net long wave loss and the line R_n is net radiation. Note that net radiation was calculated from measured fluxes of incoming short and long wave radiation, assuming that the reflectivity of the surface was 0.20 in April (e.g., vegetation) and 0.70 in January (e.g., snow). The radiative temperature of the surface was assumed equal to the measured air temperature.

S_t is smaller than L_d on every day of the year. The reflected radiation is about $0.25S_t$ except on a few days in January and February when snow increased the reflection coefficient to between 0.6 and 0.8. The net radiation R_n given by $(1 - \rho)S_t + L_d - L_u$ is shown in the top half of the graph. During summer, the ratio R_n/S was almost constant from day to day at about 0.57, but the ratio decreased during the autumn and reached zero in November. From November until the beginning of February, R_n was negative on most days. In summer, net radiation and mean air temperature were positively correlated with sunshine. In winter, the correlation was negative: sunny cloudless days were days of minimum net radiation when the mean air temperature was below average.

Staff at the University of Bergen maintained records of incoming short- and long-wave radiation for many years and overcame the problem of maintaining a standard surface by using black-body radiation at air (i.e., screen) temperature for L_u and by adopting a reflection coefficient of 0.2 (grass) or 0.7 (snow) as appropriate. (The device of referring net radiation to a surface at air temperature is convenient in *micro*climatology too (see p. 13) and deserves to be more widely adopted.) The shaded part of Figure 5.16 represents the difference between the fluxes L_d and L_u, i.e., the net loss of long-wave radiation; the bold line is R_n. In cloudless weather, the diurnal change in the two long-wave components is much smaller than the change of short-wave radiation, which follows an almost sinusoidal curve. The curve for net radiation is therefore almost parallel to the S_t curve during the day, decreases to a minimum value in the early evening, and then increases very slowly for the rest of the night (because the lower atmosphere is cooled by radiative exchange with the earth's surface). In summer, the period during which R_n is positive is usually about 2 or 3 hours shorter than the period during which S_t is positive.

Comparison of the curves for clear spring and winter days (Figures 5.16a and b) shows that the seasonal change of R_n/S_t noted in Figure 5.15 is a consequence of (i) the shorter period of daylight in winter and (ii) much smaller maximum values of S_t, in winter, unmatched by an equivalent decrease in the net long-wave loss.Figures 5.17 and 5.18 show the net radiation balance during several cloudless days over short grass [Corvallis, Oregon], and over an old growth Douglas fir/Western Hemlock forest (Wind River, Washington State) in the Pacific Northwest of the United States. By day, solar radiation is the dominant influence on the radiation balance. The reflection coefficient for solar radiation is about 0.23 over the grass and 0.07 over the forest, a consequence of the optical properties of the foliage and the canopy structure (see Chapters 6 and 8). The

Hyslop Farm
Corvallis

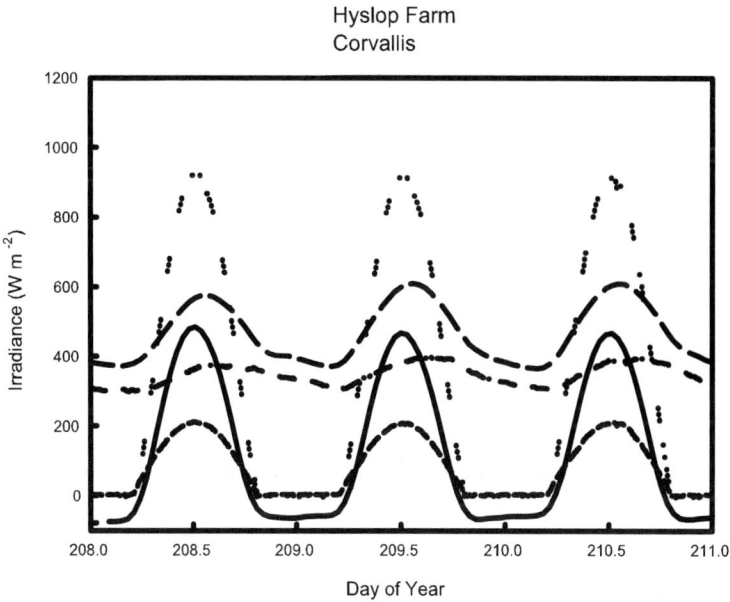

FIGURE 5.17 Components of the net radiation balance during several cloudless days over short grass near Corvallis Oregon (45°N, 123°W) $\cdots$ S_t, — — L_u, — · L_d, — R_n, — $-\rho S_t$ (data courtesy of Reina Nakamura, Oregon State University).

upward longwave radiation varies less over the forest than over the grass because the tree foliage remains close to air temperature during the day, whereas the short grass gets much warmer than the air. At night, the net radiation over both surfaces is most strongly negative shortly after sunset when the canopies are warmer than they are later in the night, and consequently lose more radiation to the cloudless sky. Unusually, net radiation at night is similar in magnitude over both canopies. More typically, heat stored in the soil would have flowed upward to keep the grass canopy warmer than the more isolated forest canopy. It is likely that the dry soil in summer did not conduct heat effectively in this example. Figures 5.17 and 5.18 demonstrate a good correlation between R_n and S_t that could be used to estimate R_n from total solar radiation records (a method that has commonly been used (e.g., Kaminsky and Dubayah, 1997)). However, the principles summarized in this chapter show that such an approach is both site-dependent and surface dependent, so that methods for estimating R_n from its components are preferable [Offerle et al, 2003].

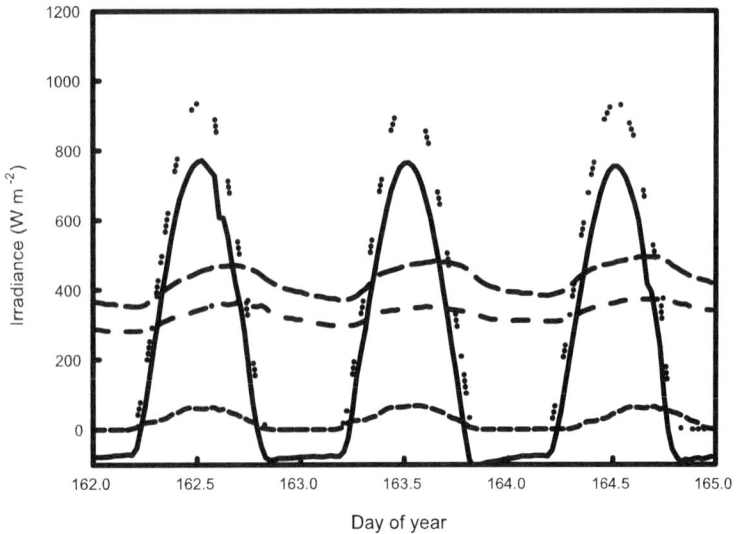

FIGURE 5.18   Components of the net radiation balance during several cloudless days over an old growth Douglas fir/Western hemlock forest (Wind River Experimental Forest, Washington, 46°N, 120°W) $\cdots$ S_t, $— —$ L_u, $— \cdot$ L_d, $—$ R_n, $- -$ ρS_t (data courtesy of Ken Bible, University of Washington).

In the material just presented we have concentrated primarily on radiation fluxes when skies are cloudless because these conditions provide maximum energy at the surface and are most amenable to analysis. In overcast weather, with low cloud, L_d becomes almost equal to σT_a^4, so the net longwave exchange is close to zero and R_n is almost zero at night. During the day under overcast skies $R_n \approx (1 - \rho) S_t$.

5.6   Problems

1. Estimate the time difference between sunset and the beginning of civil twilight and astronomical twilight in Greenwich (51.5°N, 0.0°E) on 21 June (solar declination 23.5°).

2. Use the simple model for solar radiation under cloudless skies (p. 66) to estimate the direct, diffuse, and total solar irradiance on a horizontal surface at 45°N at local solar noon on 21 June for three values of aerosol optical thickness, 0.05, 0.20, 0.40. Assume that the optical thickness for molecular attenuation is 0.30.

3. Using the results of question 2, estimate the daily insolation at 45°N at solar noon on 21 June for the three aerosol loadings assuming cloudless sky.

4. Assuming a cloudless sky, estimate the downward longwave irradiance at the ground and the atmospheric emissivity when the air temperature is 25°C. If the sky was covered by 50% cloud, estimate the longwave irradiance in this case. Why are these values only climatological approximations?

5. Four solar radiometers are exposed side by side. A and B have clear glass domes transmitting all wavelengths in the solar spectrum; C and D have domes that transmit 95% of the radiant energy from 700 to 3000 nm and no radiation below 700 nm. A and C receive global (total) radiation, whereas B and D have a shade ring that intercepts 10% of radiation from the sky and all the direct solar beam. On a cloudless summer day the instruments give the following outputs: A 11.00 mV; B 1.30 mV; C 5.30 mV; and D 0.25 mV. Assuming that all instruments have the same sensitivity, 12.0 μV per W m^{-2}, calculate (i) the ratio of diffuse to global radiation, (ii) the fraction of photosynthetically active radiation (0.4–0.7 μm) in the diffuse component, (iii) the fraction of visible radiation in the direct solar beam, and (iv) the global irradiance in the photosynthetically active waveband.

6. A farmer plans to plant seeds in a field of soil with a solar radiation reflection coefficient $\rho_s = 0.20$. She decides that the soil will be warmer (and germination more rapid) if she spreads a thin layer of black soot over it, decreasing the solar radiation reflection coefficient to $\rho'_s = 0.05$. To assess the effect of the change, she prepares adjacent plots and records the following data around noon on a cloudless day:

Incident total solar irradiance $S_t = 900$ W m^{-2} (same for both plots)

Difference in net radiation R_n(soot-covered) − R_n(bare soil) = 69 W m^{-2}

Radiative temperature of bare soil $T_s = 303$ K

Estimate the radiative temperature of the soot-covered surface, stating any assumptions you need to make.

Chapter 6

Microclimatology of Radiation (i) Absorption, Reflection, and Transmission

6.1 Radiative Properties of Natural Materials

When sunlight is intercepted by soil, by water, or by an object such as a leaf or an animal, energy is absorbed, reflected, and sometimes transmitted. This chapter describes how radiant energy is partitioned and how this process depends on wavelength. Subsequent chapters apply this information to review the interception and absorption of radiation by solid objects, vegetation canopies, and animal coats, and to discuss net radiation in terms of the radiative and geometrical characteristics of organisms.

At the short-wavelength, high-frequency end of the solar spectrum, the radiative behavior of biological materials is determined mainly by the presence of pigments absorbing radiation at wavelengths associated with specific electron transitions. For radiation between 1 and 3 μm, liquid water is an important constituent of many natural materials because water has strong absorption bands in this region; and even in the visible spectrum where absorption by water is negligible, the reflection and transmission of light by porous materials is often strongly correlated with their water content. In the long-wave spectrum beyond 3 μm, most natural surfaces behave like full radiators with absorptivities close to 100% and reflectivities close to zero.

It is important to distinguish between the *reflectivity* of a surface ρ (λ), which is the fraction of incident solar radiation reflected at a specific wavelength, and the *reflection coefficient*, which, in this context, is the average reflectivity over a specific waveband, weighted by the distribution of radiation in the solar spectrum. If this distribution is described by a function $S(\lambda)$, which is the energy per unit wavelength measured at λ, the energy in the solar spectrum between λ_1 and λ_2 is $\int_{\lambda_1}^{\lambda_2} S(\lambda)d\lambda$. The reflection coefficient of a surface exposed to solar radiation is therefore

$$\bar{\rho} = \frac{\int_{\lambda_1}^{\lambda_2} \rho\,(\lambda)\,S(\lambda)d\lambda}{\int_{\lambda_1}^{\lambda_2} S(\lambda)d\lambda} \tag{6.1}$$

For the whole solar spectrum, the integrations may be performed from 0.3 to 3 μm. The transmissivity $\tau(\lambda)$ and the transmission coefficient of a surface can be defined in the same way.

For the whole solar spectrum, the reflection coefficient of a natural surface is often called the *albedo*, a term borrowed from astronomy and derived from the Latin for "whiteness." (The alternative derivation offered by Monteith [1975], attributing the origin of the term to studies of the reflectivity of stars by an astronomer Al Bedo, is fanciful.) Because whiteness is associated with the visible spectrum, the more general term "reflection coefficient" is preferable to albedo in this context.

Gates [1980] provides a very comprehensive account of the radiative properties of plant and animal surfaces along with tables of reflection coefficient for many species. Representative values for a range of surfaces are in Table 6.1.

WATER
Reflection

When radiation is incident on clear, still water at an angle of incidence ψ less than 45°, the reflection coefficient for solar radiation is almost constant at about 5%. Beyond 45°, the coefficient increases rapidly with increasing ψ, approaching 100% at grazing incidence (Figure 6.1). The reflectivity of longwave radiation also increases as ψ increases.

Transmission

In the visible spectrum, water is usually regarded as transparent, but it has a finite absorption coefficient with a minimum in the blue-green (460 and 490 nm), which gives natural bodies of very clean water their characteristic

TABLE 6.1   Radiative properties of plant and animal surfaces .

(i) Short-wave reflection coefficients ρ (%)			
1 Leaves	Upper	Lower	Average
Maize			29
Tobacco			29
Cucumber			31
Tomato			28
Birch	30	33	32
Aspen	32	36	34
Oak	28	33	30
Elm	24	31	28

2   Vegetation at maximum ground cover

(a) Farm crops	Latitude of site	Daily mean
Grass	52	24
Sugar beet	52	26
Barley	52	23
Wheat	52	26
Beans	52	24
Maize	43	22
Tobacco	43	24
Cucumber	43	26
Tomato	43	23
Wheat	43	22
Pasture	32	25
Barley	32	26
Pineapple	22	15
Maize	7	18
Tobacco	7	19
Sorghum	7	20
Sugar cane	7	15
Cotton	7	21
Groundnuts	7	17

(b) Natural vegetation and forest		
Heather	51	14
Bracken	51	24
Gorse	51	18
Maquis, evergreen scrub	32	21
Natural pasture	32	25
Derived savanna	7	15
Guinea savanna	9	19

(c) Forests and orchards		
Deciduous woodland	51	18
Coniferous woodland	51	16
Orange orchard	32	16
Aleppo pine	32	17
Eucalyptus	32	19
Tropical rain forest	7	13
Swamp forest	7	12

3 Animal coats

(a) Mammals	Dorsal	Ventral	Average
Red squirrel	27	22	25
Gray squirrel	22	39	31
Field mouse	11	17	14
Shrew	19	26	23
Mole	19	19	19
Gray fox			34
Zulu cattle			51
Red Sussex cattle			17
Aberdeen Angus cattle			11
Sheep weathered fleece			26
newly shorn fleece			42
Man			
Eurasian			35
Negroid			18

(b) Birds	Wing	Breast	Average
Cardinal	23	40	
Bluebird	27	34	
Tree swallow	24	57	
Magpie	19	46	
Canada goose	15	35	
Mallard duck	24	36	
Mourning dove	30	39	
Starling			34
Glaucous-winged gull			52

(ii) Long-wave Emissivities ε (%)

1 Leaves	Average
Maize	94.4 ± 0.4
Tobacco	97.2 ± 0.6
French bean	93.8 ± 0.8
Cotton	96.4 ± 0.7
Sugar cane	99.5 ± 0.4
Poplar	97.7 ± 0.4
Geranium	99.2 ± 0.2
Cactus	97.7 ± 0.2

2 Animals	Dorsal	Ventral	Average
Red squirrel	95–98	97–100	
Gray squirrel	99	99	
Mole	97	—	
Deer mouse	—	94	
Gray wolf			99
Caribou			100
Snowshoe hare			99
Man			98

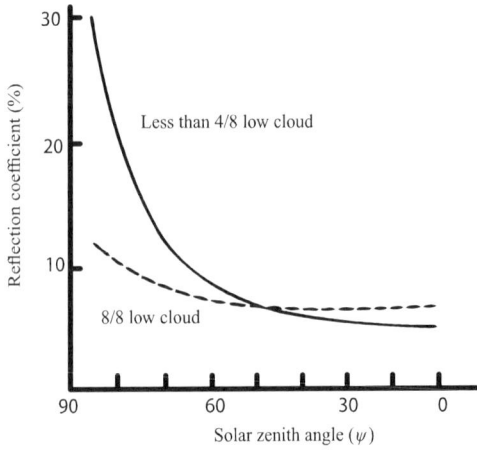

FIGURE 6.1   Reflectivity of a plane water surface as a function of solar zenith angle and cloudiness [from Deacon, 1969].

color, particularly when observed over white sand. Outside this waveband, absorption increases in both directions, but particularly towards the red end of the spectrum. Whereas 300 m of pure water are needed to reduce the transmission of blue-green radiation to 1%, the corresponding figure for red light is only 20 m. In most natural water, however, the presence of organic matter including chlorophyll depletes the blue end of the spectrum and, in some British lakes, both blue and red wavelengths are attenuated to 1% at depths less than 10 m [Spence, 1976]. Crater Lake, a National Park in Oregon, and the deepest lake in North America (600 m), is renowned for its deep aquamarine color. This is partly a consequence of the selective absorption in such deep water of wavelengths longer than that of blue light, but also because the lake is one of the clearest in the world, with very little organic material, so that ultraviolet radiation penetrates to 100 m [Hargreaves, 2003].

In the near infrared region of the spectrum, water has several absorption bands, readily identified in the transmission and reflection spectra of soil, leaves, and animal skins. The centres of the main bands are at 1.45 and 1.95 μm (Figure 6.2). Beyond 3 μm, the absorptivity and emissivity of water, measured at normal incidence, are about 0.995 but, with increasing angle of incidence, they decrease in a complementary way to the increase of shortwave reflectivity, and at 11 μm the emissivity is only about 0.7 when ψ is 80°. The angular dependence of the emissivity of water

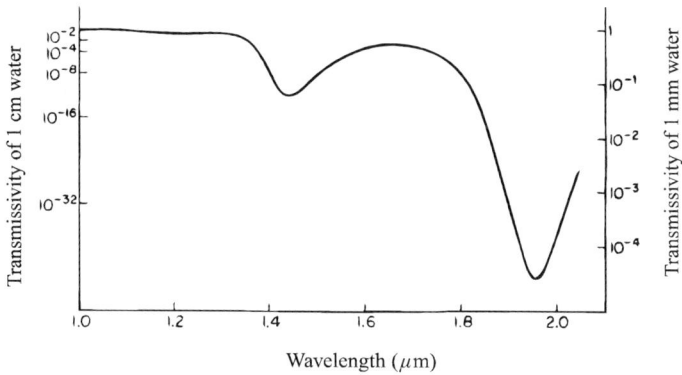

FIGURE 6.2   Transmissivity of pure water as a function of wavelength. Note logarithmic scales for 1 cm water (left-hand axis) and 1 mm (right-hand axis).

and other natural surfaces complicates the interpretation of radiometric measurements of skin temperature in human pathology [Clark, 1976] and of ground and sea surface temperature observed from aircraft or satellites [Becker, 1981, Becker et al, 1981].

SOIL

The reflectivity of soils depends mainly on their organic matter content, on water content, particle size, and angle of incidence. Reflectivity is usually very small at the blue end of the spectrum, increases with wavelength through the visible and near infrared wavebands, and reaches a maximum between 1 and 2 μm. When water is present, its absorption bands are evident at 1.45 and 1.95 μm.

Integrated over the whole solar spectrum, reflection coefficients range from about 10% for soils with a high organic matter content to about 30% for desert sand. Even a very small amount of organic matter can reduce the reflectivity of a soil. Oxidizing the organic component of a loam, which was 0.8% by weight, increased its reflectivity by a factor of two over the whole visible spectrum [Bowers and Hanks, 1965].

The reflectivity of clay minerals has been measured as a function of their particle size. Over the spectral range 0.4 to 2 μm, the reflectivity of kaolinite increases rapidly with decreasing particle diameter; e.g., from 56% for 1600 μm particles to 78% for 22 μm particles. Aggregates containing relatively large irregular particles appear to trap radiation by multiple reflection between adjacent faces, whereas finely divided powders expose a more uniform surface, trapping less radiation. Particle size also governs the *transmission* of radiation in soils. Baumgartner (1953) measured the transmission of artificial light by quartz sand. When the particle diameter was 0.2 to 0.5 mm, a depth of 1 to 2 mm of sand was enough to reduce the radiative flux by 95%, but for particles of 4 to 6 mm, a layer 10 mm deep was needed to give the same extinction. In another study, with fine seed compost, irradiance decreased by more than four orders of magnitude in a depth of 3 mm but the corresponding change of the red/far-red ratio was small—from 1.2 to 0.8 [Frankland, 1981]. The transmission of radiation by soils has had little attention from ecologists although the effects of light quality and quantity on seed germination and root development are well established.

The reflectivity of a soil sample decreases as it gets wetter, mainly because radiation is trapped by internal reflection at air-water interfaces formed by the menisci in soil pores. The dependence of reflectivity on water content is evident at all wavelengths but is strongest in the absorption band at 1.95 μm. Figure 6.3 shows that the reflectivity of a loam at 1.9 μm decreased from 60% at 1% water content to 14% at 20% water. The reflection coefficient of a stable soil can therefore be used to mon-

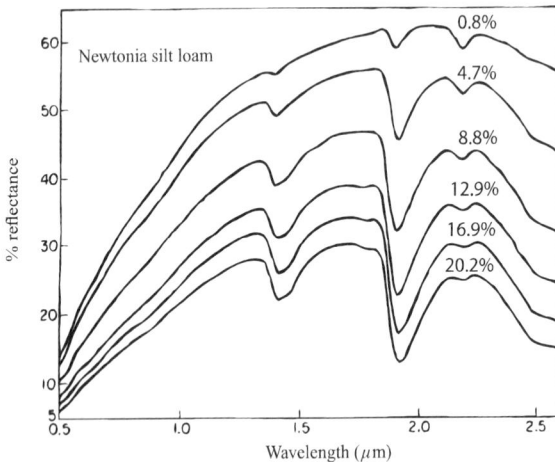

FIGURE 6.3   Reflectivity of a loam soil as a function of wavelength and water content [from Bowers and Hanks, 1965].

itor the water content or water potential of the surface layer [Idso et al, 1975; Graser and van Bavel, 1982]. Soil moisture can also be monitored by detecting emission of radiation at microwave wavelengths of about 3–6 cm but emissions at these wavelengths are sensitive to moisture only in the top few centimeters of soil [Schmugge, 1998]. Satellite observations of soil moisture using passive microwave sensing are used primarily for regional-scale studies because of their limited spatial resolution (typically about 75 km).

In the long-wave spectrum, most soils have an emissivity between 0.90 and 0.95 and the emissivities of common minerals range from 0.67 for quartz to 0.94 for marble.

LEAVES

The fractions of incident radiation transmitted and reflected by a leaf depend on the angle of incidence ψ. Tageeva and Brandt [1961] found that the reflection coefficient was almost constant for values of ψ between 0 and 50°, but as ψ increased from 50 to 90° (grazing incidence), $\rho(\lambda)$ increased sharply as a result of specular reflection. The transmission coefficient was also constant between 0 and 50° but decreased between 50 and 90°. Because changes of ρ (λ) and τ (λ) with angle were complementary, the fraction of radiation absorbed (and available for physiological processes) was almost constant for angles of incidence less than 80°.

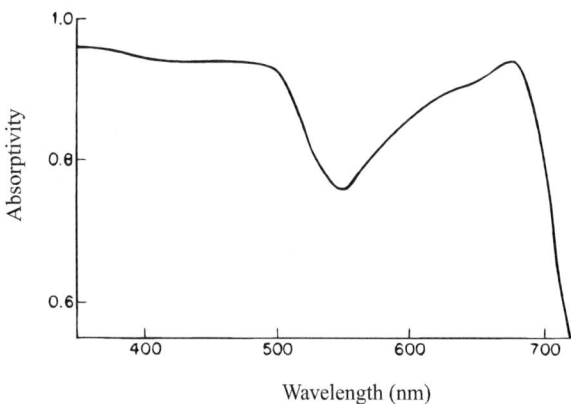

FIGURE 6.4   Average absorptivity for leaves of eight field-grown crop species [from McCree, 1972].

For the leaves of many common crop species, the absorptivity of green light at 550 nm is about 0.75 to 0.80: for blue light (400 to 460 nm) it is 0.95 and for red light (600 to 670 nm) about 0.85 to 0.95 (Figure 6.4). The very sharp decrease of absorption as wavelength increases beyond 700 nm is physiologically significant because it implies that within leaf tissue the energy per unit wavelength in an infrared waveband centred at 730 nm is much larger than the corresponding energy in the red at 660 nm. This ratio plays a major role in determining the state of the pigment phytochrome which governs many developmental processes [Smith and Morgan, 1981]. The existence of the complementary "red edge" in the reflection spectrum has been exploited in remote sensing, as discussed later.

The similarity of the transmission and reflection spectra displayed by many leaves (Figure 6.5) implies that the radiation within them is scattered in all directions by reflection and refraction at the walls of cells. The only incident radiation not scattered in this way is the component (about 10%) reflected from the surface of the cuticle without entering the mesophyll.

To derive approximate values of ρ and τ for the whole solar spectrum, the coefficients may be assumed equal at 0.1 between 0.4 and 0.7 μm and at 0.4 between 0.7 and 3 μm. Because each of these spectral bands contains about half the total radiation, the approximate coefficients for the whole spectrum are

$$\rho = \tau = (0.1 \times 0.5) + (0.4 \times 0.5) = 0.25$$

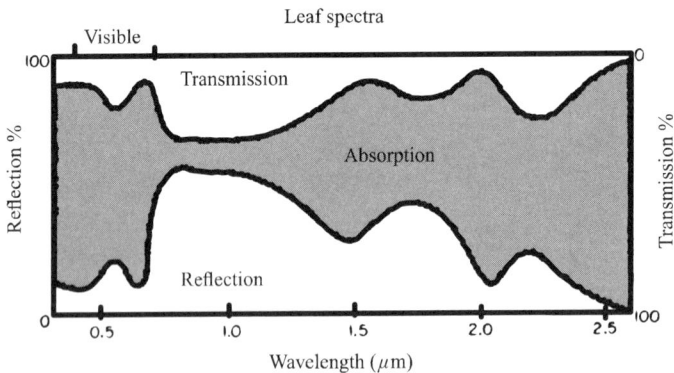

FIGURE 6.5   Idealized relation between the reflectivity, transmissivity, and absorptivity of a green leaf.

This value is consistent with measurements on leaves of a number of species recorded in Table 6.1. Because the visible spectrum makes a relatively small contribution to the reflection and transmission of solar radiation by leaves, comparisons of color are largely irrelevant in terms of the (total) reflection coefficient. However, Ehleringer and Bjorkman [1978] reported that particularly pubescent (hairy) leaves of the desert shrub *Encelia farinosa* had a reflectivity throughout the PAR region about 50% larger than closely related *Encelia* species from less arid environments, perhaps as an adaptation to large radiation loads. White pubescent leaves have reflection coefficients for total solar radiation of 0.35 to 0.40 [Jones 1992].

CANOPIES OF VEGETATION

The fractions of incident solar radiation reflected and transmitted by a dense stand of vegetation depend on two main factors: the fraction intercepted by foliage, and the scattering properties of the foliage. (In this

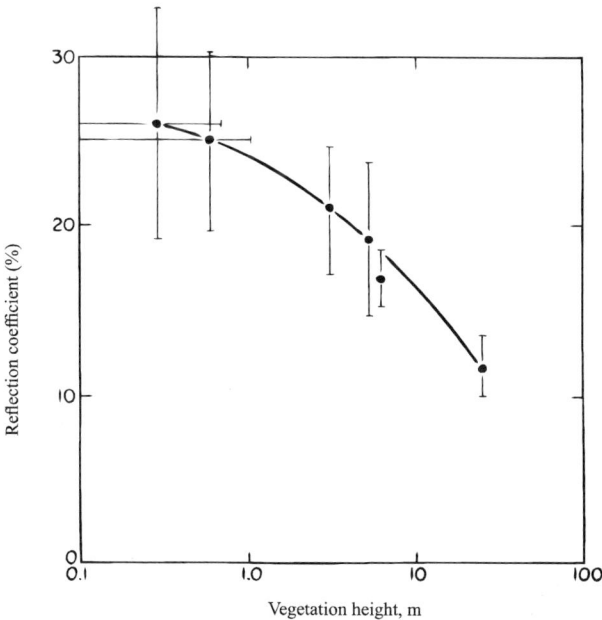

FIGURE 6.6   Relation between height of vegetation and reflection coefficient [from Stanhill, 1970].

context, "foliage" includes stems, petioles, etc., but their role in interception is often neglected.) The intercepted fraction depends, in turn, on the
area of foliage specified as a *Leaf Area Index* (plan area of leaves per unit
ground area) and on the spatial distribution of foliage with respect to the
direction of radiation. The scattered fraction depends on the optical properties of leaf cuticles, cell walls and pigments. When the foliage is not
dense enough to intercept all the incident radiation, the reflection coefficient of the stand depends to some extent on reflection from the soil and
from foliage. Detailed analysis of radiation transfer in canopies is considered in Chapter 7.

   In general, maximum values of ρ (see Table 6.1) are recorded over
relatively smooth surfaces such as closely cut lawns. For crops growing to
heights of 50 to 100 cm, ρ is usually between 0.18 and 0.25 when ground
cover is complete, but values around 0.10 have been recorded for forests
(Figure 6.6). These differences can be interpreted in terms of the trapping
of radiation by multiple reflection between adjacent leaves and stems. For
the same reason, the reflection coefficient for most types of vegetation
changes with the angle of the sun. Minimum values of ρ are recorded as
the sun approaches its zenith, and ρ increases as the sun descends to the
horizon because there is less opportunity for multiple scattering between
the elements of the canopy. The dependence of ρ on zenith angle may

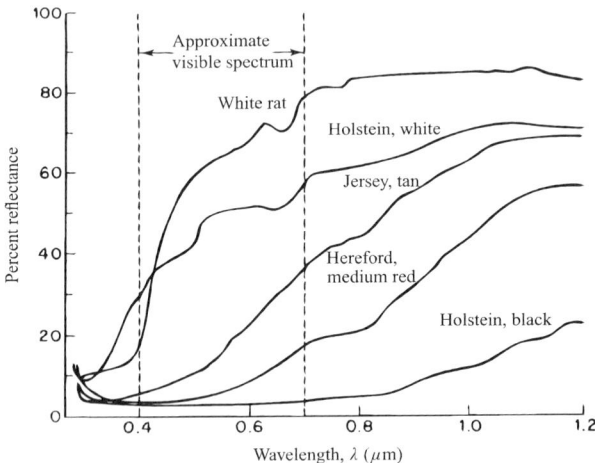

FIGURE 6.7   Reflectivity of animal coats [from Mount, 1968].

explain why reflection coefficients for vegetation measured in the tropics are usually somewhat smaller than the coefficients for similar surfaces at higher latitudes. Table 6.1 lists values of ρ for canopies of various crops and natural species.

ANIMALS

The coats and skins of animals reflect solar radiation both in the visible and in the infrared regions of the solar spectrum, and reflectivity is a function of the angle of incidence of the radiation, as for water and leaves. Figure 6.7 shows that the reflectivity of hair coats increases throughout the visible spectrum (like soil), and that intra- and inter-specific differences of reflectivity are much larger than for leaves. The maximum reflectivity of several types of coat is found between 1 and 2 μm and water is responsible for absorption at 1.45 and 1.95 μm. Beyond 3 μm, the absorptivity and emissivity of most types of coat is between 90 and 95%.

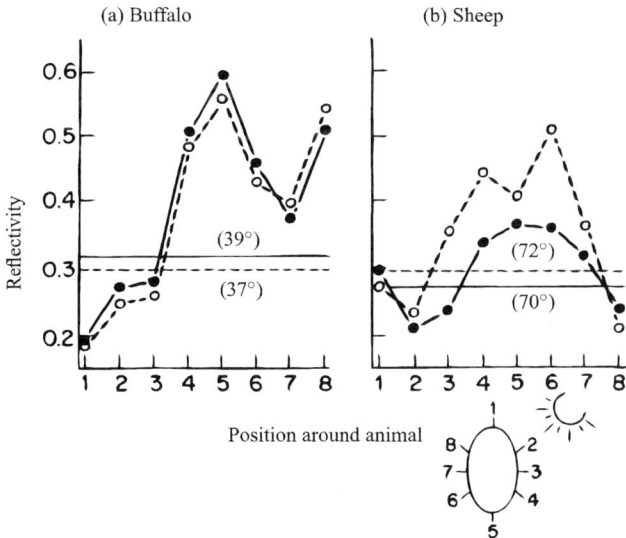

FIGURE 6.8   Reflectivity in individual positions and weighted mean reflectance (horizontal lines) (a) for swamp water buffalo calves and (b) for Merino sheep with fleeces 6 cm deep. The animals stood sideways to the sun. Numbers in brackets represent solar altitudes [from Hutchinson et al., 1975].

The measurements of coat reflectivity plotted in Figure 6.7 were made in the laboratory on samples of coats exposed to radiation at normal incidence. When Hutchinson et al [1975] used a miniature solarimeter to measure reflection from the coats of live animals standing in the field, they found a marked dependence of reflection coefficient on the local angle of incidence of direct sunlight. For water buffalo with relatively smooth coats exposed to sun at an elevation of 48° (Figure 6.8a), a minimum reflection coefficient of 0.25 was recorded when the angle of incidence was minimal but ρ was more than 0.7 at grazing incidence because of the large contribution from specular reflection. The change of reflection with angle was much smaller for Merino sheep with deep, woolly coats (Figure 6.8b). The implication of this work is that laboratory measurements of reflection, usually made at normal incidence, may be an unreliable guide to the effective reflection coefficient for an animal in the field, particularly for species with relatively smooth coats. Cena and Monteith [1975] found that forward scattering was large in white animal coats, so that, depending on coat density, the color of the skin below the coat could be important in determining overall reflection and absorption coefficients. Their analysis is discussed later (p. 133). Table 6.1 summarizes some measurements of the reflection coefficients of animal and bird coats.

Most measurements of skin reflectivity reported in the literature refer to an almost hairless animal—man. In the absence of hair, solar radiation penetrates skin to a depth that depends on pigmentation. In humans, the

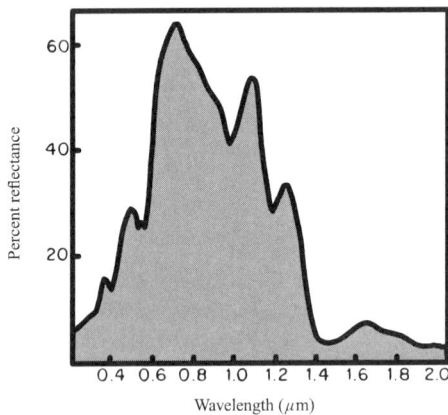

FIGURE 6.9   Spectral reflectivity of skin on an author's thumb from a recording by Dr Warren Porter in his laboratory at the University of Wisconsin on 18 April, 1969.

penetration ranges from several millimetres in Caucasian subjects with light skin to a few tenths of a millimeter in negroid subjects with a much greater concentration of melanin in the corneum. Corresponding reflection coefficients range from 20% for dark skins to 40% for light. Figure 6.9 shows that the reflectivity of white skin is maximal at about 0.7 μm and decreases to a few percent at 2 μm.

6.2 Problems

1. In clear water, 1% of incident red radiation and and 75% of blue-green radiation is transmitted to 20 m. Assuming that both wavebands are attenuated exponentially (i.e., transmitted fraction = $\exp(-kd)$, where k is an attenuation constant and d is depth), what would be the ratio of energy in the two wavebands at 100 m?

2. A leaf of a desert plant has a reflection coefficient of 0.2 in the PAR waveband and 0.4 for all wavelengths of solar radiation. What is its reflection coefficient for the waveband from 0.7 to 3μm? Speculate on the benefits these unusual reflection coefficients could have.

Chapter 7

Microclimatology of Radiation (ii) Radiation Interception by Solid Structures

In conventional problems of micrometeorology, fluxes of radiation at the earth's surface are measured and specified by the receipt or loss of energy per unit area of a horizontal plane. In this chapter, methods are described for estimating the amount of radiation intercepted by the surface of a plant or animal, for example, by multiplying the horizontal irradiance by a *shape factor* that depends on (i) the geometry of the surface and (ii) the directional properties of the radiation. To make the analysis more manageable, objects such as spheres or cylinders with a relatively simple geometry are often used to represent the more irregular shapes of plants and animals. Radiation interception by buildings, other structures, and sloping surfaces also can be analyzed using simple geometric models. Appropriate shape factors for various solid structures and slopes are derived in this chapter. In the next chapter, the principles are applied to estimate radiation interception by plants and animals.

The radiation intercepted by an organism or its analogue can be expressed as the mean flux per unit area of surface. A bar will be used to distinguish this measure of irradiance from the conventional flux on a horizontal surface; e.g., when solar irradiance is S (W per m^{-2} horizontal area), the corresponding irradiance of a sheep or a cylinder will be written $\bar{S}$ (W per m^2 total area).

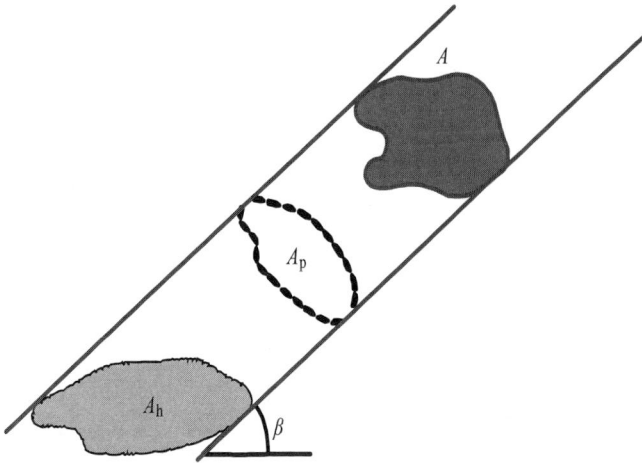

FIGURE 7.1 Area A projected on a surface at right angles to the solar beam (A_p) and on a horizontal surface (A_h).

7.1 Geometric Principles

DIRECT SOLAR RADIATION

The flux of direct solar radiation is usually measured either on a horizontal plane (S_b) or on a plane perpendicular to the sun's rays (S_p). For any solid object exposed to direct radiation at elevation β, a simple relation between the mean flux $\bar{S}_b$ and the horizontal flux S_b can be derived from the relation between the area of shadow A_h cast on a horizontal surface and the area of shadow A_p projected in the direction of the beam.

The area projected in the direction of the sun's rays is $A_p = A_h \sin \beta$ (Figure 7. 1) and the intercepted flux is

$$A_p S_p = (A_h \sin \beta) \, S_p = A_h S_b \qquad (7.1)$$

i.e. the area of shadow on a horizontal plane multiplied by the irradiance of that plane. Then if the surface area of the object is A

$$\bar{S}_b = (A_h/A) \, S_b \qquad (7.2)$$

The ratio (A_h/A), the *shape factor*, can be calculated from geometrical principles or measured directly from the area of a shadow or from a shadow photograph.

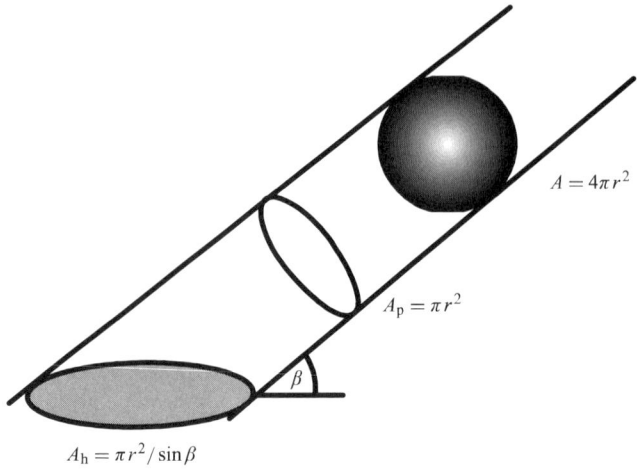

FIGURE 7.2   Geometry of a sphere projected on a horizontal surface.

Shape factors

Sphere

The shadow cast by a sphere of radius r has an area of $\pi r^2 / \sin \beta$ (Figure 7.2). The surface area of the sphere is $4\pi r^2$ so

$$\frac{A_h}{A} = \frac{\pi r^2}{4\pi r^2 \sin \beta} = 0.25 \operatorname{cosec} \beta \qquad (7.3)$$

The mean irradiance of a sphere is therefore

$$\bar{S}_b = (0.25 \operatorname{cosec} \beta)\, S_b = 0.25 S_p \qquad (7.4)$$

Ellipsoid

A sphere is a special type of ellipsoid—a solid whose cross-section is elliptical in one direction and circular about an axis of rotation at right-angles to that direction. If the elliptical cross-section through the center of the ellipsoid has a vertical semi-axis a and a horizontal semi-axis b (Figure 7.3), it can be shown that a beam at elevation β will cast a horizontal shadow with area

$$A_h = \pi b^2 \left\{ 1 + a^2 / \left(b^2 \tan^2 \beta \right) \right\}^{0.5} \qquad (7.5)$$

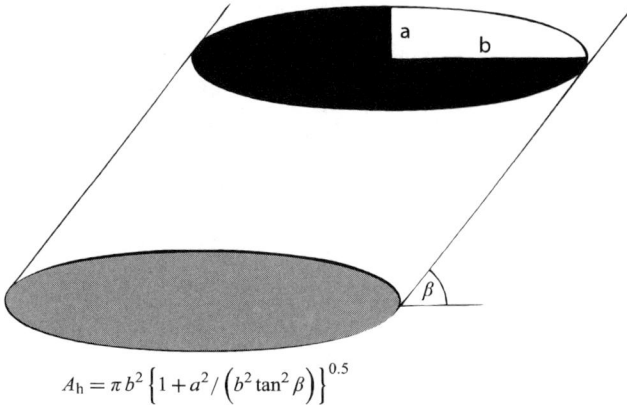

$$A_h = \pi b^2 \left\{ 1 + a^2 / \left(b^2 \tan^2 \beta \right) \right\}^{0.5}$$

FIGURE 7.3   Geometry of an ellipsoid with semi-axes a and b projected on a horizontal surface.

which reduces to $A_h = \pi b^2$ for a vertical beam and to $A_h = \pi b^2 / \sin \beta$ when $a = b$.

The surface area of an *oblate* spheroid $(b > a)$ is

$$A = 2\pi b^2 \left\{ 1 + \left(a^2 / (2b^2 \varepsilon_1) \right) \ln[(1 + \varepsilon_1)/(1 - \varepsilon_1)] \right\} \qquad (7.6)$$

where

$$\varepsilon_1 = [1 - (a^2 / b^2)]^{0.5} \qquad (7.7)$$

and the area of a *prolate* spheroid $(a > b)$ is

$$A = 2\pi b^2 \left\{ 1 + (a/(b\varepsilon_2)) \sin^{-1} \varepsilon_2 \right\} \qquad (7.8)$$

where

$$\varepsilon_2 = \left(1 - b^2 / a^2 \right)^{0.5} \qquad (7.9)$$

The ratio A_h / A can now be found by dividing Equation 7.5 by Equation 7.6 or 7.8 as appropriate (Figure 7.3).

Vertical cylinder

Figure 7.4 shows a vertical cylinder of height h and radius r suspended above the ground. The shadow area has two components: $h \cot \beta \times 2r$ from the curved surface and πr^2 from the top and bottom. The total surface area

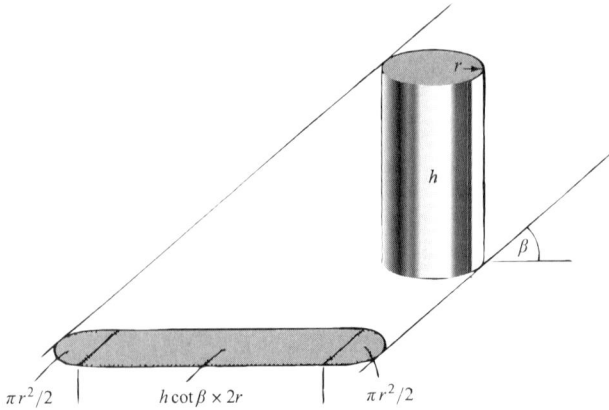

FIGURE 7.4   Geometry of a vertical cylinder projected on a horizontal surface.

of the cylinder is $2\pi rh + 2\pi r^2$ so

$$\frac{A_h}{A} = \frac{2rh\cot\beta + \pi r^2}{2\pi rh + 2\pi r^2} = \frac{(2x\cot\beta)/\pi + 1}{2x + 2} \qquad (7.10)$$

where $x = h/r$.

Underwood and Ward [1966] measured 25 male and 25 female subjects wearing slips or swim suits and photographed them from 19 different angles. (Silhouettes from 8 angles are shown in Figure 7.5.) The average projected area of all 50 subjects was found from the area of each silhouette with proper correction for parallax. When the areas presented at three azimuth angles ($0°$, $45°$, $90°$) were averaged, the mean projected body area was very close to the projected area of a cylinder with

$$h = 1.65 \text{ m}, r = 0.117 \text{ m}, x = 14.1$$

Changes of the projected area with azimuth were taken into account by fitting the measurements to the equation of a cylinder with an elliptical rather than a circular cross-section.

Values of A_h/A for vertical cylinders are plotted in Figure 7.6 as a function of β and x. At a solar elevation of $\beta = 32.5°$, $\cot\beta/\pi = 0.5$ and $A_h/A = 0.5$, independent of x. When $\beta < 60$ and $x > 10$, A_h/A is almost independent of x, so Equation 7.10 with a standard value of $x = 14$ will give a good estimate of the radiation intercepted by a wide range of human shapes from the ectomorph (slim, linear) to the endomorph (corpulent, pear-shaped) (e.g., Laurel and Hardy).

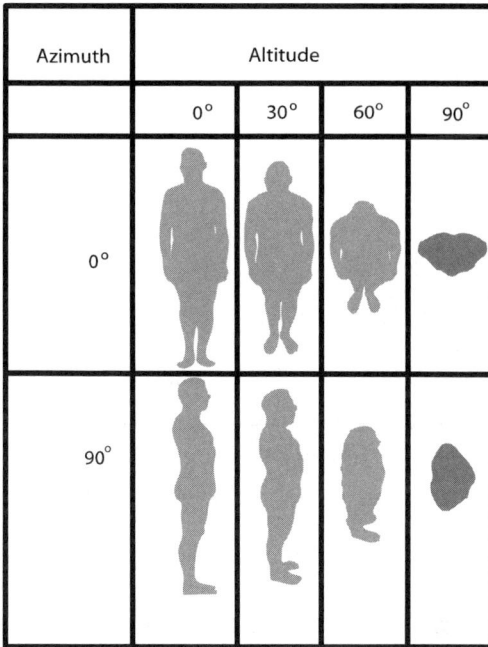

Azimuth	Altitude			
	0°	30°	60°	90°
0°				
90°				

FIGURE 7.5   The area of a male figure standing erect projected in the direction of the solar beam for different values of solar azimuth and altitude. The silhouettes were obtained photographically by Underwood and Ward [1966].

Horizontal cylinder

For a horizontal cylinder with dimensions (h, r), A_h depends on solar azimuth elevation. Defining the azimuth angle θ as the angle between the axis of the cylinder ($\theta = 0$) and the solar azimuth, it can be shown that the length projected in the direction β, θ is $h(1 - \cos^2 \beta \cos^2 \theta)$ and the projected width is $2r$, independent of β and θ. Thus for the *curved surface* only

$$A_h = A_p \mathrm{cosec}\, \beta = 2rh\, \mathrm{cosec}\, \beta \left(1 - \cos^2 \beta \cos^2 \theta\right)^{0.5} \qquad (7.11)$$

The illuminated end of the cylinder can be treated as a vertical plane (see section on inclined planes p. 109) and if α is taken as $\pi/2$ in Equation 7.16

$$A_h = \pi r^2 \cot \beta \cos \theta \ \text{(end only)} \qquad (7.12)$$

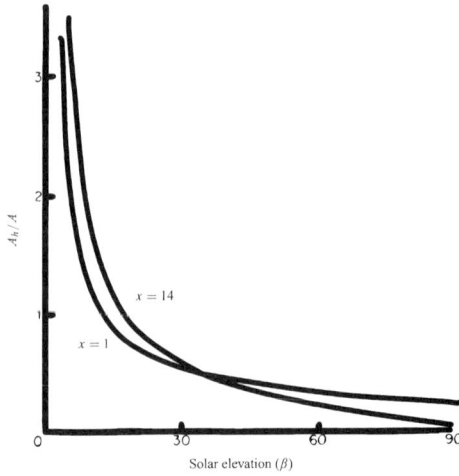

FIGURE 7.6   The ratio A_h/A for vertical cylinders with height/radius ratios (x) of 1 and 14.

The total area of the cylinder including the unlit end is $A = 2\pi r h + 2\pi r^2$, giving for the whole cylinder

$$
\begin{aligned}
\frac{A_h}{A} &= \frac{2rh \cosec \beta \left(1 - \cos^2 \beta \cos^2 \theta\right)^{0.5} + \pi r^2 \cot \beta \cos \theta}{2\pi r h + 2\pi r^2} \\
&= \frac{\cosec \beta \left\{2\pi^{-1} x \left(1 - \cos^2 \beta \cos^2 \theta\right)^{0.5} + \cos \beta \cos \theta\right\}}{2(x+1)}
\end{aligned}
\tag{7.13}
$$

where $x = h/r$.

For the special case of a cylinder at right angles to the sun's rays, $\theta = \pi/2$, and A_h/A reduces to

$$
\frac{A_h}{A} = \frac{x \cosec \beta}{\pi (x+1)} \quad \text{and} \quad \frac{A_p}{A} = \frac{x}{\pi (x+1)}
\tag{7.14}
$$

The projected area of sheep was determined photogrammetrically by Clapperton, Joyce, and Blaxter [1965] and was compared with the area of equivalent horizontal cylinders whose dimensions were determined from photographs at $\theta = 0$ (end view) and $\beta = \pi/2$ (plan view). One set of values quoted for fleeced sheep was

$$
h = 0.91 \text{ m}, r = 0.23 \text{ m}, x = 4.1
$$

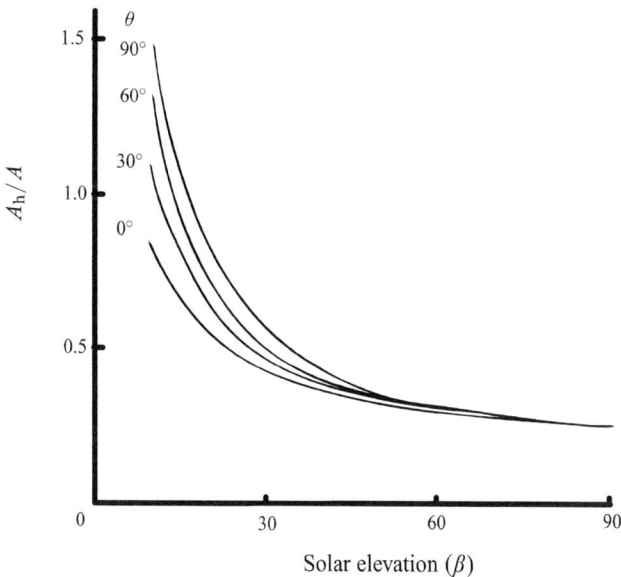

FIGURE 7.7 The ratios A_h/A for horizontal cylinders at different values of solar elevation (β) and azimuth (θ) with a length/radius ratio of $x = 4$.

With the sun at right angles to the axis, the cylindrical model under-estimated the interception of radiation by about 20% when β was less than 60°, but with the sheep facing the sun the model over-estimated interception by 10 to 30%. Agreement could probably have been improved by using Equation 7.13 instead of calculating h and r from two angles only, but for random orientation the error in using the dimensions quoted would be very small.

Figure 7.7 shows A_h/A for horizontal cylinders plotted as a function of β for four values of azimuth angle and $x = 4$. When β exceeds 40°, the shape factor is almost independent of θ and is therefore nearly proportional to cosec β. This implies that the radiation intercepted by a cylinder with $x = 4$ will be almost independent of the solar azimuth provided solar elevation exceeds 40°. For $x < 4$, the corresponding elevation angle will be less than 30°.

Cone

The interception of radiation by a cone is an interesting problem, relevant to the distribution of radiant energy when leaves on a single tree (Figure

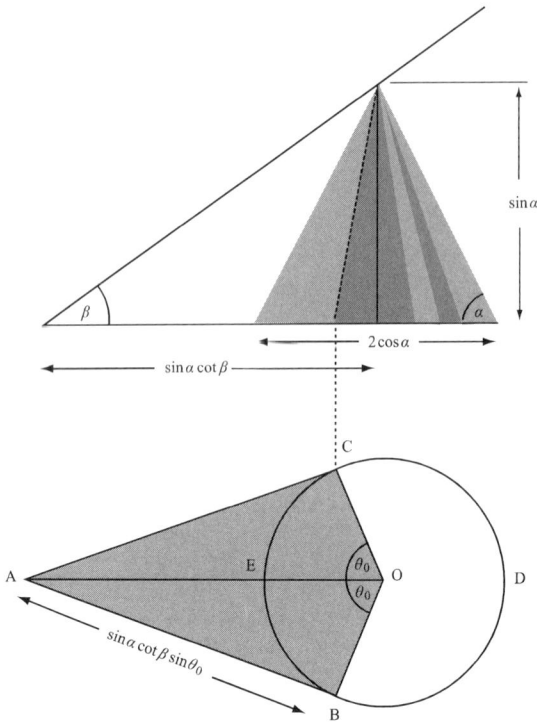

FIGURE 7.9   The geometry of a cone projected on a horizontal surface.

7.8, see color plates) or on the foliage of a crop are distributed randomly with respect to the points of the compass.

The cone in Figure 7.9 has a slant side of unit length, making an angle α with the base, so the perpendicular height is $\sin \alpha$ and the base area is $\pi \cos^2 \alpha$. For direct radiation incident at an angle $\beta > \alpha$ (not shown), the walls are fully illuminated and the shadow cast on a horizontal surface by the whole cone is simply the shadow area of the base $A_h = \pi \cos^2 \alpha$. As the area of the walls is $A = \pi \cos \alpha \times 1$, the shape factor for the walls alone is $A_h/A = \cos \alpha$. When $\beta < \alpha$, the shadow has a more complex form shown in Figure 7.9. The walls are now partly in shadow: at the base, CDB is illuminated, BEC is not, and the shadow can be delineated by projecting tangents at B and C to meet at A. The sector of the cone in shadow can now be specified by the angle AOB = AOC = θ_0. Now OB = $\cos \alpha$, AO is the horizontal projection of the axis of the cone; i.e., $\sin \alpha \cot \beta$, and as ABO is a right angle, AB = $\sin \alpha \cot \beta \sin \theta_0$. The cosine of θ_0 is OB/AO = $\cos \alpha / (\sin a \cot \beta)$. It follows that the area of the shadow is

$$ABDC = EBDC + 2ABO - CEBO$$
$$= \text{circle} + 2 \text{ triangles} - \text{sector of circle}$$
$$= \pi \cos^2 \alpha + \cos \alpha \sin \alpha \cot \beta \sin \theta_0 - \theta_0 \cos^2 \alpha$$
$$= \cos \alpha \{(\pi - \theta_0) \cos \alpha + \sin \alpha \cot \beta \sin \theta_0\}$$

As the total area of the cone is $\pi \cos \alpha (1 + \cos \alpha)$, the shape factor is

$$\frac{A_h}{A} = \frac{(\pi - \theta_0) \cos \alpha + \sin \alpha \cot \beta \sin \theta_0}{\pi (1 \cos \alpha)} \qquad (7.15)$$

where $\theta_0 = \cos^{-1} (\tan \beta \cot \alpha)$.

Inclined plane

Figure 7.10 shows the end and plan views of a square plane with sides of unit length making an angle α with the horizon XY and exposed to a beam of radiation at an elevation β at right angles to the edge **AB**.

The shadow **CEFD** has a width **EF = AB** = 1 so A_h is (**BF − BD**) × 1 or $\sin \alpha \cot \beta - \cos \alpha$.

If the beam made an angle θ with respect to the direction **AC** (Figure 7.11), the shadow would move to the position indicated by **CE'F'D** where **AE' = AE** = $\sin \alpha \cot \beta$. The shadow becomes a parallelogram with an area **CG × CD**, so A_h is [(**AE'** $\cos \theta$) −**AC**] × 1 or $\sin a \cot \beta \cos \theta - \cos \alpha$.

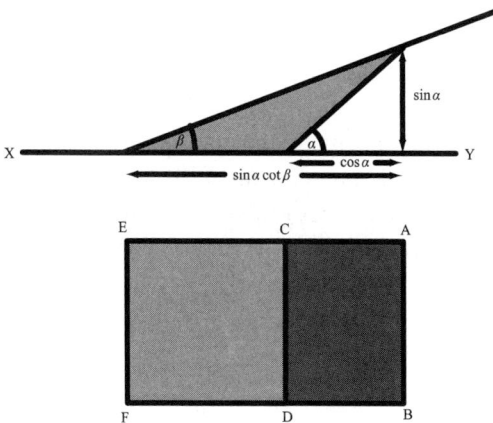

FIGURE 7.10 Geometry of a square plane, ABCD, projected on a horizontal surface when the edges AC and BD are parallel to solar beam.

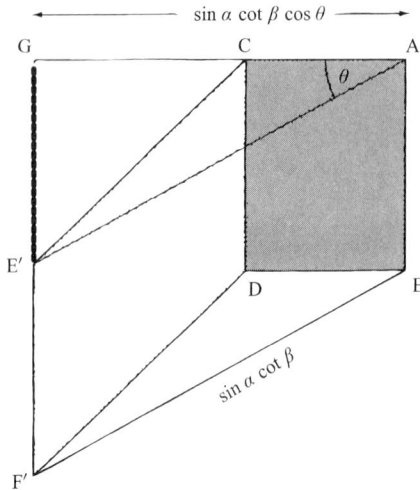

FIGURE 7.11   Geometry of a rectangle projected on a horizontal surface when the edge makes angle θ with the solar beam.

When $\beta > \alpha$, the area is $[\cos\alpha - \sin a \cot\beta \cos\theta]$ and, for all values of α and β, the projected area can be written $|\cos\alpha - \sin\alpha \cot\beta \cos\theta|$, the positive value of the function. Because any plane with area A can be subdivided into a large number of small unit squares, the shape factor of a plane is

$$\frac{A_h}{A} = |\cos\alpha - \sin\alpha \cot\beta \cos\theta| \qquad (7.16)$$

This function can be used to estimate the direct radiation on hill slopes or on the walls of houses. Then α depends on the geometry of the system, β depends on solar angle, and θ depends both on geometry and on the position of the sun. Relevant calculations and measurements can be found in a number of texts (e.g., Chauliaguet et al 1979). The example in Figure 7.12 emphasizes the large difference of direct irradiance on slopes of different aspect, which is often responsible for major differences of microclimate and plant response.

7.2   Diffuse Radiation

In addition to any incident flux of direct radiation, natural objects are exposed to four discrete streams of diffuse radiation with different directional properties.

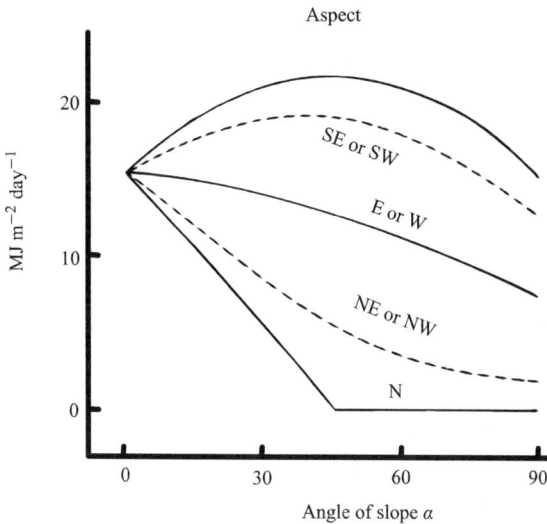

FIGURE 7.12 Daily integral of direct solar radiation on slopes at the equinoxes for latitude 45°N [from Garnier and Ohmura, 1968].

1. **Incoming short-wave diffuse radiation.**
   The spatial distribution of this flux depends on the elevation and azimuth of the sun and on the degree of cloud cover (p. 65).

2. **Incoming long-wave radiation.**
   When the sky is cloudless, the intensity of atmospheric radiation decreases by about 20 to 30% from the horizon to the zenith. Under an overcast sky, however, the flux is nearly uniform in all directions (p. 5).

3. **Reflected solar radiation.**
   The amount of radiation received by reflection from an underlying surface depends on its reflection coefficient, and the angular distribution of the flux is determined by the surface structure. Natural vegetation and agricultural crops are often composed of vertical elements that shade each other and more radiation is reflected from sunlit than from shaded areas.

4. **Long-wave radiation emitted by the underlying surface.**
   Like the reflected component of diffuse radiation, the spatial distribution of this flux depends on the disposition of sunlit (relatively warm) and shaded (relatively cool) areas.

The different angular variations of the four diffuse components are difficult to handle analytically, but for the purposes of establishing the radiation balance of a leaf or animal the angular variations can often be ignored. The following treatment deals with the interception of isotropic radiation; i.e., the intensities of the diffuse fluxes are assumed independent of angle. Additional formulae for long-wave radiation were derived by Unsworth [1975] for slopes and regular solids and by Johnson and Watson [1985] for complex shapes.

SHAPE FACTORS

Plane Surfaces

A horizontal flat plate facing upward receives diffuse fluxes of S_d and L_d in the short- and long-wave regions of the spectrum. The corresponding fluxes on a surface facing horizontally downward are ρS_t and L_u.

A flat plate at any angle α to the horizon receives radiation from all four sources on both its surfaces. To estimate how the irradiance from each of these sources depends on α, the atmosphere and the ground can be replaced by two hemispheres AOC and AO'C (Figure 7.13).

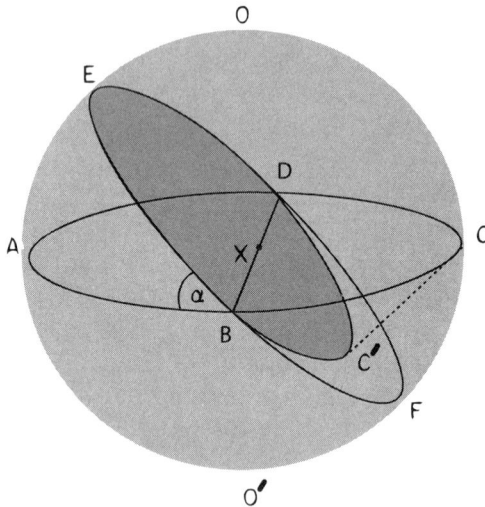

FIGURE 7.13   Diagram for calculating the diffuse irradiance at a point X in the center of a sphere from a sector of the sphere subtending an angle α at the equatorial plane.

The plane of the horizon is ABCD, and the plane of the plate is DEBF, making an angle α with the horizontal plane. The irradiance of the plate from the sector of the sky represented by DCBE could be calculated by dividing this sector into a large number of small elements dA and then integrating the product of dA and the cosine of the angle between each element and the normal to the surface of the plate. This process can be avoided by the device described on p. 4, i.e., the sector DEBC is projected onto the plane of the plate.

The area of this projection is the semi-circle DEBX plus the semi-ellipse DC'BX; i.e., $\pi/2 + (\pi \cos \alpha)/2$ if the hemisphere has unit radius. If $\mathbf{N}$ is the (uniform) radiance emitted by elements on the hemisphere, the irradiance from the sector will be $(\pi/2)(1 + \cos \alpha)\mathbf{N}$. For a horizontal surface, $\alpha = 0$ and the irradiance is $\pi \mathbf{N}$ (p. 4). The ratio of the irradiance of a surface at angle α to the irradiance of a horizontal surface is therefore $(1 + \cos \alpha)/2 = \cos^2(\alpha/2)$. For an inclined plane, the factor $(1 + \cos \alpha)/2$ for diffuse radiation is equivalent to the factor A_h/A derived for direct radiation.

If a flat leaf or other plane surface is exposed above the ground at an angle α, both surfaces will receive short- and long-wave radiation from the sky and from the ground. When the four fluxes of radiation are isotropic, they can be written as follows:

	Short-wave	*Long-wave*
Upper surface	$\cos^2(\alpha/2)\mathbf{S}_d + \sin^2(\alpha/2)\rho\mathbf{S}_t$	$\cos^2(\alpha/2)\mathbf{L}_d + \sin^2(\alpha/2)\mathbf{L}_u$
Lower surface	$\sin^2(\alpha/2)\mathbf{S}_d + \cos^2(\alpha/2)\rho\mathbf{S}_t$	$\sin^2(\alpha/2)\mathbf{L}_d + \cos^2(\alpha/2)\mathbf{L}_u$

The sum of all eight components is simply $(\mathbf{S}_d + \rho\mathbf{S}_t + \mathbf{L}_d + \mathbf{L}_u)$. The condition that the upward fluxes of radiation are approximately isotropic requires that the height of the plane above the ground should be large compared with its dimensions so its shadow can be neglected.

The total solar radiation received by plane surfaces with different slopes and aspects can be calculated by summing the direct and diffuse components. The curves in Figure 7.14 were derived by Kondratyev and Manolova [1960] who found that it was essential to allow for the spatial distribution of radiation from the blue sky (p. 5) in order to describe the diurnal change of irradiance on slopes with α exceeding $30°$. However, they also found that daily totals of radiation on slopes could be calculated accurately by assuming that the diffuse flux was isotropic, so that the daily insolation was the sum of the individual hourly values of the direct radi-

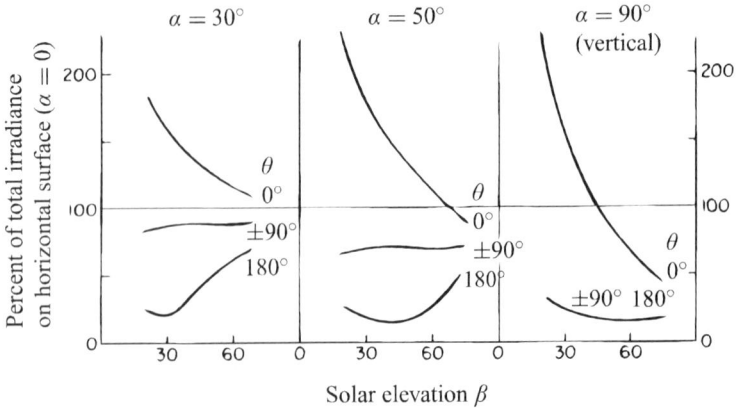

FIGURE 7.14   Irradiance of planes (direct and diffuse solar radiation) at latitude 45°N as a function of solar elevation β, elevation of the plane α, and azimuth angle θ between the solar beam and the normal to the plane. From measurements by Kondratyev and Manolova [1960].

ation on the slope, the diffuse flux from the sky, $\cos^2{(\alpha/2)}\,S_d$; and the diffuse flux from the surrounding terrain, $\sin^2{(\alpha/2)}\,\rho S_t$.

Cone

The diffuse irradiance on the walls of a cone with base angle α is equal to the irradiance of the upper surface of a plate at an elevation of α.

Vertical cylinder

For a vertical surface, $\cos\alpha = 0$, so the receipt of diffuse short-wave radiation is $(S_d + \rho S_t)/2$ and the receipt of long-wave radiation is $(L_d + L_u)/2$.

Horizontal cylinder

The components of irradiance for the upper and lower surfaces of a horizontal cylinder can be found by integrating the factors $(1 + \cos a)/2$ and $(1 - \cos a)/2$. The integration yields factors of $0.5 + \pi^{-1} = 0.82$ and $0.5 - \pi^{-1} = 0.18$. With these approximations, the components are:

The components for each half surface are the same as they would be for the upper and lower surfaces of a plane with $\alpha \approx 50°$, and the sum of all the components is simply $(S_d + \rho S_t) + (L_d + L_u)$, which is the sum for any plane surface.

	Short-wave	*Long-wave*
Upper half surface	$0.82S_d + 0.18\rho S_t$	$0.82L_d + 0.18L_u$
Lower half surface	$0.18S_d + 0.82\rho S_t$	$0.18L_d + 0.82L_u$

7.3 Problems

1. Plot a graph showing how the shape factor of an oblate spheroid, with a ratio of vertical semi-axis to horizontal semi-axis (a/b) of 0.5, varies with solar elevation β.

2. Treating a penguin as a vertical cylinder with the ratio x of height/radius = 5, calculate the mean direct radiation flux $\overline{S_b}$ incident per unit body area, when the solar elevation is $10°$ and the direct irradiance at normal incidence S_p is 800 W m^{-2}.

3. A leaf is exposed at $30°$ to the horizontal above bare soil under an overcast sky. Incident radiation on a horizontal surface is $S_d = 150$ W m^{-2} and $L_d = 290$ W m^{-2}. If the soil reflection coefficient is 0.15, emissivity is 1.0, and temperature is $15°C$, calculate the short- and long-wave irradiances of the leaf surfaces. (Assume that the diffuse radiation is isotropic and that the leaf is far enough above the surface for the shadow to be ignored).

Chapter 8

Microclimatology of Radiation (iii) Interception by Plants and Animals

The previous chapter developed equations that describe the interception of direct and diffuse radiation by solid objects and sloping planes. In this chapter, they are applied to the interception of radiation by plant canopies and animal coats. The chapter concludes with applications of the principles of the radiation balance and interception to compare the net radiation budgets of leaves, canopies, animals, and man.

8.1 Interception of Radiation by Plant Canopies

BLACK LEAVES

Direct Radiation

The principles of radiation geometry can now be applied to estimate the distribution of radiant energy within a plant canopy with horizontally uniform foliage. The amount of foliage is conventionally specified as a *leaf area index L*. For flat leaves, L is defined as the area of leaves per unit area of ground, taking only one side of each leaf into account. For the needles of conifers, which can often be assumed cylindrical, it is convenient to define L as one half the total leaf area per unit ground surface area [Chen and Black 1992]. To avoid the complications of scattering and transmission initially, the leaves are assumed to be black (close to reality for the visible spectrum at least—see p. 93).

Suppose that a thin horizontal layer of black leaves, exposed to direct solar radiation, contains a small leaf area index dL. The amount of energy intercepted by dL is the area of shadow cast by the leaves on a horizontal plane multiplied by the horizontal beam irradiance S_b (p. 101). The required shadow area is dL times the shadow cast by unit area of leaf, which is (A_h/A). The product $(A_h/A)dL$ is the area of horizontal shadow per unit ground area (a shadow area index), and the intercepted radiation can now be expressed as

$$dS_b = -(A_h/A)\,S_b dL$$
$$= -\mathcal{K}_s S_b dL \qquad (8.1)$$

where the *attenuation coefficient*, $\mathcal{K}_s$ is defined as

$$\mathcal{K}_s = A_h/A$$

Equation 8.1 implies that the mean irradiance of a leaf is $\mathcal{K}_s S_b$. The minus sign is needed if L is measured downwards from the top of the canopy. Integration of this equation gives

$$S_b(L) = S_b(0)\exp(-\mathcal{K}_s L) \qquad (8.2)$$

where $S_b(L)$ is the direct solar irradiance measured on a horizontal plane below a leaf area index L measured from the top of the canopy. This is a special case of Beer's Law (see p. 48). It may also be written as

$$\tau = \exp(-\mathcal{K}_s L) \qquad (8.3)$$

where the ratio $S_b(L)/S_b(0)$ is a *transmission coefficient* τ (i.e., the relative solar irradiance below a leaf area index L). The ratio $S_b(L)/S_b(0)$ is also the fractional area of sunflecks on a horizontal plane below L. The area of sunlit foliage between L and $(L+dL)$ is therefore $\{S_b(L)/S_b(0)\}\,dL$, and in a stand with a total leaf area of L_t the leaf area index of sunlit foliage is

$$\int_0^{L_t} \{S_b(L)/S_b(0)\}\,dL = \int_0^{L_t} \exp(-\mathcal{K}_s L)\,dL = \mathcal{K}_s^{-1}\left[1 - \exp(-\mathcal{K}_s L_t)\right]$$

which has a limiting value of $1/\mathcal{K}_s$ at large values of L_t.

For horizontal leaves, the shadow area is equal to the leaf area, independent of solar elevation, so $\mathcal{K}_s = 1$ for this special case. We now consider how values of $\mathcal{K}_s$ can be deduced for other leaf angle distributions from the area of shadows cast by cylinders, spheres, and cones as presented on pages 102 to 109.

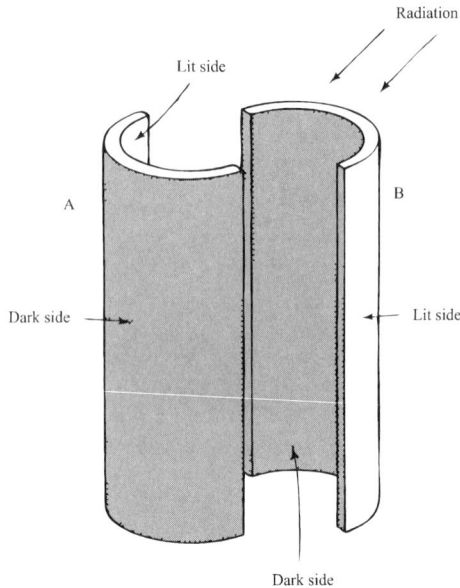

FIGURE 8.1   The distribution of radiation over two surfaces of a cylinder representing the irradiance of a large number of vertical leaves.

Vertical Leaf Distribution

If all the leaves in a canopy hung vertically facing at random with respect to azimuth or compass angle, they could be rearranged in a mosaic pattern on the curved surface of a vertical cylinder. This cylinder could be split along a central plane at right angles to the sun's rays (Figure 8.1). The convex half of the cylinder represents leaves illuminated on one side (e.g., the upper surface) and the concave half represents the leaves illuminated on the lower surface. The appropriate value of $\mathcal{K}_s = A_h/A$ is therefore twice the value derived for the curved surface of a solid cylinder, i.e.,

$$\mathcal{K}_s = 2\,(\cot\beta)\,/\pi \quad \text{(cf. p. 103)}$$

Spherical and Ellipsoidal Leaf Distribution

If the leaves in a canopy were distributed at random with respect to their angles of elevation as well as their azimuth angles, they could be rearranged on the surface of a sphere. Splitting the sphere at the equatorial plane normal to the sun's rays gives two hemispheres representing the two sides from which the leaves could be illuminated. The appropriate value

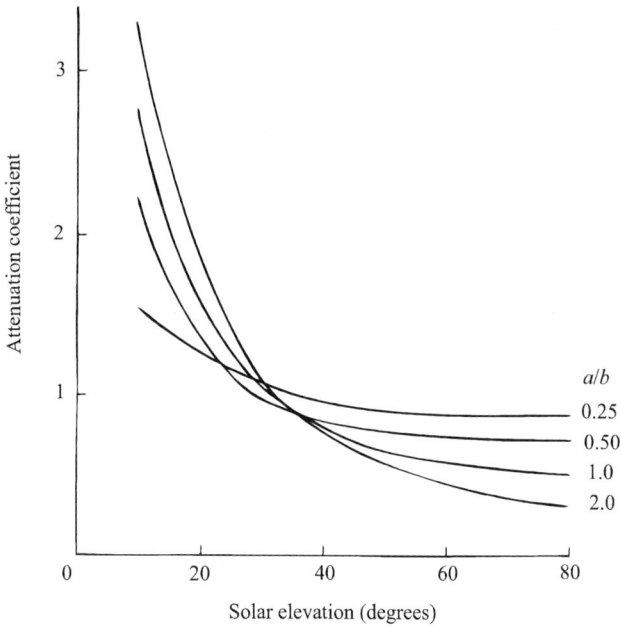

FIGURE 8.2 Dependence of attenuation coefficient for direct radiation $\mathcal{K}_s = 2A_h/A$ on solar elevation when leaf angle distribution is ellipsoidal with $a/b = $ ratio of vertical to horizontal radius (see p. 99). The case $a/b = 1$ corresponds to a spherical distribution.

of $\mathcal{K}_s$ is twice the value of the shape factor derived for a sphere, i.e.,

$$\mathcal{K}_s = 0.5 \operatorname{cosec} \beta \quad \text{(cf. p. 102)}$$

More generally, if the distribution of leaves is ellipsoidal, the value of $\mathcal{K}_s = 2A_h/A$ can be found from Equation (7.6) or (7.8) and Figure 8.2 shows the extent to which values for oblate ($a < b$) or prolate ($a > b$) spheroids depart from the spherical case $a = b$. Note that when $\beta \approx 30°$, $\mathcal{K}_s \approx 1$, almost independent of β, i.e., ellipsoids behave like planes. Campbell (1986) demonstrated that the ellipsoidal distribution describes the foliage structure of many plant species, and he derived a general expression for $\mathcal{K}_s$ when the solar elevation is β, namely

$$\mathcal{K}_s(\beta) = \frac{(x^2 + \cot^2 \beta)^{0.5}}{x + 1.774(x + 1.182)^{-0.733}} \tag{8.4}$$

where the parameter x is the ratio of the average projected areas of canopy elements on horizontal and vertical surfaces. For a spherical leaf distribution, $x = 1$; for a vertical distribution $x = 0$; and, for a canopy with horizontal leaves, x tends to infinity, giving $\mathcal{K}_s = 1$.

Conical distribution

An assembly of leaves all at an elevation of α but distributed at random with respect to their azimuth angles could be rearranged on the curved surface of a cone with a wall angle of α. For a cone exposed to a solar beam at elevation β, two cases must be considered:

(i) $\beta > \alpha$

All leaves are illuminated from above; i.e., on their upper (adaxial) surfaces. The whole curved surface of the cone is illuminated so

$$\mathcal{K}_s = \frac{A_h}{A} = \frac{\pi \cos^2 \alpha}{\pi \cos \alpha} = \cos \alpha$$

a value that is independent of solar elevation.

(ii) $\beta < \alpha$

Some of the leaves are illuminated from below; i.e., on their lower (abaxial) surfaces. In accordance with the method used for the sphere and the cylinder, the cone can be split into two parts. The relative shadow area of the convex part representing the abaxial surfaces has been calculated already (p. 109): it is

$$\cos \alpha \left\{ (\pi - \theta_0) \cos \alpha + \sin \alpha \cot \beta \sin \theta_0 \right\}.$$

The shadow area of the concave part (adaxial surfaces) is the area ACEB in Figure 7.12, i.e., the sum of two triangles less the sector of the circle or

$$\cos \alpha \left\{ \sin \alpha \cot \beta \sin \theta_0 - \theta_0 \cos \alpha \right\}.$$

The total shadow area is the sum of these expressions and as $A = \pi \cos \alpha$ for the curved surface alone,

$$\mathcal{K}_s = \frac{A_h}{A} = \pi^{-1} (\pi - 2\theta_0) \cos \alpha + 2 \sin \alpha \cot \beta \sin \theta_0$$

where

$$\theta_0 = \cos^{-1} (\tan \beta \cot \alpha).$$

A similar function, $\mathcal{K}_s \mathrm{cosec}\beta$, which is the relative shadow area on a surface normal to the sun's rays was originally tabulated by Reeve [1960] and presented graphically as a function of α and β by Anderson [1966]. Table 8.1 summarizes values of $\mathcal{K}_s$ for idealized leaf distributions and real canopies.

TABLE 8.1   Attenuation coefficients for model and real canopies (from Monteith 1969).

(a) Idealized leaf distributions	$\mathcal{K}_s$		
	solar elevation β		
	90	60	30
cylindrical	0.00	0.37	1.10
spherical	0.50	0.58	1.00
conical $\alpha = 60$	0.50	0.50	0.58
$\alpha = 30$	0.87	0.87	0.87

(b) Real canopies	$\mathcal{K}_s$
White Clover (*Trifolium repens*)	1.10
Sunflower (*Helianthus annuus*)	0.97
French Bean (*Phaseolus vulgaris*)	0.86
Kale (*Brassica acephala*)	0.94
Maize (*Zea mays*)	0.70
Barley (*Hordeum vulgare*)	0.69
Broad Bean (*Vicia faba*)	0.63
Sorghum (*Sorghum vulgare*)	0.49
Ryegrass (*Lolium perenne*)	0.43
(*Lolium rigidium*)	0.29
Gladiolus	0.20

Diffuse Radiation

The transmission of diffuse radiation through a canopy differs from that of direct radiation because diffuse radiation originates from all parts of the sky. A transmission coefficient for diffuse radiation τ_d, comparable to τ for direct radiation, can be derived by treating the diffuse flux as the sum of many beams integrated over the hemisphere. For canopies of horizontal leaves, τ is independent of solar elevation, so the diffuse and direct transmission coefficients are identical. But for other leaf distributions, where $\mathcal{K}_s$ depends on β, numerical integration reveals that τ_d does not decrease exponentially with L as it does for direct radiation (Equation 8.3). As an aid to computation, Campbell and van Evert [1994] forced an exponential equation to fit data relating τ_d to L by making the diffuse radiation attenuation coefficient $\mathcal{K}_d$ vary with L. Figure 8.3 illustrates their results for several values of leaf angle distribution. The figure, which assumes a uniform overcast sky, can be used to compare the transmission of diffuse and direct radiation. For example, a canopy with a spherical leaf distribution and $L = 3$ would have a diffuse radiation transmission coefficient of about $(\exp -0.7) \times 3 = 0.12$. For comparison, the fractional transmission

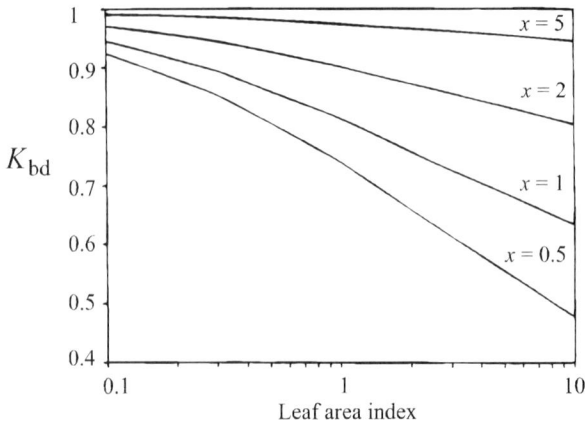

FIGURE 8.3   Apparent attenuation coefficient for diffuse radiation from a uniform overcast sky in canopies with differing leaf angle distributions (x is the ratio of averaged projected areas of leaves on vertical and horizontal surfaces, so that $x = 1$ is a spherical leaf angle distribution, $x = 0$ for a vertical distribution, and $x = \infty$ for a horizontal distribution) [from Campbell and van Evert 1994].

of direct radiation by the same canopy would be 0.12 if the solar elevation β was about 45°.

IRRADIANCE OF FOLIAGE

To estimate rates of transpiration and photosynthesis for leaves in a canopy, it is essential to calculate the irradiance of individual leaf surfaces as distinct from the irradiance of horizontal surfaces already considered. If $S_b(L)$ is the direct component of horizontal irradiance below an area index of L, the mean irradiance of foliage at this depth will be $\mathcal{K}_s S_b(L)$ from Equation 8.1. The mean irradiance can also be derived by considering an irradiance of S_b (W per m² field area) distributed over a sunlit leaf area index of $1/\mathcal{K}_s$ (m² sunlit leaves per m² field area) to give a mean irradiance of $\mathcal{K}_s S_b$ (W per m² leaf area). In the exceptional case when all leaves are facing in the same direction, the irradiance $\mathcal{K}_s S_b(L)$ will be uniform, but in general some leaves will be exposed to a stronger and some to a weaker flux of direct radiation. In the extreme, leaves parallel to the solar beam ($\alpha = \beta$) receive no direct radiation, and leaves at right angles to the beam ($\alpha = \beta + \pi/2$) receive $S_b \operatorname{cosec} \beta$.

In contrast to direct radiation, the distribution of diffuse solar radiation below a leaf area index L is relatively uniform, and its interception may be treated as independent of leaf orientation. Thus, the diffuse irradiance of foliage below a leaf area index L is the same as the diffuse irradiance on a horizontal surface at the same depth in the canopy, estimated using Figure 8.3.

The distribution of solar irradiance on foliage should be taken into account when estimating leaf temperatures precisely and is important for processes that are not proportional to the irradiance. For example, rates of leaf assimilation of carbon dioxide typically increase linearly with increasing irradiance at low irradiance, but reach an assymptote at high irradiance, giving a hyperbolic light response. Calculations of canopy assimilation can be made by estimating the sunlit and shaded leaf areas in each layer, calculating the assimilation rate for the mean leaf irradiance of each type, and summing according to the fractions of sunlit and shaded leaf area [Norman 1992]. Calculations like this can be used to show that diffuse radiation is more effective in photosynthesis than the same irradiance of direct radiation because more leaf area in the canopy is exposed to lower irradiances at which photosynthetic rates are not saturating. This implies that there is a direct effect of an increase in clouds and aerosols on the productivity of vegetation [Roderick et al, 2001], and this effect has been postulated as a partial explanation for the apparent greater uptake of atmospheric carbon dioxide by vegetation when the diffuse/total ratio in global radiation was increased by dust from a volcanic eruption [Gu et al, 2003].

Practical Aspects

In practice, values of $\mathcal{K}$ are often determined by measuring the attenuation of radiation with tube solarimeters (radiometers with an extended linear sensing surface [Szeicz et al, 1974]) or with hemispherical dome radiometers mounted on carriages that traverse on tracks within a canopy [Blancken et al, 1997]. Both types of instruments receive total solar radiation from a full hemisphere. Although estimates of $\mathcal{K}$ derived in this way (see Table 8.1) are not strictly comparable with those derived here, they cover the same range; i.e., from about 1 for stands of clover and sunflower with predominantly horizontal leaves to about 0.2 for *Allium* and *Gladiolus* with mainly vertical leaves.

The Upper Limit for Leaf Area Index
When plants produce new leaves at the top of a stand, older leaves become progressively shaded and in many species they die when their irradiance falls to a few percent of full sunlight. When production and death are bal-

anced, there is an upper limit L' to the leaf area index. Taking an arbitrary figure of 5% transmission for the level below which leaves die, L' can be related to $\mathcal{K}$ by writing

$$0.05 = \exp\left(-\mathcal{K}L'\right)$$

so that $L' = -\ln(0.05)/\mathcal{K} = 3/\mathcal{K}$.

This rough calculation is consistent with field experience that canopies with predominantly horizontal leaves usually have a maximum leaf area index of between 3 and 5, whereas values up to about 10 have been reported for cereals with a vertical leaf habit [Monteith and Elston, 1983] and old-growth needleleaved forests [Thomas and Winner 2000].

Indirect Methods for Estimating Leaf Area Index

Equations developed earlier in this section provide a basis for *indirect* optical methods of estimating leaf area index and leaf angle distribution: *direct* methods for measuring these parameters are often tedious and destructive. Campbell and Norman [1989] reviewed direct and indirect approaches, and Welles [1990] reviewed instrumentation for indirect measurements. One widely applied method is *gap fraction analysis*, which uses information on the gap distribution in canopies, derived either from hemispheric photographs taken using fish-eye lenses viewing upwards from the ground or down from above the canopy [Bonhomme and Chartier 1972, Fuchs and Stanhill 1980, Rich et al, 1993], or from measurements of the fractional length of a horizontal transect in a plant canopy that is in sunflecks. By measuring the gap fraction at several solar zenith angles, inversion of transmission equations allows both L and leaf angle distribution to be estimated using statistical methods [Campbell and Norman 1989, Chen et al, 1997].

Indirect methods described so far assume that foliage is uniformly distributed in space. This is usually appropriate for closed canopies of agricultural crops and some broadleaved forests, but is not the case for row crops, coniferous forests, and sparse natural vegetation. In coniferous forests, foliage is clumped at several scales: shoots, branches, whorls, and crowns. Branches and stems also attenuate radiation significantly, and can be described by a *wood surface area index W*, defined as half the total wood surface area per unit ground area. As a consequence of clumping, gap fraction measurements in coniferous forests often yield a measured "effective" leaf area index L_e that is smaller than the true leaf area index L of the foliage. The relation between L and L_e may be written

$$L + W = \frac{L_e \gamma_e}{\Omega_e} \tag{8.5}$$

where Ω_e is the *clumping index* of shoots, the basic foliage units that attenuate radiation, and γ_e is the ratio of needle area/shoot area . Chen et al [1997] estimated L using Equation 8.5 for several boreal forest stands. Values of Ω_e ranged between 0.70 and 0.95, and γ_e was 1.0–1.5. Values of L/L_e were typically about 1.5 for coniferous stands and 1.0 for deciduous aspen stands. Law et al [2001] estimated leaf area index directly and indirectly in a mixed age stand of open-canopied ponderosa pine forest and found that L_e was 1.3, γ_e was 1.25, and Ω_e was 0.81, giving $L + W = 2.00$. They estimated that W was about 0.3, so the foliage leaf area index L was about 1.7, about 30% larger than than the effective value.

LEAVES WITH SPECTRAL PROPERTIES
Theory and Prediction

Although the analysis of radiative transfer in canopies of black leaves is useful for revealing underlying principles, in reality leaves reflect, transmit, and absorb radiation in different proportions throughout the spectrum of short and longwave radiation (pages 93–95), and this modifies the attenuation of radiation in canopies. In addition to the radiation received from the sun at a specific zenith angle, and diffuse radiation from the atmosphere and clouds, non-black leaves are exposed to two additional streams of diffuse radiation as radiation scattered and transmitted by the foliage itself moves both upward and downward. Because, in general, $\mathcal{K}$ depends on the geometry of radiation with respect to the architecture of foliage, the value of $\mathcal{K}$ for direct radiation will usually be different from the values for diffuse fluxes. Equations for handling this complex situation have been developed from astrophysical theory [Ross, 1981]. A simplified treatment for a system in which differences in values of $\mathcal{K}$ for different fluxes may be neglected is presented here.

Assuming that the Kubelka–Munk equations (p. 49) are valid for a canopy of foliage sufficiently deep for the forward beam to be reduced to virtually zero, and putting representative values of $\rho = \tau = 0.1$ for leaves exposed to radiation in the visible spectrum, the absorption coefficient for leaves is $\alpha_p = 1 - \tau - \rho = 0.8$.

Then from Equation 4.15, assuming uniform scattering in all directions, the reflection coefficient for a deep canopy is given by

$$\rho_c^* = \left(1 - \alpha_p^{0.5}\right) / \left(1 + \alpha_p^{0.5}\right) = 0.06 \qquad (8.6)$$

Corresponding representative figures for the infrared spectrum are $\rho = \tau = 0.4$ and $\rho_c^* = 0.38$. The approximate value of ρ_c^* for the whole solar spectrum is therefore

$$\rho_c^* = (0.06 \times 0.5) + (0.38 \times 0.5) = 0.22$$

Note that multiple scattering in a deep canopy results in a canopy reflection coefficient that is smaller than for the individual leaves (for which $\rho = (0.1 + 0.4)/2 = 0.25$).

Within a canopy that is less dense, some radiation reaches the soil surface. If the fraction of radiation transmitted by the canopy is τ_c and the reflection coefficient is ρ_c, the upward flux of radiation below the canopy will be $\tau_c \rho_s$ when ρ_s is a soil reflection coefficient. To a good approximation, the fraction of incident radiation absorbed by the canopy will therefore be

$$\alpha_c = 1 + \tau_c \rho_s - \tau_c - \rho_c$$
$$= 1 - \rho_c - \tau_c \left(1 - \rho_s\right) \qquad (8.7)$$

The quantities α_c, ρ_c, and τ_c may be derived from the Kubelka–Munk equations as functions of (i) the corresponding coefficients for leaves; (ii) the leaf area index; (iii) a canopy attenuation coefficient, namely

$$\mathcal{K} = \alpha^{0.5} \mathcal{K}_b \qquad (8.8)$$

(from Equation 4.16) where $\mathcal{K}_b$ is the coefficient for black leaves ($\alpha = 1$) with the same geometry.

Strictly, the theory can be applied only to stands with horizontal leaves so that $\mathcal{K}_b = 1$. This ensures that the intercepting area of all leaves is independent of the direction of incident radiation, and therefore a single value of $\mathcal{K}$ is valid for all scattered light and for the direct beam. However, Goudriaan [1977] demonstrated that equations 8.6 and 8.7 may be used to describe the behavior of radiation in canopies where the solar elevation β was larger than the elevation angle for most of the foliage, so that the shadow area for an assembly of leaves was independent of β (see p. 120). This class includes canopies with a spherical distribution of leaves provided $\beta > 25°$. It also includes radiation from a uniform overcast sky which is transmitted by a spherical leaf distribution as if all the radiation emanated from an angle of about 45° (p. 122). For these restricted categories of foliage, the value of $\mathcal{K}_b$ to be inserted in Equation 8.8 is given by the relative shadow area of the assembly $\mathcal{K}_s$ as evaluated earlier.

After integration of the equations (4.13 and 4.14) which describe the downward and upward streams of radiation in a canopy, the reflection coefficient is given by

$$\rho_c = \frac{\rho_c^* + f \exp(-2\mathcal{K}L)}{1 + \rho_c^* f \exp(-2\mathcal{K}L)} \tag{8.9}$$

and the transmission coefficient by

$$\tau_c = \frac{\{(\rho_c^{*2} - 1) / (\rho_s \rho_c^* - 1)\} \exp(-\mathcal{K}L)}{1\rho_c^* f \exp(-2\mathcal{K}L)} \tag{8.10}$$

where

$$f = (\rho_c^* - \rho_s)/(\rho_s \rho_c^* - 1)$$

Neglecting the second-order terms ρ_c^{*2} and $\rho_c^* \rho_s$ gives approximate values of the coefficients as

$$\rho_c = \rho_c^* - (\rho_c^* - \rho_s) \exp(-2\mathcal{K}L) \tag{8.11}$$

and

$$\tau_c = \exp(-\mathcal{K}L) \tag{8.12}$$

where ρ_c clearly has limits of ρ_c^* when L is large and ρ_s when L is small. Applying Equations 8.11 and 8.12 in Equation 8.7, absorption by the canopy is

$$\alpha_c \approx 1 - \{\rho_c^* - (\rho_c^* - \rho_s) \exp(-2\mathcal{K}L)\} - (1 - \rho_s) \exp(-\mathcal{K}L) \tag{8.13}$$

The term $(\rho_c^* - \rho_s) \exp(-2\mathcal{K}L)$ is much smaller than $(1 - \rho_s) \exp(-\mathcal{K}L)$ when ρ_c^* is small and/or $\mathcal{K}L$ is large. The canopy absorption may therefore be reduced to

$$\alpha_c \approx 1 - \rho_c^* - (1 - \rho_s) \tau_c$$
$$\approx (1 - \rho_c^*)(1 - \tau_c) - \tau_c (\rho_c^* - \rho_s) \tag{8.14}$$

Some practical implications of these relations will now be considered.

Absorbed and Intercepted Radiation

In the field, the radiation intercepted by a crop or forest canopy is conveniently determined by mounting solarimeters above and below the canopy [Szeicz et al, 1964], see p. 123 herein). The fraction of the incident flux density recorded by the lower solarimeter is τ_c and the intercepted fraction is simply $1 - \tau_c$.

Empirically, the amount of total solar radiation intercepted by a canopy is often well correlated with the production of dry matter during periods when the leaf area index is increasing [Russell et al, 1989]. Theoretically, however, growth rate should depend on the *absorbed* fraction of radiation α_c. Equation 8.14 provides reassurance that, subject to the validity of the approximations used, and provided the term $\tau_c \left(\rho_s - \rho_c^* \right)$ is small, α_c is a nearly constant fraction of $(1 - \tau_c)$. The fraction is $1 - \rho_c^*$; i.e., about 0.94 for the visible spectrum and 0.77 for total solar radiation using the values derived earlier.

A further step is needed to obtain absorbed PAR from intercepted total radiation. From Equations 8.8 and 8.12

$$\tau_c \approx \exp\left(-\alpha^{0.5}\mathcal{K}_b L\right) \qquad (8.15)$$

The transmission of PAR,τ_{cP}, may therefore be estimated from the transmission of total radiation, τ_{cT}, using the identity

$$\tau_{cP} \approx \exp(-\alpha_p^{0.5}\mathcal{K}_b L) \equiv \exp(-\alpha_T^{0.5}\mathcal{K}_b L) \frac{\exp(-\alpha_p^{0.5}\mathcal{K}_b L)}{\exp(-\alpha_T^{0.5}\mathcal{K}_b L)}$$

$$= \tau_{cT} \exp\left\{ \left(\alpha_T^{0.5} - \alpha_p^{0.5}\right) \mathcal{K}_b L \right\} \qquad (8.16)$$

If $\alpha_T = 0.4$ and $\alpha_p = 0.8$, Equation 8.16 becomes

$$\tau_{cP} = \tau_{cT} \exp(-0.26\mathcal{K}_b L)$$

The fraction of PAR absorbed by a stand as given by Equation 8.14 is

$$\alpha_{cP} = 0.94 \left\{ 1 - \tau_{cT} \exp\left(-0.26\mathcal{K}_b L\right) \right\} \qquad (8.17)$$

assuming $\rho_{cP}^* = 0.06$ and neglecting the small term $\tau_{cP} \left(\rho_{sP} - \rho_{cP}^* \right)$. Finally, the fractional absorption of energy (PAR) can be converted to the equivalent fraction of absorbed quanta using a coefficient of 4.6 μmole per joule (p. 71). Figure 8.4 shows how absorbed quanta derived from Equation 8.17 are related to intercepted total radiation $(1 - \tau_c)$ derived from Equation 8.12. The figure demonstrates that the increase in absorbed PAR with leaf area index is well described by the increase in intercepted total radiation, as indicated in Equation 8.17.

Equation 8.12 may also be written

$$\tau_{cT} = \exp\left(-\mathcal{K}' L\right)$$

where $\mathcal{K}' = \alpha_T^{0.5}\mathcal{K}_b$ is the attenuation coefficient as conventionally determined by plotting $\ln \tau_{cT}$ against L to find $-\mathcal{K}'$ as a slope.

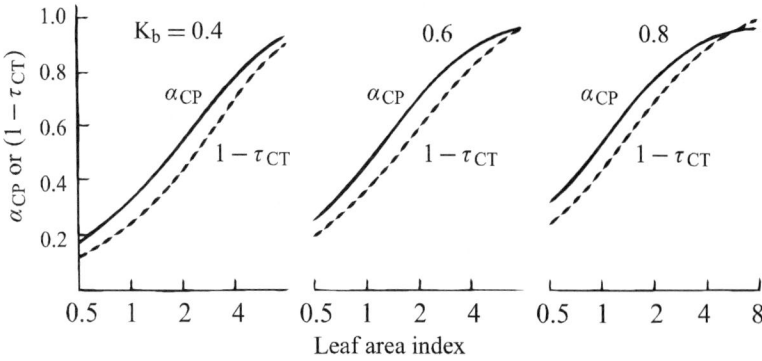

FIGURE 8.4   Fractional absorption of visible radiation (or PAR) α_{CP} (full lines), and fractional interception of total radiation $(1 - \tau_{cT})$ (dashed lines) as functions of leaf area index for three values of the attenuation coefficient for black leaves K_b (assuming $\rho_p = \tau_p = 0.05$, giving $(\alpha_p/\alpha_T)^{0.5} = 1.34$).

From Equation 8.8, the ratio of attenuation coefficients for PAR and total radiation K_P/K_T is $(\alpha_p/\alpha_T)^{0.5}$, a conservative quantity because the fraction of total radiation absorbed by a leaf is largely determined by pigments that absorb in the PAR waveband. To demonstrate this point, assume that in the PAR waveband $\rho_p = \tau_p = 0.05$ so that $\alpha_p = 0.9$, that absorptivity in the solar infrared waveband is 0.1 and that the solar spectrum contains equal amounts of energy in the two wavebands (p. 71). The absorptivity for total radiation is $\alpha_T = (0.9 + 0.1)/2 = 0.5$ and $K_P/K_T = (0.9/0.5)^{0.5} = 1.34$. Doubling ρ_p and τ_p to 0.1 makes $\alpha_p = 0.8$, $\alpha_T = 0.45$ and $K_P/K_T = (0.8/0.45)^{0.5} = 1.33$. Green [1984] found that the ratio of K for quanta (effectively the same as for PAR) and for total radiation was 1.34 for wheat grown with a range of nitrogen applications so that there was a conspicuous range of chlorophyll (the pigment absorbing PAR) between treatments.

DAILY INTEGRATION OF ABSORBED RADIATION

To relate absorbed radiation to plant productivity, it is necessary to determine or model the fraction of radiation absorbed by the canopy over periods of several days to weeks. In radiation models this requires the integration of direct and diffuse radiation over the period from sunrise to sunset. Fuchs et al [1976] suggested a much simpler solution for daily in-

tegrals. They proposed that the average fractional interception of total solar radiation (direct plus diffuse) over a whole day could be approximated
by the fractional interception of diffuse radiation, since the sun traverses
a full arc of the sky in a day (i.e., they treated the total radiation as if it
were diffuse over a whole day). Using this method, the daily fractional
interception can readily be estimated as $(1 - \tau_c)$ by replacing $\mathcal{K}_b$ in Equation 8.15 with $\mathcal{K}_d$ (from Figure 8.3) when calculating τ_c. Campbell and
Evert [1994] showed that the approximation agreed very well with daily
interception calculations made with a more complex model.

REMOTE SENSING

Because the spectrum of radiation reflected by foliage has a different shape
from the spectrum for all types of soil, the extent to which soil is covered
by vegetation can be estimated from the reflection spectrum of the area,
as recorded from an aircraft or a satellite. However, interpretation can be
very difficult if the cover is not uniform. Most information is obtained by
working near the "red edge" at 700 nm below which leaves are almost perfect absorbers and above which almost all the radiation incident on leaves
is scattered. In principle, an estimate of cover could be obtained simply
from the reflectivity ρ_i in the near infrared (say between 700 and 900 nm),
but in practice, it is often difficult to measure the incident and reflected
fluxes simultaneously and error is unavoidable if the incident flux is changing with time. The difficulty is usually overcome by measuring the ratio
x_1 of the *absolute* amounts of radiation reflected in the near infrared and
in some part of the visible spectrum, usually in the red. The corresponding ratio x_2 for the spectrum of incident radiation (which changes much
more slowly than the absolute flux at any wavelength) can be obtained
by recording the spectrum reflected from a standard white surface or by
calculation if the measurements are above an atmosphere whose radiative
behavior is known. Then, the required ratio of infra and red reflectivities
(termed the *simple ratio*) is given by

$$\frac{\rho_i}{\rho_r} = \frac{\text{reflected IR/incident IR}}{\text{reflected R/incident R}}$$
$$= \frac{\text{reflected IR/incident IR}}{\text{incident IR/incident R}} = \frac{x_1}{x_2} \qquad (8.18)$$

Many workers use a *"normalized difference vegetation index"* (NDVI) defined as

$$\frac{\rho_i - \rho_r}{\rho_i + \rho_r} = \frac{x_1 - x_2}{x_1 + x_2} \qquad (8.19)$$

The correlation between fractional ground cover and this quantity is seldom much better than the correlation with the simple ratio x_1/x_2, but the fact that the ratio has limits of -1 and $+1$ makes it convenient to handle when computing.

The main problems in using either index to estimate uniform ground cover are (i) changes in the spectral properties of foliage as a consequence of senescence, poor nutrition or disease which alter the dependence of the NDVI on leaf area [Steven, 1983] and (ii) changes in the spectrum of reflected radiation between the ground and a detector as a consequence of scattering in the atmosphere [Steven et al, 1984]. Efforts to reduce these problems have led to an *enhanced vegetation index, EVI* [Huete et al, 2002], which employs additional wavebands available from the MODIS satellite, and may be more effective in allowing for haze and in assessing dense canopies. In spite of the difficulties with NDVI, Tucker et al [1985] were able to obtain plausible distributions of ground cover month by month over the whole African continent by analyzing satellite observations of NDVI, and by integration they established a seasonal index of total intercepted radiation and therefore of biomass production. Zhou et al [2001] analyzed changes in NDVI over the northern hemisphere between 1981 and 1999. Much of the predominantly forested area between $40°$ and $70°$ N latitude in Eurasia showed a persistent increase in growing season NDVI. Over the two-decade period, the NDVI increased by 12 percent in Eurasia and by 8 percent in North America, which the authors interpreted as an indication of the influence of global climate change on the timing of leaf production and productivity of natural vegetation. The principles underlying these procedures are now outlined.

Following the example of Asrar et al [1984] but using the approximate equations already derived, a relation can be established between the observed value of ρ_i/ρ_r for a canopy and the fraction of PAR which it transmits or absorbs. First, the attenuation coefficient can be written

$$K_j = \alpha_j^{0.5} \mathcal{K}_b \qquad (8.20)$$

where j is replaced by P, r or i for the wavebands of PAR, red or near infrared radiation.

Then from Equations 8.11 and 8.12, with the same conventions for subscripts, it follows that

$$\rho_i = \rho_{ci}^* - \left(\rho_{ci}^* - \rho_{si}\right) \exp\left(-2\mathcal{K}_i L\right) \equiv \rho_{ci}^* - \left(\rho_{ci}^* - \rho_{si}\right) \exp[-\mathcal{K}_P L (2\mathcal{K}_i/\mathcal{K}_P)]$$

$$= \rho_{ci}^* - \left(\rho_{ci}^* - \rho_{si}\right) \tau_P^{2\mathcal{K}_i/\mathcal{K}_P}$$

and a similar equation can be written for ρ_r. Hence,

$$\frac{\rho_i}{\rho_r} = \frac{\rho_{ci}^* - \left(\rho_{ci}^* - \rho_{si}\right)\tau_p^{(2\mathcal{K}_i/\mathcal{K}_P)}}{\rho_{cr}^* - \left(\rho_{cr}^* - \rho_{sr}\right)\tau_p^{(2\mathcal{K}_r/\mathcal{K}_P)}} \qquad (8.21)$$

Within the limits of the approximations used, therefore, the ratio ρ_i/ρ_r should be a unique function of τ_p, independent of $\mathcal{K}$ and therefore of leaf architecture because it involves the ratio of coefficients at different wavelengths. The function also depends on the soil reflectivities ρ_{si} and ρ_{sr} and on the values of α_i, α_r, and α_p, which determine the exponents of τ_p (since $\mathcal{K}_i/\mathcal{K}_P = (\alpha_i/\alpha_p)^{0.5}$ etc.). When the second term in the denominator is small compared with the first term, ρ_i/ρ_r is a function of $\tau_p^{(2\mathcal{K}_i/\mathcal{K}_P)}$. For the special case $\alpha_i/\alpha_r = 1/4$ (e.g., $\alpha_i = 0.2$, $\alpha_r = 0.8$), $\mathcal{K}_i/\mathcal{K}_P = \sqrt{1/4} = 1/2$ and ρ_i/ρ_r is a linear function of τ_p. Figure 8.5 shows the extent to which the function departs from linearity because the approximations behind this result are not valid over the whole range of τ_p.

The simple spectral ratio ρ_i/ρ_r is therefore a valid and extremely useful index of the radiation *intercepted* by and therefore the radiation *absorbed* by a canopy. It can legitimately be used to provide an estimate of growth rate when the relation between growth and intercepted radiation is known.

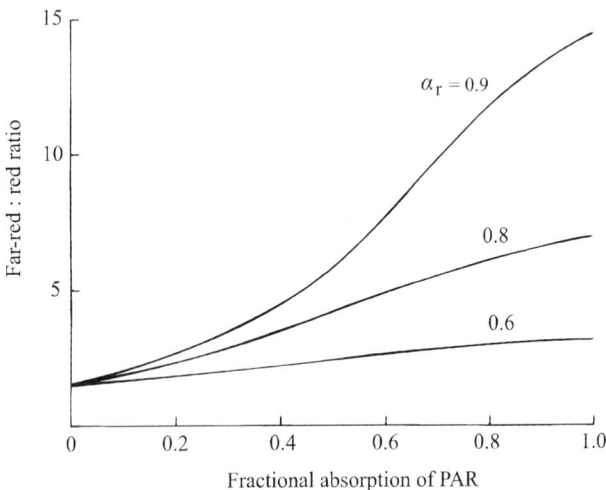

FIGURE 8.5 Relation between far-red:red reflectivity ratio for a canopy of vegetation and the amount of PAR absorbed by the canopy. Leaf absorptivity specified.

In contrast, the relation between ρ_i/ρ_r and either biomass or leaf area is strongly non-linear and depends on leaf architecture (Figure 8.4).

8.2 Interception of Radiation by Animal Coats

The transmission of radiation through animal coats was examined by Cena and Monteith [1975a] who applied the Kubelka–Munk equations (p. 50) to measurements made on samples of pelt. The parameters of their analysis were depth of coat (l), the fraction of energy $p(\theta)$ intercepted by unit depth of coat when θ is the angle of the incident flux with respect to the skin surface, the fraction of intercepted radiation reflected (ρ), transmitted (τ) or absorbed (α) by the hairs, and the absorptivity of skin (α_s). Their analysis was therefore similar to that developed earlier for canopies of leaves with spectral properties (page 125–127), treating hairs as cylinders, and calculating the interception function p from the hair diameter, length, and angle to the skin. Values of p ranged from 0.7 to about 18 cm^{-1}.

Their general solution of the Kubelka–Munk equation can be simplified for the condition $x = \alpha p l \left[1 + 2\left(\rho/\alpha\right)\right]^{0.5} > 2.7$, which is typically satisfied for many animal coats, to

$$\rho^* = \frac{(\rho/\alpha)(1 - \rho_s) + \rho_s[(1 + 2\rho/\alpha)^{0.5} - 1]}{(\rho/\alpha)(1 - \rho_s) + (1 + 2\rho/\alpha)^{0.5} + 1} \tag{8.22}$$

where ρ^* is the reflection coefficient of the coat (skin and hair together). When $\rho/\alpha \gg 1$, ρ^* tends to a value of $1 - (2\alpha/\rho)^{0.5}$; when $\rho = \alpha$, ρ^* is $1/(\sqrt{3} + 2) = 0.27$; and when $\rho/\alpha \leqslant 1$, ρ^* tends to $\rho/2\alpha$. The two limiting values and the value when $\rho = \alpha$ are all independent of skin reflection ceofficient ρ_s. It follows that, provided that $x > 2.7$, ρ^* is almost independent of ρ_s. Figure 8.6 shows the theoretical relation between coat reflection coefficient ρ^* and (α/ρ).

The fraction of radiation transmitted by a coat is given by

$$\tau^* = \eta/C \tag{8.23}$$

where

$$\eta = \alpha \left[1 + (2\rho/\alpha)\right]^{0.5}$$

and

$$C = (1 - \tau - \rho\rho_s)\sinh x + \eta \cosh x$$

with

$$x = \eta p l$$

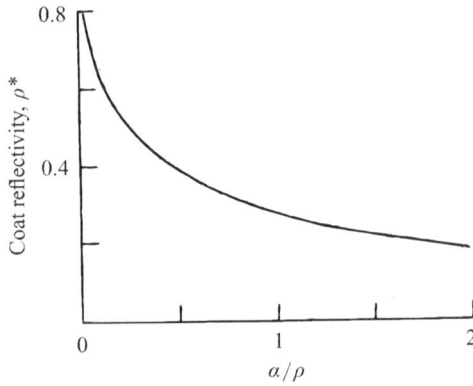

FIGURE 8.6   Theoretical relation between a coat reflection coefficient ρ^* and the ratio of absorption to reflection coefficients α/ρ for a single hair, calculated from Equation 8.22.

The fraction of radiation absorbed by the coat, α^*, is then found by noting that the fractional skin absorption is $(1 - \rho_s)\tau^*$, yielding the same identity as Equation 8.14,

$$\alpha^* = 1 - \rho^* - (1 - \rho_s)\tau^* \qquad (8.24)$$

In contrast to the situation for canopies, Cena and Monteith [1975] found that τ for short-wave radiation was larger than ρ for all species (sheep, goat, fox, fallow deer, rabbit, and cattle), especially in white coats for which α was extremely small (Table 8.2). The fact that forward scattering exceeds back scattering is probably a consequence of specular reflection when hair is nearly parallel to an incident beam of radiation.

Clean white coats of Dorset Down sheep had the largest reflection coefficient ($\rho^* = 0.80$), and the same value was found for the reflection coefficient of the skin ρ_s after removing the fleece. Using Equation 8.22 the corresponding value of α/ρ is 0.03. In contrast, the dark fleece of Welsh mountain sheep had a reflection coefficient of $\rho^* = 0.30$, giving $\alpha/\rho = 0.82$. The reflection coefficient of the skin was 0.25. To explore the implications of coat and skin color for the heat balance of coats, Table 8.3 shows calculations by Cena and Monteith for several combinations of fleece and skin color, using Equations 8.22, 8.23, and 8.24 with measured hair parameters.

The table illustrates several important points.

TABLE 8.2 Values of reflection and absorption coefficients for animal coats. ρ^* is the reflection coefficient of the hair and skin together; α/ρ is the ratio of hair absorption coefficient to reflection coefficient (from Cena and Monteith 1975).

Coat	ρ^*	α/ρ
Sheep, Dorset Down	0.79	0.03
Sheep, Clun Forest	0.60	0.13
Sheep, Welsh Mountain	0.30	0.82
Rabbit	0.81	0.02
Badger	0.48	0.28
Calf, white	0.63	0.11
Calf, red	0.35	0.60
Goat, Toggenburg	0.42	0.40
Fox	0.34	0.64
Fallow deer	0.69	0.07

TABLE 8.3 Relations between radiative properties of individual hairs and the radiation budget of whole coats (from Cena and Monteith 1975).

		Welsh Mountain clean	Dorset Down clean	Dorset Down soiled	Dorset Down soiled
	Hair parameters			light skin	dark skin
	ρ	0.240	0.066	0.240	0.240
	α	0.200	0.002	0.060	0.060
	τ	0.560	0.932	0.700	0.700
	ρ_s	0.250	0.800	0.800	0.250
Coat depth					
1 cm	ρ^*	0.30	0.80	0.51	0.50
	α^*	0.66	0.03	0.45	0.40
	$(1-\rho_s)\tau^*$	0.04	0.17	0.04	0.10
2 cm	ρ^*	0.30	0.80	0.51	0.50
	α^*	0.70	0.05	0.47	0.45
	$(1-\rho_s)\tau^*$	0	0.15	0.02	0.05
4 cm	ρ^*	0.30	0.80	0.51	0.50
	α^*	0.70	0.09	0.48	0.47
	$(1-\rho_s)\tau^*$	0	0.11	0.01	0.03

1. The values of α^* indicate that when a sheep is exposed in bright sunshine ($S_t > 1000$ W m^{-2}), the solar radiation flux intercepted by the coat will often exceed 500 W m^{-2}, and may approach 1000 W m^{-2}.

2. For the clean Dorset Down fleece, which absorbs very little radiation, the reflection coefficient is independent of fleece depth, and equal to the skin reflection coefficient; but for the soiled coat, where the hair absorption coefficient increases by a factor of 30, ρ^* is significantly less than ρ_s, although still independent of coat depth.

3. If the reflection coefficient of the skin below a soiled Dorset Down fleece were 0.25 rather than 0.80, ρ^* would decrease only very slightly, but the fractional skin absorption $((1 - \rho_s)\tau^*)$ would more than double. The fraction of incident radiation absorbed by animals with light coats (e.g., Figure 8.7b) is therefore strongly dependent on the absorptivity of skin.

4. Although the skin of the dark-fleeced Welsh Mountain sheep has a smaller reflection coefficient than that of the Dorset Down, the Welsh Mountain skin absorbs less radiation than the Dorset Down skin for all fleece lengths (clean or soiled) because transmission is less in the fleece with dark hair.

Table 8.3 helps to explain the apparent paradox that a white coat is a disadvantage for an animal in a hot sunny environment, but an advantage in a cold sunny environment, when heat absorbed at the skin surface is desirable. In principle, the thermal load on an animal with a white coat and dark skin could be larger than for a dark coat and light skin. The relevant equations have been developed and explored by Walsberg et al [1978]. For example, polar bears have the combination of a deep coat made up of transparent hollow hairs, and black skin. The coat appears white because the hollow hair core scatters and reflects visible light very effectively, much as snow does. Forward scattering of radiation allows the black skin to absorb radiative energy, and the hollow hairs are both buoyant and insulating. Table 8.3 can also be used to show the consequences for the radiation balance when a sheep with long, soiled fleece is sheared, leaving it with a clean and shorter fleece; comparing a 4 cm soiled Dorset Down fleece with a 1 cm clean fleece shows that the radiation load on the skin increases by an order of magnitude after shearing.

There have been very few definitive studies of the relation between skin color, heat stress and live weight gain in livestock. Finch et al [1984], working with Brahman and Shorthorn cattle in Queensland, were able to demonstrate a significant positive correlation between reflection coefficient and live weight gain. The largest response was observed in individuals with the thickest and woolliest coats. In the absence of heat stress, however, several workers have found a negative correlation between reflection coefficient and live weight gain for reasons that remain obscure. See Figure 8.7 (See color plates).

8.3 Net Radiation

In Chapter 5, net radiation was presented as a climatological variable with the caveat that its value depends on the temperature and reflectivity of the surface exposed to radiative exchange. Having considered how the interception of radiation by an organism is determined by its geometry, and having established characteristic values of reflectivity for natural surfaces, it is possible to compare differences in the net flux of radiant energy absorbed by contrasting surfaces exposed to the same radiation environment as specified by a flux of short-wave radiation (received from the sun and sky and reflected from the ground) and a long-wave flux (received from the atmosphere and emitted by the ground). A suitable general form of the equation describing the radiation budget is

$$\mathbf{R_n} = (1 - \rho)\,\mathbf{S_t} + \mathbf{L_d} - \mathbf{L_u} \qquad (8.25)$$

and applications of this expression are now considered for four contrasting surfaces (Figure 8.8, see color plates).

(i) **A short grass lawn**

For a continuous horizontal surface receiving radiation from above and not from below, the net radiation is simply

$$\mathbf{R_n} = (1 - \rho_1)\,\mathbf{S_t} + \mathbf{L_d} - \sigma\,\mathbf{T}_1^4 \qquad (8.26)$$

where ρ_1, T_1, are the reflection coefficient and radiative temperature of the lawn.

(ii) **A horizontal leaf**

If the leaf is assumed exposed sufficiently above the lawn for its shadow to be ignored, it receives an additional income of short-wave radiation $S_t = \rho_1 S_t$, and of long-wave radiation $L_e = \sigma T_1^4$. The net radiation is therefore

$$\mathbf{R_n} = \left\{ (1 - \rho_2)\,(1 + \rho_1)\,\mathbf{S_t} + \mathbf{L_d} + \mathbf{L_e} - 2\sigma\,\mathbf{T}_2^4 \right\} / 2 \qquad (8.27)$$

where ρ_2, T_2 are the reflection coefficient and radiative temperature of the leaf. Note that $\mathbf{R_n}$, is the net radiation per unit of *total* leaf area; i.e., twice the plan area or twice the leaf area index.

TABLE 8.4   Conditions assumed for the radiation budgets in Figure 8.8.

	1 High sun clear	2 High sun partly cloudy	3 Low sun clear	4 Overcast day	5 Clear night
Solar elevation β	60	60	10	—	—
Direct solar radiation					
$\quad S_b$ (W m^{-2})	800	800	80	—	—
Diffuse solar radiation					
$\quad S_d$ (W m^{-2})	100	250	30	250	—
Downward long wave radiation					
$\quad L_d$ (W m^{-2})	320	370	310	380	270
Surface temperature (°C)					
$\quad$ air	20	20	18	15	10
$\quad$ lawn	24	24	15	15	6
$\quad$ leaf	24	25	15	15	4
$\quad$ sheep	33	36	15	20	10
$\quad$ man	38	39	15	20	10
Reflectivities					
$\quad$ lawn	0.23	0.23	0.25	0.23	—
$\quad$ leaf	0.25	0.25	0.35	0.25	—
$\quad$ sheep	0.40	0.40	0.40	0.40	—
$\quad$ man	0.15	0.15	0.15	0.15	—

(iii) **A sheep**

The sheep is assumed standing on the lawn and so receives radiation reflected and emitted by the surface below it. When the area of shadow is ignored, the net radiation for the sheep is

$$R_n = \left(1 - \bar{\rho}_3\right)\left(1 + \rho_1\right)\bar{S}_t + \bar{L}_d + \bar{L}_e - \sigma T_3^4 \qquad (8.28)$$

where the bars indicate averaging over the exposed surface. The sheep is further assumed to be a horizontal cylinder with its axis at right angles to the sun's rays, reflectivity is ρ_3, and mean surface temperature is T_3.

(iv) **A man**

For a man standing on the lawn, the radiation balance is formally identical to Equation 8.28 with ρ_4, T_4 replacing ρ_3, T_3.

Values assumed for the radiation fluxes and for ρ and T are given in Table 8.4. Long-wave emissivities are assumed to be unity.

It is instructive to compare the net radiation received by different surfaces at the same time, noting the effects of geometry and reflection coefficients, or to compare the same surface at different times to see the influence of solar elevation and cloudiness. Salient features of these comparisons are as follows.

(a) During the day, the lawn absorbs more *net* radiation than any of the other surfaces, including the isolated leaf. The leaf receives less *short-wave* radiation than the lawn $((1 + \rho_1)S_t/2$ compared with $S_t)$ and absorbs more *long-wave* radiation $((L_d + \sigma T_1^4)/2$ compared with $L_d)$.

(b) The sheep absorbs less *net* radiation than the other surfaces. This is partly a consequence of the relatively large reflection coefficient (0.4) and partly a consequence of geometry.

(c) The geometry of the man ensures that R_n is large in relation to the other surfaces when the sun is low.

(d) For all surfaces, the *net* radiation is greatest when the sun is shining between clouds and is larger under an overcast sky than it is when the sun is near the horizon.

(e) At night, the leaf, sheep, and man receive *long-wave* radiation from the lawn and from the sky, so their *net loss of long-wave* radiation is less than the *net loss* from the lawn.

NET RADIATION MEASUREMENT

Measuring the net exchange of radiation at the surface of a uniform plane poses no problems: a net radiometer is exposed with its sensing surfaces parallel to the plane. For more complex surfaces, the net flux can be derived from an application of Green's theorem, which states that the flux of any quantity received within or lost by a defined element of space is the integral of the flux evaluated at right angles to the envelope that defines the space. This principle has been applied to measure the net radiant exchange of apple trees with tubular net radiometers defining the surface of a cylinder and with thermopile surfaces parallel to the surface of the cylinder [Thorpe, 1978]. Similarly, Funk [1964] measured the net radiant flux received by a man standing inside a vertical cylindrical framework over which a single net radiometer moved on spiral guides, always pointing toward the axis of the cylinder.

8.4 Problems

1. A canopy has a vertical (cylindrical) leaf angle distribution and leaf area index of 4. When the solar elevation is $45°$ and S_b above the

canopy is 500 W m^{-2}, calculate, for surfaces at the base of the canopy, (i) the mean direct solar irradiance per unit surface area, (ii) the fractional area of sunflecks, (iii) the mean irradiance of sunlit leaves.

2. A forest canopy with a leaf area index of 3.0 intercepted about 80% of the incident direct solar radiation when the solar elevation was 60°. Determine whether the leaf angle distribution function was spherical or vertical (cylindrical). What assumptions about the foliage properties have you needed to make, and how might they influence interception in a "real" situation?

3. The daily insolation on a horizontal surface was 32 MJ m^{-2} on a cloudless summer day.

   a) Estimate the fraction τ of this daily energy receipt that would be measured at the bottom of a crop canopy with a leaf area index of 4.0.

   b) Hence, estimate the fractional interception of solar radiation by the canopy.

   c) Find the leaf area index that would intercept 95% of the daily radiation.
   (Assume a spherical leaf distribution for all parts of the question.)

4. A canopy of a shrub species with hairy leaves is sufficiently dense and deep for the assumptions of isotropic multiple scattering of solar radiation to apply. Assume that the individual leaves have reflection (ρ) and transmission (τ) coefficients both of 0.15 for PAR.

   a) What is the absorption coefficient (α_p) for PAR of the leaves?

   b) What is the reflection coefficient (ρ_c^*) of the canopy for PAR?

   c) Repeat the calculations of (a) and (b) for near infrared (NIR) radiation assuming leaf properties $\rho = \tau = 0.40$ for this waveband.

   d) If the PAR and NIR bands each contain half the energy in the solar spectrum at the surface, what is the value of ρ_c^* for the whole solar spectrum?

5. Consider a sparse forest canopy with the following characteristics:

Leaf area index	$L = 1.0$
Theoretical attenuation coefficient	$\mathcal{K}_b = 1.0$
	(assuming black leaves)
Soil reflection coefficient	$\rho_s = 0.15$
Leaf reflection coefficient (PAR)	$\rho_p = 0.10$
Leaf transmission coefficient (PAR)	$\tau_p = 0.10$

Calculate:

a) The canopy attenuation coefficient $\mathcal{K}$ for PAR.

b) The canopy reflection and transmission coefficients (ρ_c and τ_c) for PAR, assuming second order terms can be ignored.

c) The absorption coefficient of the canopy (α_c) for PAR.

d) The ratio of absorption (α_c) to interception ($1 - \tau_c$) of PAR by the canopy.

Chapter 9

Momentum Transfer

When plants or animals are exposed to radiation, the energy they absorb can be used in three ways: for heating, for the evaporation of water, and for photochemical reactions (e.g., photosynthesis). Heating of the organism itself or of its environment implies a transfer of heat by conduction or by convection; evaporation involves a transfer of water vapor molecules in the system; and photosynthesis involves a similar transfer of carbon dioxide molecules. At the surface of an organism, heat and mass transfer are sustained by molecular diffusion through a thin skin of air known as a *boundary layer* in contact with the surface. The behaviour of this layer depends on the viscous properties of air and on the transfer of momentum associated with viscous forces. Similar principles apply to organisms in water; further details may be found in Leyton [1975] and Ellington and Pedley [1995]. A discussion of momentum transfer is therefore needed as background to the next three chapters, which consider different aspects of exchange between organisms and their environment.

9.1 Boundary Layers

Figure 9.1 shows the development of a boundary layer over a smooth flat surface immersed in a moving fluid (i.e., a gas or liquid). When the streamlines of flow are almost parallel to the surface, the layer is said to be *laminar*, and the flow of momentum across it takes place by the momentum exchange between individual molecules discussed on p. 29. The thickness of a laminar boundary layer cannot increase indefinitely because the flow becomes unstable and breaks down to a chaotic pattern of swirling motions called a *turbulent boundary layer.* A second laminar layer of restricted depth—*the laminar sublayer*—forms immediately above the surface and below the turbulent layer.

The transition from laminar to turbulent flow depends on the relative magnitudes of inertial forces associated with the horizontal movement of the fluid and of viscous forces generated by inter-molecular attraction (sometimes referred to as "internal friction"). The ratio of inertial to viscous forces is known as the *Reynolds number* (Re) after the physicist who first showed that this ratio determined the onset of turbulence when a liquid flowed through a pipe. When Re is small, viscous forces predominate so that the flow tends to remain laminar, but when the ratio increases beyond a critical value Re_c, inertial forces dominate and the flow becomes turbulent. (See Figure 9.1).

The general form of Re is Vd/v where V is the fluid velocity, d is an appropriate dimension of the system, and v is the coefficient of kinematic viscosity of the fluid. For flow over a flat plate, d is the distance from the leading edge. When a very smooth flat plate is exposed to a parallel flow of air virtually free from turbulence, Re_c is of the order of 10^6 but the engineering literature quotes values of about 2×10^4 observed in less rigorous conditions.

Both in laminar and in turbulent boundary layers, velocity increases from zero at the surface to the free-stream value V at the top of the bound-

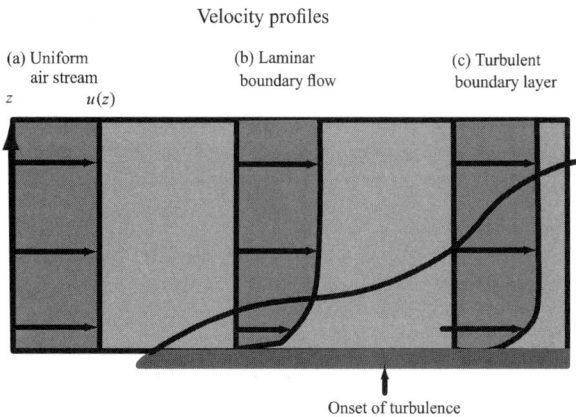

Velocity profiles

(a) Uniform air stream   (b) Laminar boundary flow   (c) Turbulent boundary layer

z $u(z)$

Onset of turbulence

FIGURE 9.1 Development of laminar and turbulent boundary layers over a smooth flat plate (the vertical scale is greatly exaggerated).

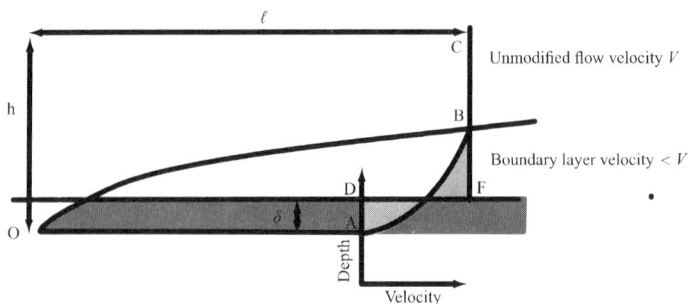

FIGURE 9.2   Boundary layer OB, displacement boundary layer (gray) and wind profile
(CBA) over a smooth flat plate exposed to an airstream with uniform velocity V.

ary layer, but definitions of boundary layer depth are necessarily arbitrary.
One definition is the streamline where the velocity is $0.99V$, but in problems of momentum transfer it is more convenient to work with an average
boundary layer depth.

In Figure 9.2, the flow of air reaching a flat plate across a vertical cross-
section of depth h is proportional to the velocity V times h. At a distance
l from the edge of the plate, the velocity profile is represented by the line
ABC, and the flow through the cross section at C will be less than Vh
because the velocity in the boundary layer is less than V. The same reduction in flow would be produced by a layer of completely *still* air with
thickness δ (shaded) above which the air moves with a uniform velocity V.
The velocity profile in this equivalent system is represented by ADFC. The
depth δ, known as the "*displacement boundary layer*," can be regarded as
an average depth of the boundary layer between the leading edge of the
plate and the cross section at C.

By applying the principle of momentum conservation to flow within a
laminar boundary layer over a smooth plate, it can be shown that the depth
of the displacement boundary layer expressed as a fraction of the distance
l from the leading edge is

$$\delta/l = 1.72 \ (\mathrm{Re})^{-0.5} \tag{9.1}$$

implying that δ increases with $l^{0.5}$ and $\nu^{0.5}$, and decreases with $V^{0.5}$. (For
an estimate of the actual boundary layer depth at l as distinct from an
average depth over a distance l, the numerical factor in this equation can
be replaced by 5.) The depth of a turbulent boundary layer increases with
$l^{0.8}$.

SKIN FRICTION

The force that air exerts on a surface in the direction of the flow is a direct consequence of momentum transfer through the boundary layer and is known as *skin friction*. To establish analogies with heat and mass transfer later, the transfer of momentum is conveniently treated as a process of diffusion (see p. 29). If t is a diffusion path length for momentum transfer from air moving with velocity V to a surface where velocity is zero, then the frictional force is

$$\tau = \upsilon \rho V / t = \rho V / r_M \qquad (9.2)$$

when $r_M = t/\upsilon$ is a resistance for momentum transfer. From theoretical analysis for the flow over a smooth plate, the drag per unit surface area is proportional to $V^{3/2}$ and is given by

$$\tau = 0.66 \rho V \, (V \upsilon / l)^{0.5} \qquad (9.3)$$

Comparison of Equations 9.2 and 9.3 then gives the resistance as

$$r_M = 1.5 \, (l / V \upsilon)^{0.5} = 1.5 V^{-1} \, \text{Re}^{0.5} \qquad (9.4)$$

For example, if $V = 1$ m s^{-1} and $l = 0.05$ m, then $r_M = 90$ s m^{-1}, establishing an order of magnitude relevant to many micrometeorological problems.

Over natural surfaces, the flow of air is usually much more complex than Figure 9.1 suggests, but for a poplar leaf parallel to a laminar flow of air in a wind-tunnel (Figure 9.3) the profiles of windspeed developed on the upper surface were equivalent to those in Figure 9.1. At the lower surface, profiles were distorted by curvature of the leading edge, which produced shelter but also generated turbulence in its wake. When V was 1.5 m s^{-1} or with minimum turbulence in the airstream, windspeed within a few millimeters of the upper surface upwind of the mid-rib was close to the theoretical value for a flat plate. The value of Re$_c$ for the onset of turbulence was about 9×10^3. Increasing the velocity or the level of turbulence in the airstream increased the velocity at a fixed point in the boundary layer relative to V and decreased Re$_c$ to 1.9×10^3. These observations are relevant to the exchange of momentum and related forces on the surfaces of leaves and to the forces responsible for detaching fungal spores [Grace and Collins, 1976; Aylor, 1975, 1990].

FORM DRAG

In addition to the force exerted by skin friction, a consequence of momentum transfer to a surface across the streamlines of flow, bodies immersed

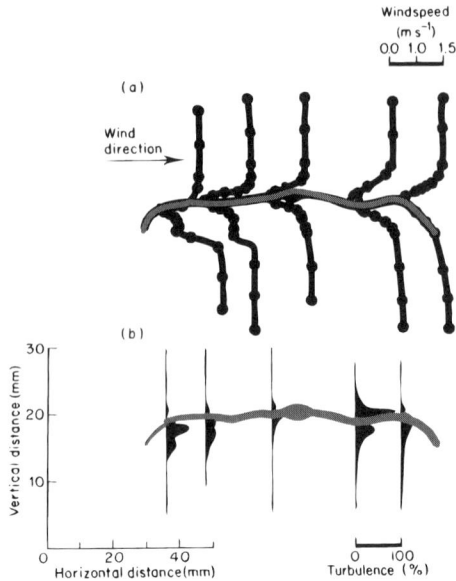

FIGURE 9.3   Profiles of mean windspeed (a) and turbulence intensity i (b) around a *Populus* leaf shown in transverse section in a laminar free stream (from Grace and Wilson, 1976).

in moving fluids experience a force in the direction of the flow as a result of the deceleration of fluid. This force is known as *form drag* because it depends on the shape and orientation of the body. Maximum form drag is experienced by surfaces at right angles to the fluid flow, and the force can be estimated by assuming that there is a point on the surface where the fluid is instantaneously brought to rest after being uniformly decelerated from a velocity V. If the initial momentum per unit volume of fluid is ρV and the mean velocity during deceleration is $V/2$, the rate at which momentum is lost from the fluid is $\rho V \times V/2 = 0.5\rho V^2$. This is the maximum rate at which momentum can be transferred to unit area on the upstream surface of a bluff body, and it is therefore the maximum pressure excess a fluid can exert in contributing to the total form drag over the body. In practice, fluid tends to slip around the sides of a bluff body so that a force smaller than $0.5\rho V^2$ is exerted on the upstream face. However, in the wake that forms downstream of a bluff body, the fluid pressure is less than in the free stream, and the associated suction force often makes an important contribution to the total form drag on a body. (As examples, gulls diving for fish assume a "streamlined" shape that minimizes the wake to maximize their fall speed, and in the sporting world, golf ball manufacturers adopt complex patterns for the dimples on golf balls to reduce the wake and consequently to reduce the drag.) The total drag force on unit area is conveniently expressed as $c_f \times 0.5\rho V^2$ where c_f is a drag coefficient.

In most problems, it is appropriate to combine skin friction and form drag to give a total force τ, usually the force on a unit area projected in the direction of the flow; i.e., $2rl$ for a cylinder of radius r and length l in cross-flow and πr^2 for a sphere. The ratio $\tau/(0.5\rho V^2)$ then defines the total *drag coefficient* c_d, which for spheres and for cylinders at right angles to the flow lies between 0.4 and 1.2 in the range of Reynolds numbers between 10^2 and 10^5. Manufacturers of energy-efficient cars boast drag coefficients of about 0.3.

As background for later discussion of mass and heat transfer, it should be noted that the diffusion of momentum in skin friction is analogous to the diffusion of gas molecules and of heat provided the surface is *parallel* to the airstream. For such a surface, close relations may be expected between the transfer resistances r_M, r_H, and r_V. For a surface at *right angles* to the air stream however, there is no frictional force in the direction of flow. Friction will operate in all directions at right angles to the flow, but the net sum of all these (vector) forces must be zero. In contrast, the net flux of heat or mass, which are scalar quantities, must be finite in the plane of the surface. In this case, r_V and r_H may be similar to each other but will be unrelated to the value of r_M.

9.2 Momentum Transfer to Natural Surfaces

The atmosphere in motion exerts forces on all natural surfaces—individual leaves, plants, trees, crops, animals, bare soil, and open water. Conversely, every object or surface exposed to the force of the wind imposes an equal and opposite force on the atmosphere proportional to the rate of momentum transfer between the air and the surface. Momentum transfer is always associated with wind "shear": the windspeed is zero at the surface of the object and increases with distance from the surface through a boundary layer of retarded air.

Isolated objects such as single plants or trees tend to have very irregular boundary layers, and disturb the motion of the atmosphere by setting up a train of eddies in their wake rather like the eddies formed downstream from the piers of a bridge. Surfaces such as bare soil and uniform vegetation also generate eddies in the air moving over them because the drag that they exert on the air is incompatible with laminar flow. Vogel

[1994] provides a comprehensive survey of fluid dynamics applied to air- and water-dwelling organisms.

DRAG ON LEAVES

To avoid the fluctuations of windspeed that are characteristic of the atmosphere, the drag on natural objects can be measured in a wind tunnel where the flow is steady and controlled. Thom [1968] studied the force on a replica of a "leaf" made of thin aluminium sheet. Figure 9.4 shows the dimensions of the replica and Figure 9.5 shows how the drag coefficient c_d changed with windspeed and direction. Note that the quantity c_d shown in Figure 9.5 was calculated by dividing the force per unit plan area of the leaf (τ/A) by ρV^2 which is numerically the same as the force per unit of total surface area $(\tau/2A)$ divided by $0.5\rho V^2$. This departure from an aero-

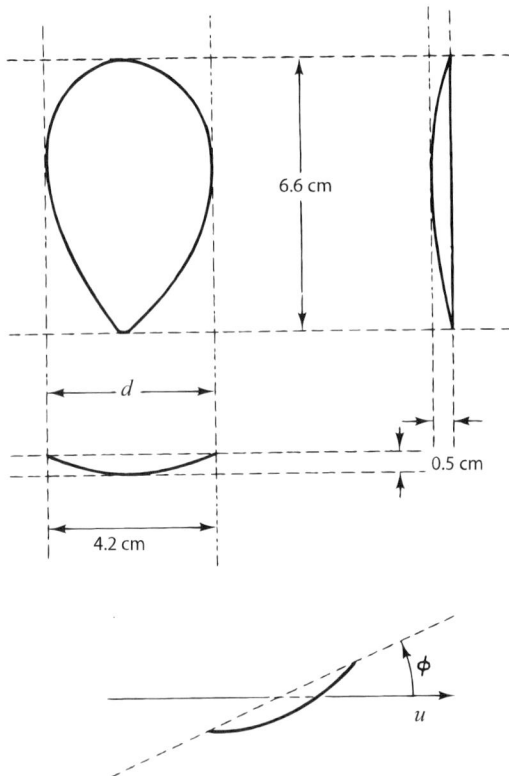

FIGURE 9.4   Shape and dimensions of model leaf used by Thom [1968].

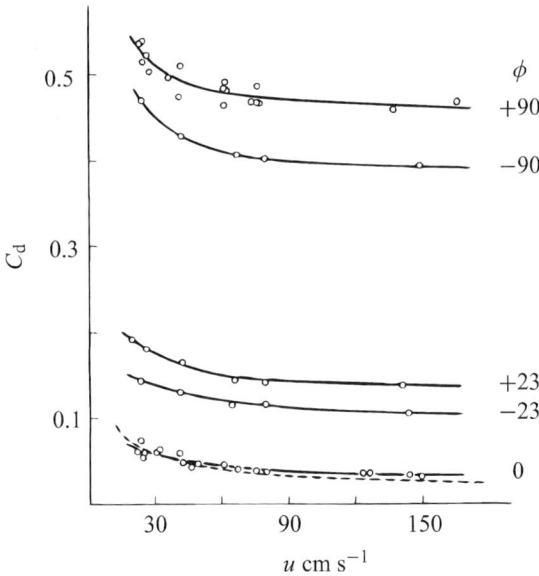

FIGURE 9.5 Drag coefficient of a model leaf (full lines) as a function of windspeed u and angle between leaf and airstream (see Figure 9.4). The broken curve is a theoretical relationship for a thin flat leaf at $\phi = 0$ [from Thom, 1968].

dynamic convention (which uses a projected area) allows the resistance to momentum transfer for the whole leaf to be written as $r_M = 1/(V c_d)$, equivalent to the combination in parallel of the two resistances of $2/(V c_d)$ for each surface separately.

When the leaf was oriented in the direction of the airstream ($\varphi = 0$) the drag was minimal and r_M was about half the value derived from Equation 9.4 which gives the resistance to momentum transfer for one side of a plate only (i.e., the resistance of the leaf was consistent with theory for a two-sided flat plate). When the concave or convex surfaces were facing the airstream ($\varphi = +90$ or $\varphi = -90$, respectively), form drag was much larger than skin friction.

The general form of the curves in Figure 9.5 is consistent with a combination of form drag proportional to V^2 and skin friction proportional to $V^{1.5}$. The total drag coefficient c_d can be expressed as the sum of a form drag component c_f and a frictional component $n V^{-0.5}$ where n is a constant; i.e.,

$$c_d = c_f + n V^{-0.5}$$

The relevance of wind tunnel measurements of leaf drag to momentum transfer in a crop is a matter for debate. In the first place, turbulence is usually suppressed in wind tunnels to achieve laminar flow, whereas the movement of air in canopies is nearly always turbulent. Turbulent eddies of an appropriate size can increase rates of momentum transfer by disturbing flow in the laminar boundary layer, and the disturbance is likely to be accentuated if the turbulence is strong enough to make the leaves flutter. Measurements by Rashke indicate that the characteristic diameter of eddies shed by cylinders of diameter d is about $5d$ when Re exceeds 200. Within plant canopies, the eddies shed by leaves and stems decay to form smaller and smaller eddies drifting downwind and penetrating the boundary layer of other leaves and stems. Above plant canopies or bare soil, turbulent eddies generated at the surface increase with height. Mitchell [1976] measured heat transfer from spheres to the atmosphere at various heights above the ground, and found that when the ratio of height to sphere diameter was between 2 and 10, the resistance to transfer was around 0.75 of the theoretical (laminar flow) value.

The drag on real leaves may also be increased by the roughness of the cuticle and by the presence of hairs. Sunderland [1968] found that the drag on an aluminium replica of a wheat leaf increased when a real leaf was attached to the metal surface. The increase was about 20% at 1.5 m s^{-1} growing to 50% at 0.5 m s^{-1} because of the increasing importance of skin friction at low windspeeds.

For calculations of momentum transfer between single leaves and the atmosphere, it is reasonable to reduce resistances to momentum transfer calculated from Equation 9.4 by a factor of about 1.5 to allow for the effects of turbulence, but it should be recognized that the factor depends on the actual scale of turbulence in any particular situation.

In the real world, leaves rarely exist in aerodynamic isolation, and the drag coefficient for foliage therefore depends on foliage density and on windspeed. A convenient parameter specifying density in this context is the ratio of the total surface area of laminae, petioles, etc., in a specimen, divided by the plan area facing the wind. In the canopies of arable crops or of a deciduous tree, most leaves are exposed to turbulent air in the wake of their upwind neighbors and in coniferous trees there is similar interference between needles. The extent to which the drag on individual elements of foliage is reduced by the presence of neighbors was expressed by Thom [1971] in terms of a *shelter factor*, the ratio of the actual drag coefficient observed to the coefficient measured (or estimated) for the same element in isolation. Shelter factors depend on foliage density and on windspeed. Representative values found in the literature range from 1.2 to 1.5 for shoots of apple trees, to 3.5 for a stand of field beans and for a pine forest.

Even without the complications of turbulence and shelter, the dependence of drag on windspeed is very complex because of the interaction between aerodynamic forces on leaves and elastic forces opposing them. As windspeed increases from zero, fluttering begins, even at a constant windspeed, if the two sets of forces cannot achieve equilibrium, and the consequence is a sudden increase of momentum transfer and drag. With a further increase of windspeed, many types of leaf tend to bend into a streamlined position [Vogel 1993], reducing drag and eliminating flutter when the windspeed is constant. In nature, however, wind fluctuates continuously both in speed and in direction so that leaves, stems, and small branches oscillate even in quite light winds.

WIND PROFILES AND DRAG ON EXTENSIVE SURFACES

The wind also exerts a drag force on extensive surfaces such as bare soil and canopies of uniform vegetation as a consequence of wind shear. This force can be described as the rate of momentum transfer (momentum flux) between the atmosphere and the surface. In Chapter 3, it was shown that the vertical flux of an entity s from the atmosphere to an extended surface can be represented by the mean value of the product $\rho w's'$, which is finite when fluctuations of vertical velocity w' are correlated, either positively or negatively, with simultaneous fluctuations of the entity s'. When the entity is horizontal momentum, and coordinates are aligned so that the x-axis points in the direction of the mean flow, s' becomes the fluctuation of the horizontal velocity u', and the vertical flux of horizontal momentum from the atmosphere to the surface is given by $\overline{\rho u'w'}$. Provided that the vertical flux is constant with height, this quantity can be identified as the drag force per unit ground area, otherwise known as the shearing or Reynolds stress (τ); i.e.,

$$\tau = \overline{\rho u'w'} \tag{9.5}$$

Based on this relationship, a velocity known as the friction velocity u_* is defined as

$$u_* = \overline{(u'w')}^{0.5} = (\tau/\overline{\rho})^{0.5} \tag{9.6}$$

so that

$$\tau = \overline{\rho}u_*^2 \tag{9.7}$$

Equation 9.6 implies that the friction velocity is related to the magnitude of turbulent fluctuations. The shearing stress may also be written as

$$\tau = \rho K_M du/dz \tag{9.8}$$

where K_M is a *turbulent transfer coefficient* for momentum with dimensions $L^2 T^{-1}$. Equation 9.8 is similar to Equation 3.6 which related momentum flux in laminar boundary layers to molecular viscosity, but, in the atmosphere, both windspeed and turbulent mixing normally increase with height, so K_M is height dependent. In Chapter 16, simple dimensional arguments are used to obtain the functional relations between windspeed and height.

DRAG ON PARTICLES

Particles in the atmosphere experience drag forces when there is relative motion between the air and the particle. Such forces allow air movements to detach pollen from stamens or spores of pathogens from surfaces, and partly determine whether particles in an airstream are captured at the surfaces of vegetation (see Chapter 12).

When a particle is in steady motion, the drag force exerted on it by the air must balance other external forces such as gravity, electrical attraction, etc. The origins of the drag forces on particles can be conveniently treated as three special cases applied to spherical particles:

(i) Particles with radius r much smaller than the mean free path λ of gas molecules.

In this case, particles behave like giant gas molecules, and the drag force is the result of more molecules impinging on the surface of a moving particle from in front than from behind.

Provided that the mass of a particle is much larger than the mass m_g of the gas molecules and assuming perfect reflection in collisions, kinetic theory can be used to show that the drag force on a particle moving at velocity V in a gas is

$$F = \frac{4}{3}\pi n m_g \bar{c} r^2 V \tag{9.9}$$

where n is the number of gas molecules per unit volume, and $\bar{c}$ is their mean velocity. Thus, the drag force is proportional to particle velocity and to the surface area (r^2).
(At 20°C and 101kPa, n is about 3×10^{25} molecules m^{-3}, the average value of m_g for air is about 5×10^{-26} kg, and $\bar{c}$ is about 500 m s^{-1}.

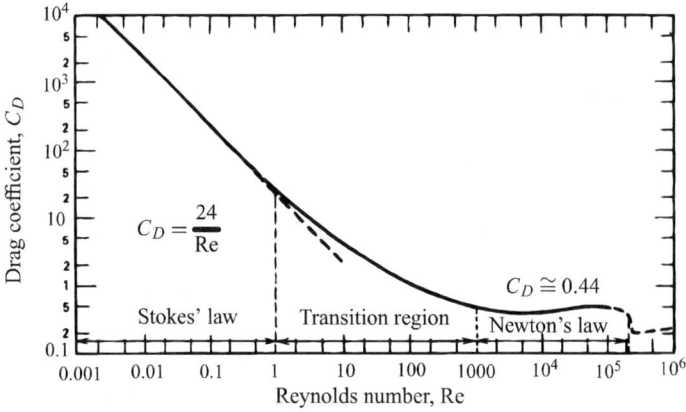

FIGURE 9.6 Relation between the drag coefficient for a sphere and Reynolds number. The dashed line is based on Stokes' Law. The discontinuity at Re $\simeq 3 \times 10^5$ corresponds to the transition from a laminar to a turbulent boundary layer [Hinds 1999].

The mean free path of molecules in air is about 0.066 μm, so that this case applies only to very small particles with radii less than 0.01 μm.)

(ii) Particles with radius r larger than λ, but where the Reynolds number Re$_P$ for relative motion between gas and particle is small (i.e., $2rV/\upsilon$ is less than about 0.1).

In this case, the viscous forces arising from skin friction between the particle and the gas dominate the drag. Stokes showed that for this case

$$F = 6\pi \rho_g \upsilon r V \qquad (9.10)$$

where ρ_g is the density of the gas. Thus, the drag force is proportional to particle velocity and to radius. (The kinematic viscosity υ of air is about 15×10^{-6} m^2 s^{-1}, and so Stokes' Law applies to 10 μm radius particles typical of pollen grains when $V \times 2 \times 10 \times 10^{-6}/(15 \times 10^{-6}) < 0.1$; i.e., $V < 0.08$ m s^{-1}.)

(iii) Particles with $r > \lambda$ and Re$_P > 1$

In this case, the inertial forces corresponding to the form drag as the gas flows around the particle dominate the drag. The drag force is given by

$$F = c_d \times 0.5\rho_g V^2 A \qquad (9.11)$$

where A is conventionally taken as the cross-sectional area of the particle and c_d is a drag coefficient. Equation 9.11 was first derived by Isaac Newton as part of an analysis of the motion of cannonballs. He thought that c_d was independent of velocity for a given shape, but this is only a good approximation for $Re_p > 1000$ (relevant for cannonballs, but not for most natural aerosols), and the drag coefficient of a sphere increases as Re decreases, as shown in Figure 9.6. At small values of Re_p, combining Equations 9.6 and 9.7 shows that

$$c_d = \frac{6\pi \upsilon \rho r V}{0.5\rho V^2 \pi r^2} = \frac{12\upsilon}{Vr} = \frac{24}{Re_p} \qquad (9.12)$$

and this reciprocal relation gives the linear section on Figure 9.6 when $Re_p < 0.1$. The empirical expression

$$c_d = (24/Re_p)\left(1 + 0.17\ Re_p^{0.66}\right) \qquad (9.13)$$

[Fuchs, 1964] is accurate within a few percent for $1 < Re_p < 400$; it overestimates c_d by about 7% when $Re_p \simeq 0.5$. In the region when $Re_p > 10^3$, c_d is constant and so drag force is proportional to V^2 and to cross-sectional area. Further details of particle motion in the atmosphere are in Hinds [1999].

Aylor [1975] used these principles to calculate the force required to detach spores of the pathogen *Helminthosporium maydis* from infected leaves of maize. This fungal pathogen produces roughly cylindrical spores (diameter about 20 μm and projected area about 4×10^{-10} m^2), which grow on stalks projecting 150 μm from the leaf surface. To find the minimum force necessary to detach a spore from its stalk, Aylor observed spores through a microscope while dry air was blown on them from a narrow tube. Figure 9.7 shows that 50% of the spores were removed when the (steady) windspeed was about 10 m s^{-1}. The corresponding drag force may be found from Equation 9.7 which, taking $\rho_g = 1.2$ kg m^{-3}, $A = 4 \times 10^{-10}$ m^2 and $c_d = 4$ for a cylinder at $Re_p \simeq 10$ (similar to the value for a sphere, Figure 9.6), yields $F \simeq 1 \times 10^{-7}$ newtons. Aylor also used a centrifugal method of detaching spores, and this gave excellent agreement with his estimates of F. Spores are detached at much lower *mean* windspeeds than this in the field, indicating the importance of brief gusts in breaking down the leaf boundary layer and dispersing pathogens into the atmosphere.

9.3 Lodging and Windthrow

The forces exerted by the wind on trees, crops and isolated plants are sometimes sufficient to cause them to be broken or uprooted. The principles underlying lodging of crops and windthrow of trees are similar. Figure 9.8 shows the forces involved.

The drag force on a plant is the resultant of the drag on many individual leaves exposed to differing windspeeds and turbulence. The drag produces a turning moment about a rotation point where the stem enters the soil. As the plant is displaced, turning moments also occur due to the mass of the tree and the soil attached to the roots. The force of the wind is opposed by tension in roots and by the resistance of the soil to shear. Baker [1995] developed a theoretical model for windthrow and applied it to predict conditions in which cereal crops and forest would be damaged. His model related mean and turbulent wind velocities to the mean and fluctuating displacements of the plant. These displacements generate forces and moments acting at or near the stem base. When the bending moment at the base exceeds the strength of the stem or of the soil-root system, the plant is assumed to fail. The model demonstrated that the natural frequency of oscillation is an important factor determining stability. Oscillations are damped by interference between adjacent plants in canopies, aerodynamic forces, and energy dissipation within the plant itself.

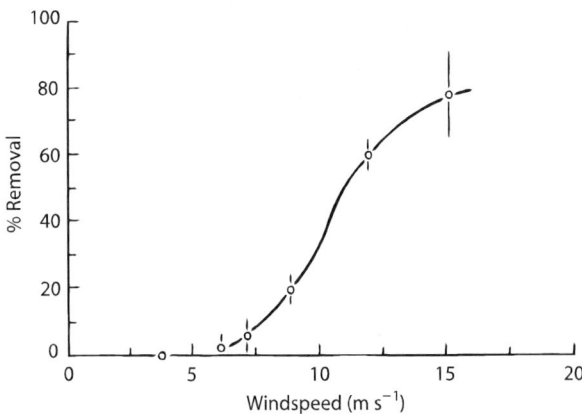

FIGURE 9.7 Percentage of spores removed by blowing for 15 s on spores reared on dried plant material (from Aylor, 1975). Vertical bars represent standard deviations.

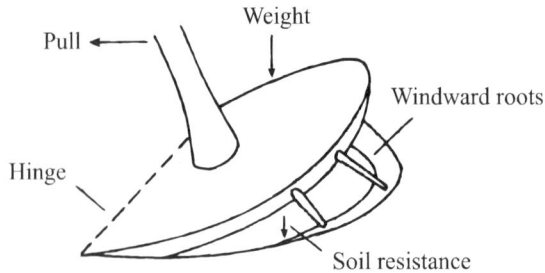

FIGURE 9.8   Forces acting on an isolated shallow-rooted tree exposed to the wind, showing the four components of anchorage that resist overturning (after Stokes et al. 1995).

LODGING OF CROPS

Lodging of grass and cereal crops occurs as they near maturity because their tall stems are pushed sideways by the wind. Failure may occur when the stem buckles (stem lodging) or when the root system rotates in, or is pulled out of, the soil (root lodging, Ennos 1991). Plant breeders have developed cultivars with shorter, stiff stems to resist stem lodging, and farmers also may apply growth regulating chemicals to reduce stem elongation, but these strategies cause greater forces to be transmitted to root systems and may make root lodging more prevalent. Lodging is most likely to occur when the weight of the upper parts of plants is increased by the interception of rain, when the lower parts of stems are weakened by disease or by a heavy application of nitrogenous fertilizer, or when the shearing strength of the soil around the roots is weakened by rainfall. Tani [1963] investigated the forces needed to bend the stems of rice beyond their elastic limit. Ennos [1991] measured forces associated with stem bending and root anchorage in mature spring wheat (*Triticum aestivum*) grown in the laboratory.

Tani found that when mature rice plants were exposed to a uniform wind in the laboratory, the stems broke when the forces on the plants produced a moment of about 0.2 N m on the base of the stems. Plants growing to a height of about 0.84 m in the field were found to lodge when the windspeed exceeded 20 m s^{-1}. At this speed, the moment on the base of the stem had two components: a moment of 0.034 N m induced by the force of the wind (mainly form drag), and a moment of 0.023 N m induced by the force of gravity acting because the top of the stem was displaced by about 40 cm. The total moment needed to break the stem under field conditions was therefore 0.057 N m, about a quarter of the laboratory value. The discrepancy between the figures may be explained by (a) the much larger forces exerted in the field during strong gusts when the instantaneous windspeed can be two or three times larger than the mean, or (b) a resonance set up between the natural period of oscillation of the plants (about 1 s) and the dominant period of turbulent eddies at the top of the canopy, or (c) the effect of disease on field-grown plants.

Ennos [1991] concluded that the resistance of wheat stems to bending was about 30% greater than the resistance of the root anchorage, so root systems would fail before stems. His measured values for the moments necessary to buckle stems were about 0.2 N m, comparable to the laboratory values of Tani. Ennos found that the stiffness of roots and their resistance to axial movement (i.e., being pulled out of soil along their axes) contributed about equally to the anchorage strength, but this equality is likely to change in soils with different shear strengths.

DRAG ON TREES

Trees may be uprooted or broken at the stems by strong winds. Understanding why most trees uproot but some trees break requires knowledge of the dynamic forces exerted by the wind and the physical properties of the tree and soil structure [Wood, 1995]. As trees grow, they adapt to their wind exposure by developing root and branch structure sufficient to resist forces that have occurred during their lifetime. They are thus more vulnerable to wind damage if they become more exposed (e.g., if neighboring trees die or are harvested).

Wind forces on an isolated tree

To measure drag, several investigators have put small trees or parts of larger trees in wind tunnels and measured the drag forces on them. Fraser [1962], Mayhead [1973] and Mitchell [Rudnicki et al 2004, Vollsinger et al 2005] exposed small conifers, broadleaves, and tops of larger trees in wind tunnels (Figure 9.9a–d, see color plates) and found that the relation between drag force F and windspeed V was strongly affected by a decrease in the effective cross-section of the crown as the wind got stronger, a result of streamlining by individual leaves and by whole branches. Whereas, for rigid objects, the drag at high windspeeds is almost proportional to the square of velocity V (p. 145), the drag on tree specimens increased more slowly with V, an important result of streamlining.

Wood [1995] related the drag force F on individual trees to the windspeed V by defining a drag coefficient c_d such that

$$c_d = \frac{F}{0.5\rho V^2 h^2} \qquad (9.14)$$

where h is the tree height (m). (Note that Equation 9.14 uses tree height rather than area as on p. 145). With F measured in newtons, Wood showed that Mayhead's data for Sitka spruce (*Picea sitchensis*) fitted the relationship

$$c_d = a \left[\frac{M}{h^3} \right]^{0.67} \exp(-bV^2) \qquad (9.15)$$

where M is the mass of branches (kg), V is windspeed (m s^{-1}), and the constants a and b had the values 0.71 and 9.8×10^{-4}, respectively. The exponential term indicates the effect of streamlining, which decreases the drag coefficient as windspeed increases. Equations 9.14 and 9.15 together imply that the drag force F is proportional to $V^2 M^{0.67} \exp(-V^2)$. For a typical British-grown Sitka spruce with height 15 m and branch mass 50 kg, Equation 9.15 shows that the drag coefficient is halved between 1 and 27 m s^{-1}, so that over this range of windspeed F increases approximately proportionately to $V^{1.8}$ rather than V^2. Michell and his colleagues (Rudnicki et al, 2004] also found that Equation 9.15 provided a good description of drag on three conifer species exposed in a wind tunnel, and speculated that a low drag force per unit branch mass could be a factor accounting for the lack of damage to some tree species in wind storms.

Wind Forces on Trees in Forests

A difficulty in applying Equation 9.14 to predict drag forces on trees in forests is that V is assumed uniform over the height of a tree. In forests, windspeed often decreases exponentially from the top to the bottom of the canopy. To investigate how the drag on a tree was modified by canopy shelter, Stacey et al [1994] measured forces on a model tree in a wind tunnel, first in isolation with uniform flow to match Mayhead's data, and then with the tree among a large array of similar tree models (Figure 9.10, see color plates). The mean drag on a tree sheltered by many others was only 6–8% of the fully exposed value at the same windspeed.

Studies in which known forces are applied to trees using winches and cables [e.g., Coutts 1986, Milne 1991] demonstrate that trees blow down at mean windspeeds considerably lower than those predicted from static pulling tests. Possible explanations are that damage is caused by resonant swaying movements set up by the turbulent wind, or by large coherent gusts (the "Honami" waves seen moving over cereal crops on windy days). Gardiner [1995] studied the swaying of trees in relation to turbulence, and concluded that trees respond to intermittent impulses from gusts by behaving like damped harmonic oscillators. The frequency of gust arrival increases with windspeed, making it more likely that gusts occur in phase with the motion of the trees. The process of coherent gust generation at the top of aerodynamically rough canopies is discussed in Chapter 16.

9.4 Problems

1. Lycopodium spores with diameter 4.2 μm have a sedimentation velocity in air of 0.50 mm s^{-1}. Estimate their drag coefficient.

2. Using the empirical relationship in Equation 9.15, plot a graph showing how the drag force on a Sitka spruce tree 15 m high with a branch mass of 50 kg would vary with windspeed V up to 25 m s^{-1}. Hence, show that the drag force increases approximately as V^n over this range of windspeed, and estimate the value of n.

► Chapter 10

Heat Transfer

Three mechanisms of heat transfer are important in the environment of plants and animals: *radiation*, governed by principles already considered in Chapter 3; *convection*, which is the transfer of heat by moving air or fluid; and *conduction* in solids, and still gases and fluids, which depends on the exchange of kinetic energy between molecules. Two types of convection are important in micrometeorology: "forced" convection or transfer through the boundary layer of a surface exposed to an airstream, proceeding at a rate that depends on the velocity of the flow—a process analogous to skin friction; and "free" convection, which depends on the ascent of warm air above heated surfaces or the descent of cold air toward or beneath cooled surfaces, both a result of differences in air density.

All these mechanisms of transfer are exploited in domestic heating systems. Fan heaters distribute hot air by forced convection; convector heaters and hot water "radiators" circulate warm air by free convection; underfloor heating depends on the conduction of heat from cables buried below the floor; and the conventional bar-type radiator loses heat both by convection and radiation. Convection, conduction, and systems where mixed heat transfer occurs will be considered in turn.

10.1 Convection

The analysis of convection is greatly simplified by using non-dimensional groups of quantities, and a short description of these groups is needed to

introduce a comparison of the convective heat loss from objects of different size and shape.

NON-DIMENSIONAL GROUPS

When a surface immersed in a fluid loses heat through a laminar boundary
layer of uniform thickness δ, the heat transfer per unit area ($\mathbf{C}$) can be
written as

$$\mathbf{C} = k\,(T_s - T)\,/\delta \qquad (10.1)$$

where k is the thermal conductivity of the fluid, T_s is surface temperature,
and T is fluid temperature. The same equation can be used in a purely
formal way to describe the heat loss by forced or free convection from
any object with a mean surface temperature of T_s surrounded by fluid at
T, even though the boundary layer is neither laminar nor uniformly thick.
In this case, δ is the thickness of an *equivalent* rather than a real laminar
layer. It is determined by the size and geometry of the surface and by the
way in which fluid circulates over it. A more useful form of Equation 10.1
can be derived by substituting a characteristic dimension d of the body
for the equivalent boundary layer thickness which cannot be measured
directly. For a sphere or cylinder with air blowing across it, the diameter
is a logical choice for d, and for a rectangular plate d is the length in the
direction of the wind. The equation then becomes

$$\mathbf{C} = \left(\frac{d}{\delta}\right) k\,(T_s - T)\,/d \qquad (10.2)$$

The ratio d/δ is called the *Nusselt number* (Nu) after the German engineer Wilmhelm Nusselt, who first proposed the dimensionless number and
made several important contributions to the theory of heat transfer. Just as
the Reynolds number is a convenient way of comparing the forces associated with momentum transfer to geometrically similar bodies immersed in
a moving fluid, the Nusselt number provides a basis for comparing rates
of convective heat loss from similar-shaped bodies of different scales exposed to different windspeeds.

The rate of convective heat transfer in air can be written as

$$\mathbf{C} = \rho c_p\,(T_s - T)\,/r_H \qquad (10.3)$$

where r_H is a resistance to heat transfer (p. 33). Comparison of Equations
10.2 and 10.3 gives

$$r_H = \frac{\rho c_p d}{k\mathrm{Nu}} = \frac{d}{\kappa\mathrm{Nu}} \qquad (10.4)$$

where κ is the thermal diffusivity of air. Thus, resistances to convective
heat transfer are inversely proportional to the Nusselt number. Much of

this chapter explores how Nusselt numbers can be estimated, and consequently heat transfer resistances calculated, for objects of different shapes in forced and free convection.

FORCED CONVECTION

In forced convection, the Nusselt number depends on the rate of heat transfer through a boundary layer from a surface hotter or cooler than the air passing over it, a process analogous to the transfer of momentum by skin friction. The Nusselt number is therefore expected to be a function of the Reynolds number (specifying the boundary layer thickness for momentum) modified by the ratio of boundary layer thicknesses for heat (t_H) and for momentum (t_M). The ratio t_M/t_H is a function of the *Prandtl number* defined as (v/κ) (Ludwig Prandtl, a German physicist, is credited with the "discovery" of the boundary layer and with many developments in aerodynamic theory).

Measurements of heat loss by forced convection from planes, cylinders and spheres can be described by the general relation

$$\text{Nu} \propto \text{Re}^n \, \text{Pr}^m \qquad (10.5)$$

where m and n are constants and $t_M/t_H = \text{Pr}^m$.

Writing resistance as (boundary layer thickness)/ diffusivity, the ratio of resistances for heat and momentum transfer can be expressed as

$$\frac{r_H}{r_M} = \frac{t_H/\kappa}{t_M/v} = \left(\frac{v}{\kappa}\right) \bigg/ \left(\frac{t_M}{t_H}\right) = \text{Pr}^{1-m} \qquad (10.6)$$

For forced convection over plates, m = 0.33 and for air Pr = 0.71 so that $r_H/r_M = 0.89$.

Because the Prandtl number is independent of temperature (p. 32) and because micrometeorologists are rarely concerned with heat transfer in any gas except air, Pr^m can be taken as constant in order to reduce Equation 10.5 to

$$\text{Nu} = A\,\text{Re}^n \qquad (10.7)$$

Values of the constants A and n for different types of geometry are given in Appendix A.5(a). In environmental physics, objects such as leaves, stems, and animals are often treated as planes, cylinders, and spheres, as appropriate, for calculations of heat and mass transfer. Over the range of Reynolds numbers likely to be encountered in natural environments, values of Nu vary by about $\pm 20\%$ for the different geometric shapes, so for approximate calculations it is often adequate to use a single pair of values of A and n, e.g., for a flat plate.

FREE CONVECTION

In free convection, heat transfer depends on the circulation of fluid over and around an object, maintained by gradients of temperature that create gradients of density. In this case, the Nusselt number is a function of another non-dimensional group, the *Grashof number* Gr and of the Prandtl number Pr. (The Grashof number is apparently named after Franz Grashof, an eminent German engineer, who surprisingly, unlike most scientists whose names are attached to units, does not appear to have made contributions to free convection research.) The Grashof number is determined by the temperature difference between the hot or cold object and the surrounding fluid $(T_s - T)$, the characteristic dimension of the object d, the coefficient of thermal expansion of the fluid a, the kinematic viscosity of the fluid v, and the acceleration of gravity g. Physically, the Grashof number is the ratio of a buoyancy force times an inertial force to the square of a viscous force. Numerically, it is calculated from

$$Gr = agd^3(T_s - T)/v^2 \qquad (10.8)$$

In a system with a large Grashof number, free convection is vigorous because buoyancy and inertial forces that promote the circulation of air are much larger than the viscous forces that tend to inhibit circulation. The Grashof number has a similar role to that of the Reynolds number in forced convection, and is the primary criterion determining the transition from a turbulent to a laminar boundary layer in free convection.

The onset of free convection occurs when the product Gr Pr (the *Rayleigh number*) exceeds a critical value, around 1100 for the free atmosphere. Most leaves and animals in natural environments would therefore be expected to exchange heat by free convection in still conditions.

The Nusselt number for free convection in a gas is proportional to (Gr Pr)^m and can therefore be written as

$$Nu = BGr^m \qquad (10.9)$$

for a specific gas such as air (Pr $= 0.71$). The numerical constants B and m which depend on geometry are tabulated in Appendix A.5(b). As with forced convection, it is often adequate to use the flat plate values as approximations for all geometries.

When appropriate values of a and v for air at $20°C$ are inserted in Equation 10.8, it can be shown that

$$Gr\, Pr = 112d^3 (T_s - T) \qquad (10.10)$$

and

$$Gr = 158d^3(T_s - T) \qquad (10.11)$$

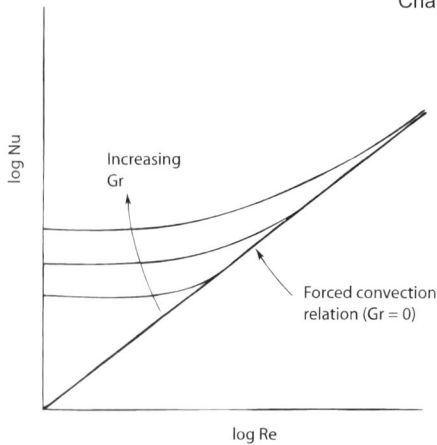

FIGURE 10.1 Influence of Grashof number (Gr) on relation between Nusselt (Nu) and Reynolds number (Re). (Adapted from measurements using a heated sphere exposed to horizontal wind by Yuge, 1960.)

where d is the characteristic dimension in centimetres (Note the use of centimeters in this relationship).

For laminar free convection, m $= 1/4$ irrespective of the shape of the object losing heat. In this case, Equation 10.2 demonstrates that the rate of convective heat loss is proportional to $(T_s - T)^{5/4}$, the so-called five-fourths power law of cooling.

MIXED CONVECTION

In many natural systems, convection is a very complex process because of continuous changes in windspeed and direction, often coupled with movement of the surface losing heat. During strong gusts, the loss of heat from a leaf or an animal will usually be determined by forced convection, but during lulls free convection may be dominant. The convection regime may therefore be described as "mixed" in the sense that both modes of convection are present although their relative importance changes with time. Because this situation is too complex to handle either experimentally or theoretically, heat transfer in the environment is usually calculated from a mean windspeed when forced convection is thought to be dominant or from a temperature difference if the windspeed is very light.

Figure 10.1, based on measurements by Yuge [1960], provides a general illustration of the variation of the measured Nusselt number Nu for heated spheres in a crosswind as the Grashof number increases.

The straight line is the forced convection case (Gr = 0), where Nu $\propto$ Re^n. Nu was determined almost solely by the Grashof number when Re was less than about 16, but approached the value for forced convection as Re increased. The three curves in Figure 10.1 represent the relationship between Nu and Re as the Grashof number increased. Yuge found that when Gr $= 400$, (the lowest curve), there was a relatively sharp transition from free to forced convection. When Gr was about 1000, the transition was still apparent, but occurred at larger Re. But with Gr $= 1800$, there was a wide range of Reynolds numbers over which the Nusselt number had a value substantially greater than the value for forced or free convection separately but substantially less than the sum of the two Nusselt numbers. This is what is usually understood as a regime of "mixed" convection.

As a rough criterion for distinguishing the two regimes, the Grashof number may be compared with the square of the Reynolds number. As Gr depends on

$$\frac{\text{buoyancy} \times \text{inertial forces}}{(\text{viscous forces})^2}$$

and Re^2 depends on $(\text{inertial forces})^2/(\text{viscous forces})^2$, the ratio Gr/Re^2 is proportional to the ratio of buoyancy to inertial forces. When Gr is much larger than Re^2, buoyancy forces are much larger than inertial forces and heat transfer is governed by free convection. When Gr is much less than Re^2 buoyancy forces are negligible and forced convection is the dominant mode of heat transfer. Based on Schuepp's [1993] extensive review of leaf boundary layers, mixed convection is probable when $0.1 < \text{Gr}/\text{Re}^2 < 10$.

For example, when a leaf with $d = 5$ cm is $5°C$ warmer than the surrounding air, its Grashof number is about 10^5, whereas Re^2 is about $10^7 V^2$ when V is in m s^{-1}. A regime of forced convection is expected when V exceeds 1 m s^{-1}, but at windspeeds between 0.1 and 0.5 m s^{-1}, which are often found in crop canopies, both forced and free convection will be active mechanisms of heat transfer.

A cow with $d = 0.5$ m and a surface temperature of $20°C$ above the ambient air has Gr $= 4 \times 10^8$ and $\text{Re}^2 = V^2 \times 10^9$ when V is in m s^{-1}. In this case, free convection will be the dominant form of heat transfer when the animal is exposed to a light draught indoors, but at windspeeds of the order 1 m s^{-1} in the field, the convection regime will again be mixed.

LAMINAR AND TURBULENT FLOW

Both in forced and in free convection, the size of the Nusselt number depends on the degree of turbulence in the boundary layer. In turn, this depends partly on the turbulence in the airstream and partly on the roughness of the surface which tends to generate turbulence. When a *smooth* flat plate exchanges heat by forced convection in an airstream effectively free from turbulence, the transition from laminar to turbulent flow in the boundary layer occurs at a Reynolds number of the order of 10^6, but in a *turbulent* airstream the critical Reynolds number decreases to an extent that depends partly on the amplitude of the velocity fluctuations and partly on their frequency, and was reported by Grace [1978] to be about 4×10^3 for a poplar leaf. In micrometeorological problems involving leaves or other plant organs, Re is usually between 10^3 and 10^4 but it has never been clearly demonstrated whether the boundary layer of a leaf in a crop canopy, for example, should be regarded as laminar or turbulent. At a Reynolds number of 10^4, the Nusselt number for laminar forced convection from a flat plate is $0.60 \times (10^4)^{0.5}$ or 60, compared with $0.032 \times (10^4)^{0.8}$ or 51 for a turbulent boundary layer; and at $\mathrm{Re} = 4 \times 10^4$, the corresponding numbers are 120 and 150. Thus for values of Re in the range of micrometeorological interest, there will usually be little difference between the conventional Nusselt numbers for laminar and turbulent boundary layers. It does not necessarily follow that the same Nusselt numbers will be valid when the airstream itself is turbulent or when the surface is rough. When the air is free from turbulence, elements of surface roughness with a height of less than 1% of the characteristic dimension can increase the Nusselt number for a cylinder by a factor of about 2 and can reduce the critical Reynolds number for the transition to a turbulent boundary layer by an order of magnitude [Achenbach, 1977]. A more detailed discussion of these matters was given by Gates [1980].

The onset of turbulence in free convection occurs when the Grashof number exceeds about 10^8, an unusual situation in micrometeorology. For example, the mean surface temperature of a sheep or a man would need to be at least $30°\mathrm{C}$ above the temperature of the ambient air to achieve $\mathrm{Gr} = 10^8$. The assumption of laminar boundary layer flow will therefore usually be valid in cases of free convection and in forced convection.

10.2 Measurements of Convection

PLANE SURFACES

When the boundary layer over a plane surface is laminar, the rate of heat transfer between the surface and the airstream can be calculated from first principles for two discrete cases. First, if the *temperature* is uniform over the whole surface the Nusselt number is

$$\mathrm{Nu} = 0.66\,\mathrm{Re}^{0.5}\,\mathrm{Pr}^{\,0.33} \tag{10.12}$$

and this relation is quoted in engineering texts that are concerned mainly with the heat transfer from metal surfaces with high thermal conductivity.

In the second case, which is more biologically relevant, the *heat flux* per unit area is constant over the whole surface. This condition should be valid for poor thermal conductors exposed to a uniform flux of radiation, e.g., leaf laminae in sunshine. The uniformity of heat flux from a leaf surface has not been established experimentally, but it is clear from radiometric measurements of leaf temperature (see Figure 10.3) that it is not legitimate to treat sunlit leaves as isothermal surfaces. According to Parlange et al, [1971], the assumption of a uniform heat flux leads to the prediction that the excess leaf temperature θ (= $T_s - T$) should increase with the square root of the distance from the leading edge x (cf. the uniform temperature case in which the flux decreases with the square root of x). It is convenient to incorporate x in a local Reynolds number $\mathrm{Re}_x = (Vx/\upsilon)$. In laminar flow, the excess temperature θ then becomes

$$\theta = 2.21\mathbf{C}\frac{x}{k}\,(\mathrm{Re}_x)^{-0.5}\,\mathrm{Pr}^{-0.33} \tag{10.13}$$

and the mean temperature excess over a plate of length d is

$$\bar{\theta} = \frac{\displaystyle\int_0^d (T_s - T)\,dx}{\displaystyle\int_0^d dx} = 1.47\mathbf{C}\frac{d}{k}\,(\mathrm{Re})^{-0.5}\,\mathrm{Pr}^{-0.33} \tag{10.14}$$

The mean Nusselt number, defined as $\overline{\mathrm{Nu}} = \mathbf{C}d/k\bar{\theta}$ is

$$\overline{\mathrm{Nu}} = 0.68\,\mathrm{Re}^{0.5}\,\mathrm{Pr}^{0.33} \tag{10.15}$$

which is only a few percent larger than the Nusselt number for the uniform temperature case.

A Nusselt number for a plate of irregular length (e.g., serrated or compound leaves) can be calculated from an appropriate mean length in the direction of the airstream. If W is the width of a leaf at right angles to the flow, the leaf area can be expressed as $\int_0^W y\,dx$ (Figure 10.2).When the Nusselt number is $A\mathrm{Re}^n$, an effective mean length $\bar{y}$ can be defined by writing the total heat loss from the leaf as

$$\mathbf{C} = A\left(\frac{V\bar{y}}{\upsilon}\right)^n \frac{k}{\bar{y}}\theta\int_0^W y\,dx \tag{10.16}$$

But the total heat loss can also be written in the form

$$\mathbf{C} = \int_0^W A\left(\frac{Vy}{\upsilon}\right)^n k\theta\,dx \tag{10.17}$$

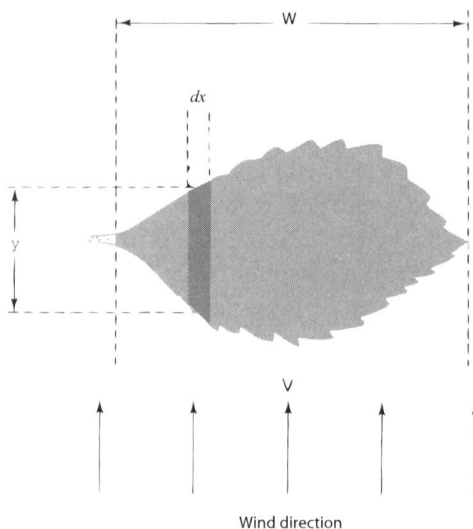

FIGURE 10.2   Coordinates for integrating heat loss over the surface of a leaf of irregular shape.

and by equating these expressions the mean length is given by

$$\bar{y} = \left\{ \int_0^W y^n \, dx \Big/ \int_0^W y \, dx \right\}^{1/(n-1)} \tag{10.18}$$

For laminar forced convection, $n = 0.5$. Equation 10.18 is also valid for free convection with $n = 0.75$. Parkhurst et al [1968] measured the heat loss and excess temperature from a series of metal leaf replicas with a wide range of shapes. Almost all the Nusselt numbers based on the mean dimension $\bar{y}$ lay between the values $0.60 \, \mathrm{Re}^{0.5}$ and $0.80 \, \mathrm{Re}^{0.5}$.

LEAVES

Many attempts have been made to determine the Nusselt number of leaves or of heated metal replicas of leaves as a function of size, shape, windspeed, intensity of turbulence, degree of fluttering, etc, and very comprehensive reviews were provided by Gates [1980] and Schuepp [1993]. Values of Nu reported by some workers have been close to the so-called engineering values summarized in Appendix A.5 but some are larger by a factor between 1 and 2.5. The apparent excess loss of heat from leaves has sometimes been a consequence of undetected free convection at low windspeeds or of turbulence at high windspeeds but the main reason probably lies in the difference between the uniform boundary layer which develops downwind from the edge of a wide, flat plate in a wind tunnel and the very irregular and unstable boundary layer which must exist over a relatively narrow leaf with irregular edges, sometimes curled, and with protrusions in the form of veins.

Figure 10.3 (see color plates, from Wigley and Clark, 1974) clearly illustrates the difference in the thermal behavior of a leaf-shaped plate exposed to a uniform airstream and a real leaf exposed in a comparable environment. Since the imposed heat flux was uniform in both cases, the temperature distributions give a striking impression of how the boundary layer thickness changed across the two contrasting laminae. With this contrast in mind, it is surprising that convection from leaves does not differ much more from predictions based on plates. For comparisons, most workers use the ratio β of the observed Nusselt number for a leaf (or group of leaves) to the corresponding value of Nu for a smooth plate of comparable size at the same windspeed. From Equation 10.4, β is also the ratio of boundary layer conductance $(1/r_H)$ for both types of objects.

For the simple leaf replica shown in Figure 9.4, heated electrically and exposed in a wind-tunnel, β was about 1.1 and the ratio of resistances for heat and momentum transfer was almost exactly the value predicted from Equation 10.6 [Thom 1968]. In a field experiment where heat loss was measured from leaves attached to apple trees growing in rows [Thorpe and Butler, 1977], the relation between Nu and Re was very scattered but the line of best fit gave $\beta \approx 1$, possibly because turbulence compensated for the decrease of windspeed between the alleys (where windspeed was measured) and in the trees (where leaves were exposed). Schuepp [1993] concluded that the consensus from many studies of heat transfer from single real and model leaves when boundary layer flow was laminar was that β lay between 1.4 and 1.5, but there are a few reports of β exceeding 2.5. For example, Grace and Wilson [1976] reported values of $\beta = 2.5$ for poplar leaves exposed in a turbulent flow, and Parlange et al [1971] found similarly large values for tobacco leaves in turbulent wind tunnel flow. Measurements derived from replicas of *Phaseolus* leaves (as in Figure 10.3) gave $\beta = 1.1$ for laminar flow (turbulent intensity $i = 0.01$ to 0.02) and for Re up to 2×10^4 [Wigley and Clark, 1974]. For turbulent flow ($i = 0.3$ to 0.4), $A = 0.04$ and $n = 0.84$ in Equation 10.7, Nu exceeded the laminar flow value above (but not below) Re $= 10^3$, and β was about 2.5 for Re $= 10^4$.

Relatively little work has been published on free convection from leaves despite its significance for heat transfer in canopies where windspeeds be-

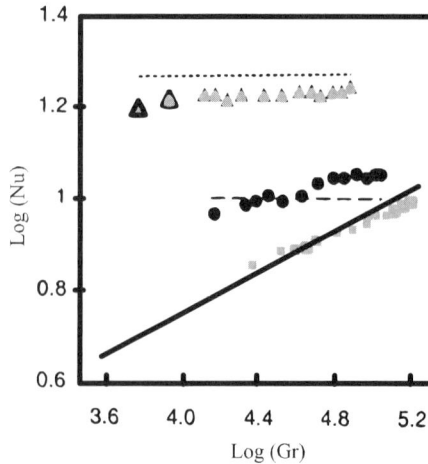

FIGURE 10.4   Measured experimental values of Nusselt number for heat transfer from artificial leaves, compared to calculated values for forced convection (Equation 10.7) and free convection (Eq.10.9). Symbols: □ measured at Re = 0; ● measured at Re = 296; △ measured at Re = 1014; — Eq.10.9; - - - Eq.10.7, Re = 296; ..... Equation 10.7, Re = 1014 (from Bailey and Meneses 1995).

low 0.5 m s^{-1} are common. In one study with leaves of *Acer* and *Quercus*, Nu was close to the prediction from Equation 10.9 when GrPr was 10^6, but β increased as GrPr decreased below this figure and was about 2 at GrPr $= 10^4$. Roth–Nebelsick [2001] explored free and mixed convection from leaves using numerical modeling, and concluded that free convection was unlikely to be a significant mode of leaf heat loss except in very still indoor environments. Buoyancy did not appear to enhance heat transfer from small leaves in mixed convection, perhaps because the buoyant plume was displaced to the end of the leaf surface, and did not disturb the main boundary layer. Bailey and Meneses [1995] studied heat transfer from two artificial leaves exposed vertically in a wind tunnel under conditions generating free, forced and mixed convection. Figure 10.4 summarizes their findings. In free convection, measured Nusselt numbers agreed closely with values calculated from Equation 10.9 as leaf-air temperature differences were increased from 0.3 to 11.5 K. At a wind speed of 0.10 m s^{-1}, corresponding to Re = 296, Nusselt numbers were closer to the value for forced convection calculated from Equation 10.7, but showed some increase with increasing Gr, indicating that mixed convection was occurring. At a higher wind speed (0.43 m s^{-1}, Re = 1014), Nu was virtually independent of Gr. Bailey and Meneses concluded that even at wind speeds as low as 0.1 m s^{-1}, typical of non-ventilated glasshouses, heat transfer was by mixed convection, not by free convection alone.

Influence of Leaf Shape and Leaf Hairs

The dissipation of heat from a set of copper plates exposed to low windspeed was studied by Vogel (1970). All the plates had the same area but their shapes ranged from a circle and a regular 6-point star to replicas of oak leaves with characteristic lobes (Figure 10.5). The amount of electrical energy needed to keep each plate 15°C warmer than the surrounding air was recorded at windspeeds from 0 to 0. 3 m s^{-1} and at different orientations. This energy is proportional to Nu and inversely proportional to the resistance r_H under the conditions of the experiment. The main conclusions were:

(i) increasing airflow from 0 to 0.3 m s^{-1} decreased r_H by 30 to 95%: the decrease was greater for the leaf models than for the stellate shapes;

(ii) the resistances of all the stellate and lobed plates were smaller than the resistance of the round plate and were less sensitive to orientation;

(iii) a deeply lobed model simulating a sun leaf of oak always had a smaller resistance than a shade leaf with smaller lobes;

(iv) the resistance of the leaf models was least when the surface was oblique to the airstream;

(v) serrations about 5 mm deep on the periphery of the circular plate had no perceptible effect on its thermal resistance;

(vi) the measurements could not be correlated using a simple Nusselt number based on a weighted mean width as described on p. 167.

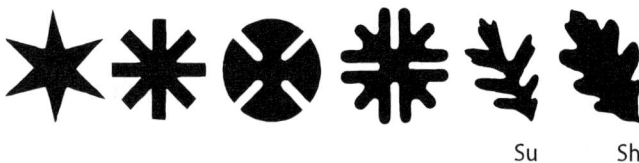

Su   Sh

FIGURE 10.5   Metal replicas of leaves used by Vogel [1970] to study heat losses in free convection. The shapes Su and Sh represent sun and shade leaves of white oak.

Ecologists have long been intrigued by the role of leaf hairs in modifying leaf temperature and gas exchange. Parkhurst [1976] re-analyzed data of Wuenscher who compared pubescent and shaved leaves of common mullein (*Verbascum thapsus*). Pubescence increased the boundary layer resistance for heat transfer r_H by a factor of about two. Meinzer and Goldstein [1985] studied the energy balance of the giant rosette plant *Espeletia timotensis* that grows up to 3 m tall in the Andes at elevations above about 4000 m. They estimated that, at a windspeed of 2 m s^{-1}, the thick leaf pubescence (up to 3 mm) would increase r_H by almost a factor of six (17 s m^{-1} to 95 s m^{-1}) compared to a "hairless" leaf. (Note that the increase is only about half the resistance that would result from a layer of still air 3 mm thick). As a result of their greater aerodynamic resistance, leaf temperatures of pubescent leaves would be about 5°C or more warmer than those of smooth leaves in bright sunshine, probably increasing photosynthetic efficiency in an environment where air temperature is often sub-optimal.

These measurements support the hypothesis favored by many ecologists that the shapes and surface structure of leaves may represent an adaptation to their thermal environment. Because the natural environment is very variable, and because physiological responses to changes of leaf tissue temperatures are complex, it is difficult to obtain conclusive experimental proof.

To summarize, non-dimensional groups provide a useful way of summarizing and comparing heat loss from objects with similar geometry but different size exposed to different wind regimes. Standard values of these groups should be regarded as a useful yardstick for estimating heat transfer in the field, allowing mean surface temperature to be estimated within the precision needed for most ecological studies. When precision is essential, appropriate values of Nu must be obtained experimentally, preferably by measurements on real leaves at the relevant site.

CYLINDERS AND SPHERES

The flow of air over cylindrical and spherical objects is more complex than the flow over plates because of the separation of the boundary layer that occurs toward the rear, and the wake generated by this separation. In forced convection the Nusselt number can again be related to the Reynolds number by the expression $Nu = A Re^n Pr^{0.33}$ but, unlike the corresponding constants for flat plates, both A and n change with the value of Re (Appendix A.5a). Both for spheres and for cylinders exposed to cross-diameter flow, the obvious quantity to adopt for the characteristic dimension is the diameter, but for the irregular bodies of animals, the cube root of the volume may be more appropriate.

Mammals

McArthur measured the heat loss from a horizontal cylinder electrically heated and with a diameter of 0.33 m to simulate the bare trunk of a sheep [McArthur and Monteith, 1980a]. In a second set of measurements, the cylinder was covered with the fleece of a sheep.

For the bare cylinder, estimates of Nu fitted the relation $Nu = A\,Re^n$ with $A = 0.095$ and $n = 0.68$ in the region $2 \times 10^4 < Re < 3 \times 10^5$, cf. $A = 0.17$ and $n = 0.62$ in Table A.5(a).

To avoid the difficulty of comparing equations with different exponents, Figure 10.6 shows the heat transfer (aerodynamic) resistance of cylinders and cylindrically-shaped animals obtained from various sources. Consistency is reassuring, and as the aerodynamic resistance is often small compared with the coat resistance of an animal, the degree of uncertainty implied by Figure 10.6 is immaterial for most calculations.

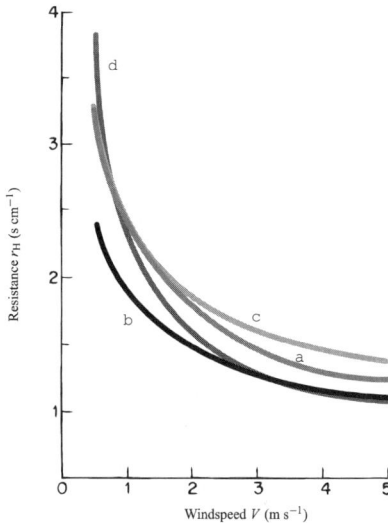

FIGURE 10.6 Relations between boundary layer resistance r_H and windspeed V for cylindrical bodies. Line a, smooth isothermal cylinder (McAdams, 1954); line b, cattle (Wiersma and Nelson, 1967); line c, sheep (Monteith, 1975); line d, sheep (McArthur and Monteith, 1980a).

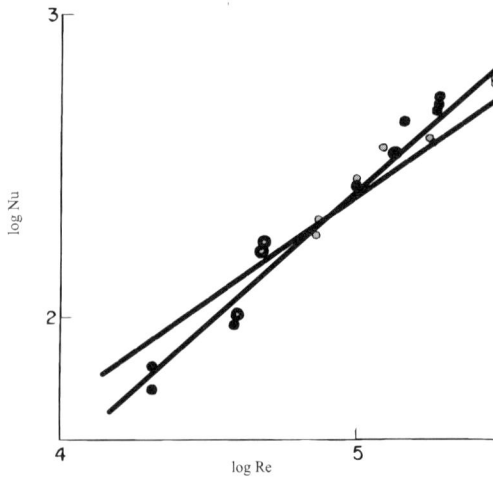

FIGURE 10.7   The relation between Nu and Re for model sheep without fleece—dashed line; the relation for the same model with fleece—full line and points (from McArthur and Monteith, 1980a).

Nusselt numbers for hair surfaces are harder to establish because of the problem of measuring the temperature at the tips of the hairs. Using a radiation thermometer to measure this quantity, McArthur found $A = 0.112$ and $n = 0.88$. Comparison of the relation between Nu and Re for smooth and hairy cylinders (Figure 10.7) suggests that in a light wind (< 1 m s^{-1}), the effective boundary layer depth ($\propto 1/$Nu) was thicker when the surface was composed of hairs, presumably because the effective surface "seen" by the radiometer was some distance below the physical hair tips. But when Re exceeded 10^5, the boundary layer depth was smaller with hair present, suggesting that wind penetrated the coat, possibly generating turbulence.

When measurements of this kind are used to estimate heat loss from mammals, Nusselt numbers need to be obtained for appendages too: legs and tails can be treated as cylinders, and heads as spheres. Rapp [1970] reviewed several sets of measurements of the loss of heat from nude human subjects in different postures and exposed to a range of environments. Agreement between measured and standard Nusselt numbers was excellent except when the convection regime was mixed. In one experiment, the heat loss from vertical subjects standing in a horizontal airstream was $8.5V^{0.5}$ W m^{-2} K^{-1} (V in m s^{-1}), and for a man with a characteristic dimension of 33 cm, this is equivalent to $A = 0.78$, $n = 0.5$ in Equation

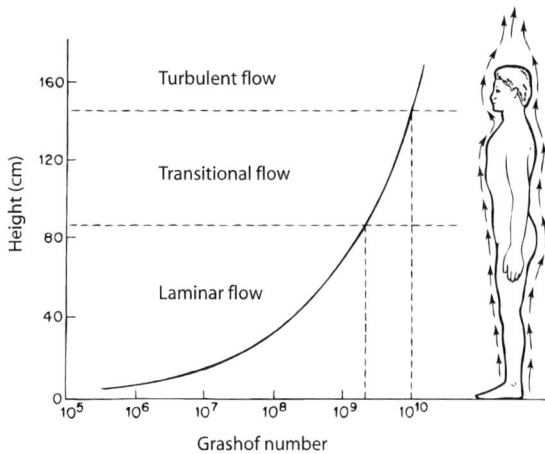

FIGURE 10.8   Variation of Grashof number with vertical height over the body surface for a mean skin temperature of 33°C and an ambient temperature of 25°C [from Clark and Toy, 1975].

10.7. Although the corresponding values in Appendix A.5 are $A = 0.24$, $n = 0.6$, the two sets of coefficients give similar values of Nu for a relevant range of Re (10^4 to 10^5). Similarity with values of Nu already quoted for horizontal cylinders illustrates the usefulness of non-dimensional groups for comparing heat loss from different systems.

In a light wind, the heat loss from sheep and other animals may be governed by free convection, particularly when there is a large difference of temperature between the coat and the surrounding air. When Merino sheep were exposed to strong sunshine in Australia, fleece tip temperatures reached 85°C when the air temperature was 45°C. With $d = 30$ cm, the corresponding Grashof number is about 2×10^8, so free and forced convection would be of comparable importance when $\mathrm{Re}^2 = 2 \times 10^8$, i.e. when windspeed was about 0.7 m s^{-1}. To calculate fleece temperatures in similar conditions, Priestley [1957] used a graphical method to allow for the transition from free to forced convection with increasing windspeed (see p. 164).

For humans standing erect, heat loss by free convection is complex because the Grashof number depends on the cube of the characteristic dimension. The nature of the convection regime therefore changes with height above the ground as shown in Figure 10.8. The movement of air associated with free convection from the head and limbs has been demon-

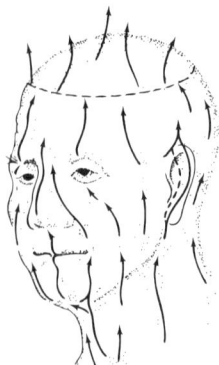

FIGURE 10.9   Routes taken by the flow of naturally convected air over the human head [from Schlieren photography by Lewis et al, 1969].

strated by Lewis and his colleagues using the technique of Schlieren photography (Figure 10.9). The technique has also been used to study free convection from wheat ears and a rabbit (Figure 10.10, see color plates). The way in which air ascending over the face of humans is deflected from the nostrils may be important in preventing the inhalation of bacteria and other pathogens. Similarly, the posture adopted by dogs and other animals when "sniffing" the air extends the nose beyond the layer of air ascending the body, thus increasing the sensitivity for external odor detection.Figure 10.11 shows that the velocity and temperature profiles close to a bare leg are characteristic of the flow in free convection from vertical surfaces.

The application of conventional heat transfer analysis is much more difficult for non-cooperative subjects like piglets, particularly when the animal changes its posture and orientation with respect to windspeed in response to heat or cold stress.

Birds

Several workers have measured the heat loss from replicas of birds constructed from appropriate combinations of metal spheres and cylinders. The model of a domestic fowl used by Wathes and Clark [1981a] consisted of a copper sphere with a diameter of 265 mm to which cylinders of 15 mm diameter were soldered at points corresponding to the head and legs. The Nusselt number for forced convection was

$$Nu = 2 + 0.79 \ Re^{0.48}$$

Scale of ordinates   0 10 20 30 cm s^{-1}

Scale of abscissae   0 1 2 3 cm

Scale of leg   0 2 4 6 cm

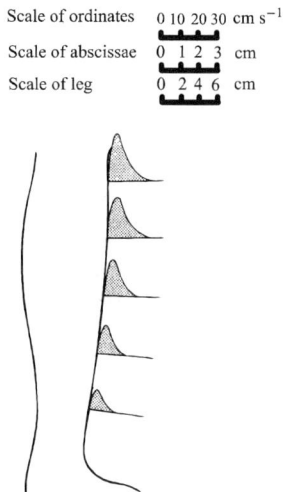

FIGURE 10.11   Development of velocity profiles on the front of the leg measured with a hot wire anemometer (from Lewis et al, 1969).

giving values of Nu similar to those from the standard relation for a simple sphere within the experimental range of Reynolds numbers (3×10^3 to 10^4).

Insects

The convective heat loss from insect species was measured by Digby [1955], who mounted dead specimens in a wind tunnel with transparent walls. Heating was provided by radiation from an external 2 kW lamp. The difference between the temperature of the insects and the ambient air was (i) directly proportional to incident radiant energy, and (ii) inversely proportional to the square root of the windspeed, as predicted for cylinders in the appropriate range of Reynolds numbers.

   Evidence for the variation of excess temperature with body size was more difficult to interpret. When the maximum breadth of the thorax was taken as an index of body size d, the excess temperature was expected to change with a power of d between 0.5 (cylinders) and 0.4 (spheres). For the locusts *Schistocerca gregaria* and *Carausius morosus*, the exponent of d was 0.4, but for 20 species of Diptera and Hymenoptera, Digby's measurements support an exponent of unity; i.e., excess temperature was directly proportional to the linear dimensions of the insect. This anomalous result may be a consequence of considerable scatter in the measurements or of changes in the distribution of absorbed radiant energy with body size.

In a similar series of measurements, Church [1960] used microwave radiation to heat specimens of bees and moths. He argued that the amount of energy produced by an insect in flight would be roughly proportional to body weight, and therefore used radiant flux densities that were proportional to the cube of the thorax diameter. When the insects were shaved to remove hair, the excess temperature was proportional to $d^{1.4}$. As the heat loss per unit area was proportional to d, the Nusselt number was proportional to $d^{0.6}$, close to expectation for a sphere. The excess temperature of insects covered with hair was about twice the excess for denuded insects, showing that the thermal resistances of the hair layer and the boundary layer were of similar size.

If the proportionality between Nu and $d^{0.6}$ is consistent with Nu $=$ $A \, \mathrm{Re}^n$, the index n must be 0.6 and Nu should be proportional to $V^{0.6}$. In fact, Church found that the excess temperatures of a shaved *Bombus* (bumble bee) specimen was proportional to $V^{0.4}$, but he explained this discrepancy in terms of the difference between temperatures at the surface of and within the thorax. If all the experimental results are assumed consistent with $n = 0.6$, the measurements on the *Bombus* specimen give a Nusselt number of $0.28 \, \mathrm{Re}^{0.6}$, close to the standard relation for a sphere, Nu $= 0.34 \, \mathrm{Re}^{0.6}$. The values of Nu from this equation are close to the values predicted from either of the relationships in Appendix A.5 within the appropriate range of Reynolds numbers. Ecologists have noted that bumble bee foragers have a wide range of body sizes, and have speculated that this may be related to adaptation to climates where temperatures fluctuate through a wide range [Peat et al, 2005].

Conifer Leaves

Many coniferous trees have needle-like leaves that can be treated as cylinders in order to estimate heat losses, though their cross section is seldom circular. Because the needles are closely spaced they shield each other from the wind and do not behave like isolated cylinders (see also p. 150). Engineers have developed empirical formulae for the heat loss from banks of cylinders in regular arrays and their relevance to coniferous branches has been examined by Tibbals et al [1964].

To avoid the difficulty of measuring the surface temperature of real needles, branches of pine, spruce, and fir were invested with dental compound and then burnt to leave a void that was filled with molten silver. The compound was then removed leaving a silver replica of the branch— a modern version of the Midas touch! For Blue spruce (*Picea pungens*), the Nusselt numbers based on the diameter of single needles were about one-third to one-half of Nu for isolated cylinders at the same Reynolds number, a measure of mutual sheltering. For White fir (*Abies concolor*), the Nusselt number in transverse flow (across four rows of needles) was about 60% greater than in longitudinal flow (across 20 or 30 rows of needles). When an average of the two modes was taken, Nu was close to the experimental values for banks of tubes and similar agreement was obtained for *Pinus ponderosa*. In still air, the Nusselt number for all three species was close to an experimental value for horizontal cylinders at the same Grashof number.

10.3 Conduction

Conduction is a form of heat transfer produced by sharing of momentum between colliding molecules in a fluid, by the movement of free electrons in a metal, or by the action of inter-molecular forces in an insulator. In micrometeorology, conduction is important for heat transfer in the soil and through the coats of animals, but not in the free atmosphere where the effects of molecular diffusion are trivial in relation to mixing by turbulence.

If the temperature gradient in a solid or motionless fluid is dT/dz, the rate of conduction of heat per unit area (**G**) is proportional to the gradient, and the constant of proportionality is called the thermal conductivity of the material k'. In symbols

$$\mathbf{G} = -k' \, (dT/dz) \tag{10.19}$$

where the negative sign is a reminder that heat moves in the direction of decreasing temperature. For a steady flow of heat between two parallel surfaces at T_2 and T_1, separated by a uniform slab of material with thickness t, integration of Equation 10.19 gives

$$\mathbf{G} = -k' \, (T_2 - T_1) / t \tag{10.20}$$

In general, layers of motionless gas provide excellent insulation, a fact exploited by the thick coats of hair with which many mammals trap air and by man's design of clothing and double glazing. The conductivity of still air is four orders of magnitude less than the values for copper and silver.

Equation 10.19 is used to describe the vertical flow of heat in soil as discussed in Chapter 15. When it is applied to conduction in cylinders and spheres, the area through which a fixed amount of heat is conducted changes with position, and the integration of Equation 10.19 then yields expressions somewhat more complex than Equation 10.20. For a hollow cylinder with interior and exterior radii r_1 and r_2 at temperatures T_1 and T_2 respectively, the heat flux per unit area of the outer surface is

$$G = \frac{k'\,(T_1 - T_2)}{r_2 \ln(r_2/r_1)} \tag{10.21}$$

and for a hollow sphere the corresponding flux density is

$$G = \frac{k'\,(T_1 - T_2)}{r_2\,(r_2/r_1 - 1)} \tag{10.22}$$

Equations 10.21 and 10.22 have been used to calculate heat transfer through parts of animals and birds that are approximately cylindrical or spherical.

The form of Equation 10.21 has important implications for the efficiency of insulation surrounding a cylinder or any object that is approximately cylindrical such as the trunk of a sheep or a human finger. The equation predicts that, for a fixed value of the temperature difference across the insulation, the heat flow per unit length of the cylinder $2\pi r_2 G$ will be inversely proportional to $\ln(r_2/r_1)$. In the steady state, assuming that radiative exchange is negligible, the rate of conduction must be equal to the convective heat loss from the outer surface of the insulation. If the air is at a temperature T_3, the convective heat loss will be

$$2\pi r_2 \mathrm{Nu}(k/2r_2)(T_2 - T_3) = 2\pi r_2 G = 2\pi k'(T_1 - T_2)/\ln(r_2/r_1)$$

where k is the thermal conductivity of air and Nu is the Nusselt number. By rearranging terms it can be shown that the heat loss per unit length of cylinder is

$$2\pi r_2 G = \frac{2\pi k(T_1 - T_3)}{\{(k/k')\ln(r_2/r_1)\} + \{2/\mathrm{Nu}\}} \tag{10.23}$$

The Nusselt number is proportional to $(r_2)^n$ where n is 0.5 for forced and 0.75 for free convection. When r_2 increases, the heat loss $2\pi r_2 G$ will therefore increase or decrease depending on which of the terms in curly brackets dominates the denominator. To find the break-even point where $2\pi r_2 G$ is independent of r_2, Equation 10.23 can be differentiated with respect to r_2 to give

$$\mathrm{Nu} = 2nk'/k$$

When the Nusselt number is greater than this critical value, the heat loss from the insulation increases as r_2 increases; when it is smaller than the critical value, the heat loss decreases with increasing r_2.

If the conductivity of air is assumed constant and $n = 0.6$ is an approximate mean value for mixed convection, the critical value of Nu depends only on k', the conductivity of the insulating material. For animal coats and for clothing, $k \approx k'$ so the critical Nu is of the order of unity. In nature, Nusselt numbers of this size may be relevant to a furry caterpillar under a cabbage leaf on a calm night or to an animal in a burrow, but for most organisms freely exposed to the atmosphere, Nu will exceed 10. In general, therefore, the thermal insulation of animals will increase with the thickness of their hair or clothing.

The conductivity of fatty tissue on the other hand is about 12 times the conductivity of still air, so the critical Nusselt number for insulation by subcutaneous fat is about 14. When Nu is larger than 14, fat provides insulation in the conventional sense, but in an environment where Nu is less than this critical value, a naked ape suffering from middle-aged spread might have increasing difficulty keeping warm as his girth increased.

10.4 Insulation

Insulation is defined as the temperature difference per unit heat flux and per unit area. It is equivalent to the term $r/\rho c_p$ where r is a thermal insulation resistance and ρc_p is a volumetric specific heat. To convert from units of insulation (e.g., $K\ m^2\ W^{-1}$) to units of resistance (e.g., $s\ cm^{-1}$) and for comparison with the resistance of the boundary layer r_H, it is necessary to choose an arbitrary value of ρc_p for air (e.g., $1.22 \times 10^3\ J\ K^{-1}\ m^{-3}$, which is the value at $20°C$). On this basis, a resistance $r = 1\ s\ cm^{-1}$ is equal to an insulation $r/\rho c_p = 0.082\ K\ m^2\ W^{-1}$.

A unit of insulation found in human studies is the "*clo*," equal to 0.155 $K\ m^2\ W^{-1}$, and therefore equivalent to a resistance of $1.86\ s\ cm^{-1}$ or 0.39 cm of still air (putting $r_H = t/\kappa$). The clo was originally conceived as the insulation that would maintain a resting man, whose metabolism is 50 kcal $m^{-2}\ h^{-1}$ (about 60 W m^{-2}), indefinitely comfortable in an environment of $21°C$, relative humidity less than 50%, and air movement 20 ft min^{-1}. Insulation of clothing is commonly reported in clo. A unit used in Europe is the "*tog*," defined as an insulation of $0.1\ K\ m^2\ W^{-1}$, so that 1 clo = 1.55 tog. Tog units are often quoted for the insulation of duvets (comforters) and some insulated outdoor wear. Specialized units of this type have few merits and tend to separate a subject from other related branches of science. Table 10.1 summarizes insulation values for clothing, sleeping bags, and duvets in specialized and SI units.

TABLE 10.1 Insulation of still air, human clothing, sleeping bags and duvets.

Insulation material	clo	tog	$\mathbf{K\ m^{-2}\ W^{-1}}$	$\mathbf{r\ (s\ cm^{-1})}$
Still air (1 cm at 20°C)	2.6	4.0	0.4	4.8
Naked human	0	0	0	0
Tropical clothing	0.3	0.5	0.05	0.6
Light summer clothing	0.5	0.8	0.08	1.0
Indoor winter clothing	1.0	1.6	0.16	2.0
Heavy weight business suit	1.5	2.3	0.23	2.8
Maximum insulation outdoor clothing	4.0	6.2	0.62	7.6
Summer weight sleeping bag (8–15°C)	2–3	3–5	0.3–0.5	3.7–6.1
Winter weight sleeping bag (−3–+10°C)	4.5–6.5	7–10	0.7–1.0	8.5–12.2
Winter weight duvet	8.4	13	1.3	15.9

(Values of resistance are based on an arbitrary value for ρc_p of 1.22×10^3 J K^{-1} m^{-3})

INSULATION OF ANIMALS

The insulation of animals has three components: a layer of tissue, fat, and skin across which temperature drops from deep body temperature to mean skin temperature; a layer of relatively still air trapped within a coat of fur, fleece, feathers, or clothing; and an outer boundary layer whose resistance is given by $d/(\kappa \text{Nu})$ (p. 161). A comprehensive analysis of heat exchange in mammals would need to consider separately the amount of heat lost from the trunk, legs, head, etc. Because these appendages are usually less well insulated and are smaller than the trunk, they are capable of losing more heat per unit area. In practice, the loss of heat from appendages is usually small compared with the total loss from the rest of the body, although some animals subject to heat stress are believed to dissipate large amounts of heat through their ears or tails. For example, African elephants have large ears (the surface area of both sides of the ears is about 20% of the surface area of the animal) that are flapped in hot conditions and are well supplied with blood vessels that can be dilated to lose heat. There is some evidence that the rate of ear flapping by African elephants increases with temperature [Buss and Estes, 1971]. Phillips and Heath [1992] estimated that in some conditions a large African elephant could achieve most of its heat loss requirements by flapping its ears and modulating the blood supply to them. In a follow-up theoretical study, Phillips and Heath [2001] concluded that the excessively large ears of the cartoon elephant Dumbo not only allowed Disney's baby elephant to fly, but would allow him to dissipate the large quantity of metabolic heat produced during aerobatics.

TABLE 10.2 Thermal resistances of animals. Peripheral tissue and coats.

Tissue (Blaxter, 1967)	s cm^{-1}	
	Vaso-constricted	Dilated
Steer	1.7	0.5
Man	1.2	0.3
Calf	1.1	0.5
Pig (3 months)	1.0	0.6
Down sheep	0.9	0.3
Coats (Blaxter, 1967; Hammel, 1955)	s cm^{-1} per cm depth	per cent of still air
Air	4.7	100
Red fox	3.3	70
Lynx	3.1	65
Skunk	3.0	64
Husky dog	2.9	62
Merino sheep	2.8	60
Down sheep	1.9	40
Blackfaced sheep	1.5	32
Cheviot sheep	1.5	32
Ayrshire cattle: flat coat	1.2	26
erect coat	0.8	
Galloway cattle	0.9	19

Average values of insulation for specific species can be determined from measurements of metabolic heat production, external heat load, and the relevant mean temperature gradients.

TISSUE

The insulation of tissue is defined as the temperature difference per unit heat flux density between deep body temperature and the skin surface. Insulation is strongly affected by the circulation of blood beneath the skin, and the constriction and dilation of blood vessels can change resistance by a factor of 2 to 3. For comparison with the thermal resistances of hair and air, values of tissue insulation found in the literature have been multiplied by the volumetric heat capacity of air at 20°C. Table 10.2 shows values ranging from a minimum of 0.3 s cm^{-1} for dilated tissue to between 1 and 2 s cm^{-1} for vaso-constricted tissue.

The empirical equation

$$r_{max} = 0.155 W^{0.33} \qquad (10.24)$$

relates maximum tissue resistance r_{max} (s cm^{-1}) to body mass W (kg) and provides reasonable estimates for several animal species ranging from baby pigs weighing 1.5 kg to cattle weighing 450 kg [Bruce and Clark 1979, Turnpenny et al 2000, Berman 2004].

Taken as an average over the whole body, the thermal conductivity of human skin during vaso-constriction appears to be about an order of magnitude greater than the conductivity of still air; i.e., about 0.2 to 0.3 W m^{-1} °K^{-1}, and measurements on fingers showed that the conductivity increases linearly with the rate of blood flow. The effective mean thickness of the skin during vaso-constriction is equivalent to about 2.5 cm of tissue or 0.25 cm of still air. These figures must conceal large local differences of insulation of the limbs and appendages depending on the thickness and nature of subcutaneous tissue and the degree of curvature.

COATS—MIXED REGIMES

There are many systems both in the field and within buildings where several modes of heat transfer operate simultaneously, but one mode usually dominates and a good approximation to the rate of transfer can then be obtained by neglecting the others. In the coats of animals and in clothing, however, molecular conduction, radiation, free and forced convection all play a significant role, and evidence for this is now considered.

The thermal resistance of animal coats has been measured by a number of workers and reported in units such as K m^2 W^{-1} or clo/inch. These units obscure an important physical fact; the thermal conductivity of hair, fleece and clothing is the same order of magnitude as the conductivity of still air. (A distinction made by others between "still" air and "dead air space" is based mainly on an arithmetical error. The two are physically identical by definition although in practice it is difficult to achieve a temperature gradient across a layer of still air without setting up a circulation of air by convection which increases in thermal conductivity.)

The thermal conductivity of air is 2.5×10^{-2} W m^{-1} K^{-1} at 20°C, equivalent to a resistance of 4.8 s cm^{-1} for a layer of 1 cm; or 2.58 clo/cm; or 4.0 tog/cm; or 6.6 clo/inch. Scholander et al [1950] showed that the insulation per unit thickness of coat was remarkably uniform for a wide range of wild animals from shrews to bears (Figure 10.12). The average insulation derived from his measurements is often quoted as 4 clo/inch meaning that 1 inch of fur had the same insulation as 4/6.6 or 0.6 inches of still air. In terms of still air, the efficiency of insulation is therefore 60%. A simple calculation suggests that conduction along hair fibers was trivial, but at least two other modes of heat transfer may have been responsible for the failure of the coats to behave like still air:

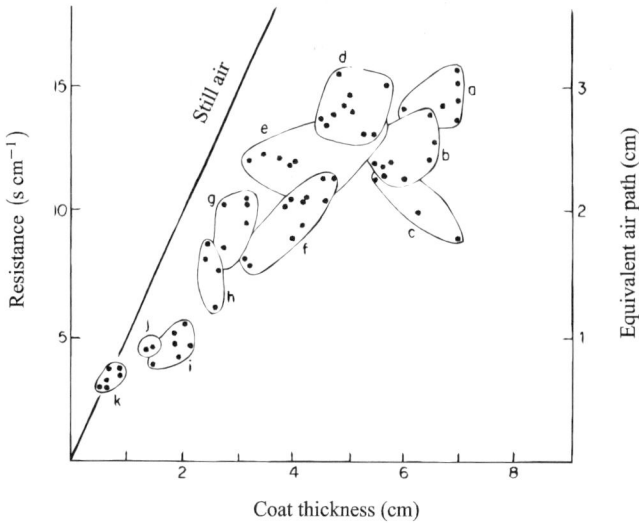

FIGURE 10.12 Thermal resistance of animal coats as a function of their thickness (after Scholander et al, 1950). a, dall sheep; b, wolf, grizzly bear; c, polar bear; d, white fox; e, reindeer, caribou; f, fox, dog, beaver; g, rabbit; h, marten; i, lemming; j, squirrel; k, shrew.

(1) there was significant radiative transfer from warmer to cooler layers of hair;

(2) buoyancy generated by temperature gradients was responsible for free convection.

Starting with the first possibility, Cena and Monteith [1975a] used Beer's Law to analyze the change of radiative flux with depth in samples of coat exposed to radiation. Depth within the coat was specified by an interception parameter (p), as described on p. 133. For long-wave radiation in an isothermal coat, the assumption that hairs behaved like black-bodies gave estimates of p close to those obtained by measurements of hair geometry, ranging from 4 to 5 cm for sheep to 18 cm for calf and deer with much smoother coats.

For a non-isothermal coat with a uniform temperature gradient between the skin and the external surface, the heat flux was found to be proportional to the gradient, and the equivalent thermal conductivity was

$$k_R = (4/3)\, 4\sigma\, T^3 / \rho \qquad (10.25)$$

Assuming $4\sigma T^3 = 6.3$ W m^{-2} K^{-1} ($T = 293$ K), k_R ranged from 0.004 W m^{-1} K^{-1} for calf hair to 0.02 W m^{-1} K^{-1} for sheeps' fleece. These figures imply that the effective thermal conductivity of animal coats, determined by combining the conductivities for molecular conduction and radiation in parallel, can never be as small as the value for still air and may be nearly twice that value for fleece and coats with similar structure. Corresponding values of resistance per unit depth of coat are obtained by writing $r_R = \rho c_p / k_R$ where ρc_p has some arbitrary value (e.g., 1.22 kJ m^{-3} K^{-1} for 20°C).

Even allowing for radiative transfer, the conductivity of several types of coat was substantially larger than the value predicted on the basis of transfer by molecular conduction and radiation, and the discrepancy increased with the size of the temperature gradient across the hairs. This behavior suggested that free convection must be implicated, since the effective conductivity for this mode of heat transfer is proportional to the 0.25 power of a temperature difference. The rates of heat transfer assigned to free convection were found to be consistent with air velocities within the coat of the order of 1 cm s^{-1}.

Molecular conduction, radiation, and free convection therefore account for the range of values of thermal resistance in Table 10.2. Few attempts have been made to relate the insulation of an animal coat to the structure and weight of coat elements. Wathes and Clark [1981b] were able to show that the pelt resistance of the domestic fowl (i.e., feathers and skin) increased with feather mass per unit area W_f (kg m^{-2}) according to the relation

$$r = 6.5W_f + 1.5 \qquad (10.26)$$

where r is in s cm^{-1}. A range of W_f from 0.05 to 0.8 kg m^{-2} was obtained from pelts damaged by abrasion or pecking when birds were housed in battery cages. The corresponding range of resistance from 2 to 7 s cm^{-1} suggests that, in this type of system, substantial losses of heat and therefore of productivity may be caused by poor management. In undamaged coats, the average plumage resistance of 6 s cm^{-1} was similar to values reported for other avian species, and the resistivity of 1.8 s cm^{-1} per cm coat depth was comparable with figures for sheep. However Ward et al [2001] compared the insulation of chicken pelts from housed and free-range birds, and found surprisingly little difference between the two environments. They concluded that behavior (e.g., huddling, or choosing sheltered microclimates) is probably more important than differences in pelt insulation in determining the heat balance of housed and free-range poultry.

In the natural environment, forced convection must also play a significant role within coats to an extent determined by windspeed and direction with respect to an animal's body. The literature contains accounts of many experiments purporting to show that resistance of coats and clothing is reduced by a quantity proportional to the square root of V. However, careful re-analysis of the data showed that the conductance of a coat, which is the reciprocal of resistance, increased linearly with V; i.e.,

$$r(V)^{-1} = r(0)^{-1}(1 + bV) \qquad (10.27)$$

where $r(V)$ is resistance at windspeed V and b is a constant that depends on the wind permeability of the coat [Campbell, McArthur and Monteith, 1980]. Values of b for dense coats were around 0.1 s m^{-1}, so that a dense coat with a resistance of 10 s m^{-1} in still air would have only half this resistance at a windspeed of 10 m s^{-1}. For a range of coat structures, b ranged between 0.03 and 0.23 s m^{-1}.

The simplest interpretation of Equation 10.27 is that wind completely destroys the insulation of a depth of fleece t given by

$$t = lbV/(1 + bV) \qquad (10.28)$$

where l is total fleece depth.

Figure 10.13 unites all four modes of heat transfer, and emphasizes the thermal complexity of coats and clothing.

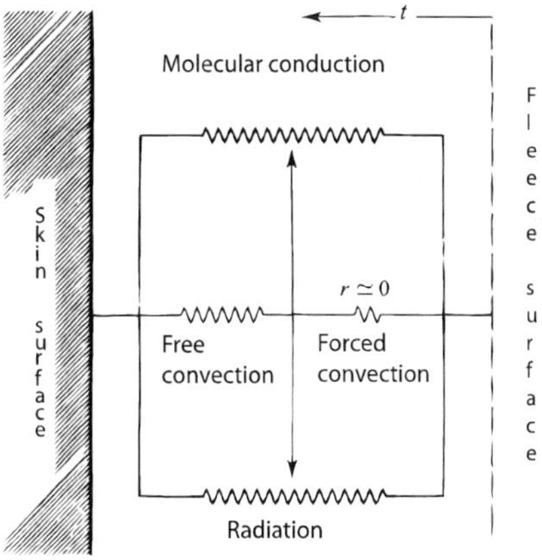

FIGURE 10.13 Electrical analogue for heat transfer through animal coat with wind penetration to depth t below surface (from McArthur and Monteith, 1980b).

10.5   Problems

1. Estimate the Nusselt number for a flat plate with characteristic dimension 50mm exposed to a windspeed of 2.0 m s^{-1}. Assume now that a real leaf can be treated as a flat plate with the same characteristic dimension, but that its Nusselt number is twice the value calculated earlier (i.e., $\beta = 2$, p. 169). Estimate the rate of convective heat transfer from the leaf at a wind speed of 2.0 m s^{-1} when the difference between leaf and air temperature is 1.5°C.

2. A silk worm, diameter 4mm and length 30 mm, is suspended vertically on a thread in a sun fleck in a plant canopy where the windspeed is 0.1 m s^{-1}. If the caterpillar's temperature is 5°C greater than the air temperature, estimate its rate of convective heat loss.

3. A thermometer element, reflection coefficient 0.40, is exposed to a mean irradiance of 300 W m^{-2} of solar radiation in an environment where effective radiative temperature and air temperature are 20.0°C and windspeed is 1.0 m s^{-1}. If the resistance r_H to convective heat transfer of the thermometer is 80 s m^{-1}, what temperature will the thermometer indicate? Two methods of improving the thermometer were tested (i) coating the element with white paint, reflection coefficient 0.90, and (ii) surrounding the element with a radiation shield, emissivity unity, but which reduced the wind speed u around the bulb to 0.5 m s^{-1}. If the radiation shield reduced the mean solar irradiance of the bulb to 90 W m^{-2}, calculate which method gave the smallest temperature error, assuming that r_H was proportional to $u^{-0.5}$.

4. An isolated flower bud can be treated as a sphere, diameter 10mm. The bud is exposed to a cloudless night sky with $L_d = 230$ W m^{-2}. Ground temperature is 273 K and air temperature is 275 K. The bud is susceptible to damage if its temperature falls below 273 K. Calculate the extra heat flux that would be necessary to supply to the bud to maintain its temperature at 273 K. Suggest ways in which the microclimate could be modified to protect the bud.

5. The fleece temperature of a sheep (treated as a cylinder, diameter 1.0 m) was 40°C greater than air temperature in bright sunshine when the windspeed was 0.5 m s^{-1}. Would its convective heat loss be best treated as forced or free convection? Would the boundary layer be most likely laminar or turbulent? Estimate the rate of convective heat loss.

6. A young animal is roughly cylindrical in shape, diameter 20 cm. Its core (diameter 16 cm) temperature is $37°C$, skin temperature is $30°C$, and the thermal conductivity of the peripheral tissue is 0.60 W m^{-1} K^{-1}. Estimate the heat flux per unit area of the skin surface.

Chapter 11

Mass Transfer (Gases and Water Vapor)

Two modes of diffusion are responsible for exchanges between organisms and the air surrounding them. *Molecular diffusion* operates within organisms (e.g., in the lungs of an animal or in the substomatal cavities of a leaf) and in a thin skin of air forming the boundary layer that surrounds the whole organism. *Turbulent diffusion* is the dominant transfer process in the free atmosphere, although molecular diffusion continues to operate and is responsible for the ultimate degradation of turbulent energy into heat.

Turbulence is ubiquitous in the atmosphere except close to the earth's surface on calm clear nights. The turbulent transfer of water vapor and carbon dioxide is of paramount importance for all higher forms of life. As a measure of effectiveness, the amount of carbon dioxide absorbed by a healthy green crop in one day is equivalent to all the CO_2 between the canopy and a height of 30 m. In practice, although the concentration of carbon dioxide in the atmosphere decreases between sunrise and sunset as a result of photosynthesis, this depletion rarely exceeds 15% of the diurnal mean concentration near the ground. These figures imply that turbulent transfer enables vegetation to extract CO_2 from a substantial depth of the mixed planetary boundary layer. For example, Bakwin et al [1998] reported that the CO_2 concentrations measured on tall television towers in Wisconsin and North Carolina on summer afternoons were about 2–3 ppm less than the daily mean at 400–500 m, and up to 40 ppm less than the daily mean in the lowest 20 m.

The process of mass transfer will now be described in terms of diffusion across boundary layers, through porous septa and within the free atmosphere.

11.1 Non-Dimensional Groups

Mass transfer to or from objects suspended in a moving airstream is analogous to heat transfer by convection and is conveniently related to a non-dimensional parameter similar to the Nusselt number of heat transfer theory. This is the *Sherwood number* Sh defined by the equation

$$\mathbf{F} = \mathrm{Sh}D(\chi_s - \chi)/d \tag{11.1}$$

where $\mathbf{F}$ = mass flux of a gas per unit surface area (e.g., g m^{-2} s^{-1}); χ_s, χ = mean concentration of gas at the surface and in the free atmosphere (e.g., g m^{-3}) ; D = molecular diffusivity of the gas in air (e.g., m^2 s^{-1}). (Thomas Sherwood was an American chemical engineer who made substantial advances in the analysis of interactions between mass transfer and flow.) As

$$\mathrm{Sh} = \frac{\mathbf{F}}{D(\chi_s - \chi)/d} \tag{11.2}$$

the Sherwood number can be defined as the ratio of actual mass transfer $\mathbf{F}$ to the rate of transfer that would occur if the same concentration difference were established across a layer of still air of thickness d. The corresponding resistance to mass transfer is derived by comparing Equation 11.1 with

$$\mathbf{F} = (\chi_s - \chi)/r \tag{11.3}$$

giving $r = d/(D\mathrm{Sh})$ (cf. $r_\mathrm{H} = d/(\kappa\mathrm{Nu})$). Resistances and diffusion coefficients for water vapor and carbon dioxide will be distinguished by subscripts and are related by $r_\mathrm{V} = d/(D_\mathrm{V}\mathrm{Sh})$ and $r_\mathrm{C} = d/(D_\mathrm{C}\mathrm{Sh})$.

FORCED CONVECTION

Just as the Nusselt number for forced convection is a function of Vd/υ (Reynolds number) and υ/κ (Prandtl number), the Sherwood number is the same function of Vd/υ and the ratio υ/D , known as the *Schmidt number* Sc. (Ernst Schmidt, a German engineer, drew attention to the similarity between heat and mass transfer, and is also credited with first proposing the use of aluminium foil for radiation shielding.) As an example of the analogy between mass and heat transfer, the Sherwood number for mass exchange at the surface of a flat plate is

$$Sh = 0.66\ Re^{0.5}\ Sc^{0.33} \tag{11.4}$$

cf.

$$Nu = 0.66\ Re^{0.5}\ Pr^{\ 0.33} \tag{11.5}$$

The presence of the term $0.66\ Re^{0.5}$ in both expressions is a consequence of a fundamental similarity between the molecular diffusion of heat, mass, and momentum in laminar boundary layers, and the numbers $Sc^{0.33}$ and $Pr^{0.33}$ take account of differences in the effective thickness of the boundary layers for mass and heat.

For any system in which heat transfer is dominated by forced convection, the relation between Sh and Nu is given by dividing Equation 11.4 by 11.5:

$$Sh = Nu(Sc/Pr)^{0.33} = Nu(\kappa/D)^{0.33} \tag{11.6}$$

The ratio κ/D is sometimes referred to as a *Lewis number* Le. In air at $20°C$, $(\kappa/D)^{0.33}$ is 0.96 for water vapor and 1.14 for CO_2 (see Appendix A.2). The corresponding ratios of resistances (cf. p. 162) are

$$r_V/r_H = (\kappa/D_V)^{0.67} = 0.93$$
$$r_C/r_H = (\kappa/D_C)^{0.67} = 1.32$$

FREE CONVECTION

In free convection, the circulation of air around a hot or cold object is determined by differences of air density produced by temperature gradients, by vapor concentration gradients, or by a combination of both. If the Nusselt number is related to the Grashof and Prandtl numbers by $Nu = BGr^n Pr^m$, the Sherwood number will be $Sh = BGr^n Sc^m = NuLe^m$ where m is 1/4 in the laminar regime and 1/3 in the turbulent regime. To calculate the Grashof number, it is convenient to replace the difference between the surface and air temperature $T_0 - T$ by the difference of virtual temperature (p. 16). If e_0 and e are vapor pressures at the surface and in the air and p is air pressure, the gradient of virtual temperature is

$$T_{v_0} - T_v = T_0\,(1 + 0.38e_0/p) - T\,(1 + 0.38e/p) \tag{11.7}$$
$$= (T_0 - T) + 0.38\,(e_0 T_0 - eT)\,/p$$

where temperatures are expressed in K. The importance of the vapor pressure term when T is close to T_0 can be illustrated for the case of a man

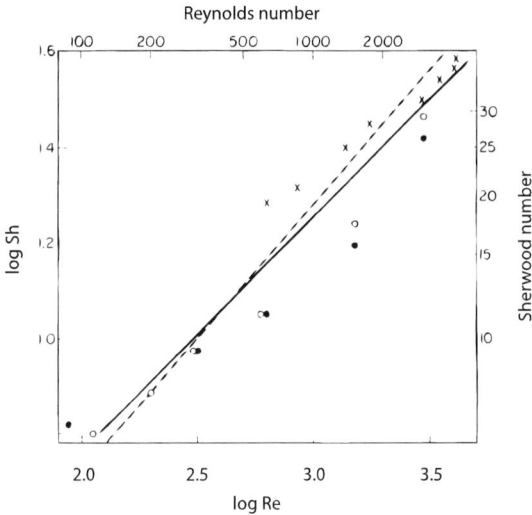

FIGURE 11.1 Relation between Sherwood and Reynolds numbers for plates parallel to the airstream. Continuous line, standard relation Sh $= 0.57\,\mathrm{Re}^{0.5}$; dashed line, measurements on discs by Powell [1940]; X, measurements on model bean leaf by Thom [1968]; O, measurements on replicas of alfalfa and Cocksfoot leaves by Impens [1965].

covered with sweat at 33°C and surrounded by still air at 30°C and 20% relative humidity. Then $e_0 = 5.03$ kPa and $e = 0.85$ kPa. The term $T_0 - T$ is 3 K, $0.38\,(e_0 T_0 - eT)\,/p$ is 4.9 K and the difference of virtual temperature is 7.9 K. The size of the Grashof number allowing for the difference in vapor pressure is 2.6 times the number calculated from the temperature difference alone. The corresponding error in calculating a Nusselt or Sherwood number (proportional to $\mathrm{Gr}^{0.25}$) is about -27%.

The same type of calculation is used to determine atmospheric stability when temperature and water vapor concentration are both functions of height (see p. 326).

11.2 Measurements of Mass Transfer

PLANE SURFACES

For laminar flow over smooth flat plates, the Sherwood number for water vapor, $0.57\,\mathrm{Re}^{0.5}$, is shown by the continuous line in Figure 11.1. Powell [1940] obtained a similar relation Sh $= 0.41\,\mathrm{Re}^{0.56}$ for circular discs with diameters between 5 and 22 cm parallel to the wind (dashed line). Thom

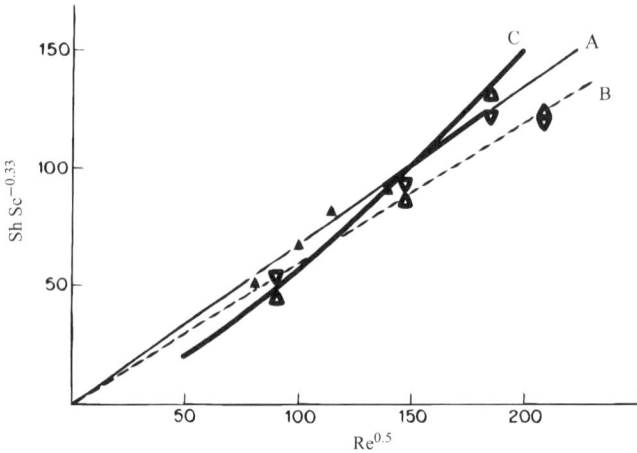

FIGURE 11.2 Comparison of measurements of mass transfer between air flow and flat surfaces parallel to the flow. The non-dimensional group $Sh\ Sc^{-0.33}$ is plotted as a function of $Re^{0.5}$ to give the constant of Equation 9.2 as a slope. ▲ small leaves; △ large leaves; ∇ large leaves (central strip only)—Chamberlain [1974]; A—Pohlhausen [1921]; B—Thom [1968]; C—Powell [1940]: (from Chamberlain, 1974).

[1968] measured evaporation from filter paper attached to the model bean leaf described on p. 148 and used bromobenzene and methylsalicylate and water to get a range of diffusion coefficients from 5.4×10^{-6} to 24×10^{-6} $m^2\ s^{-1}$. At windspeeds exceeding $1\ m\ s^{-1}$, the mass transfer of all three vapors was described by $Sh = 0.7\ Re^{0.5}Sc^{0.33}$ i.e., within a few percent of the predicted value (Equation 11.2). The measurements for water vapor plotted in Figure 11.1 show that when the windspeed was less than $1\ m\ s^{-1}$ (Re < 2800), the Sherwood numbers were larger than the predicted value, possibly because the rate of mass transfer was increased by differences of density in the air surrounding the leaf.

By using radioactive lead vapor as a tracer, perfectly absorbed by the surfaces exposed to it, Chamberlain [1974] was able to measure rates of mass transfer as a function of position on model bean leaves about 11×11.5 cm. Figure 11.2 shows that when the samples were parallel to the flow in a wind tunnel, measurements of the Sherwood number were consistent with the standard relation and with Thom's observations. With leaves at an angle to the flow (not shown), the local boundary layer re-

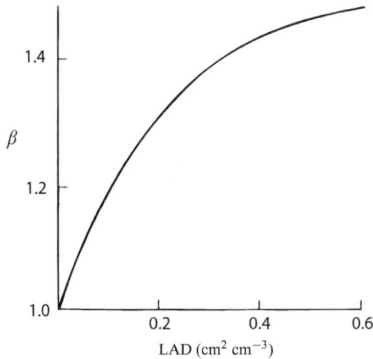

FIGURE 11.3 Ratio β of boundary layer resistance for an isolated *Citrus* leaf to resistance measured within a canopy of leaves of specified leaf area density (LAD); (redrawn from Haseba, 1973).

sistance of the surface tilted in the upwind direction increased by a factor of about four from the leading edge to the trailing edge. In contrast, the resistance over the downwind surface was greatest just behind the leading edge—a shelter effect—and then decreased with increasing distance, presumably because the surface was exposed to eddies forming in the lee of the edge. The mean resistance for the whole area of both surfaces appeared to be almost independent of exposure angle.

For real leaves in the field, wind direction is rarely constant and differences in local boundary resistance will usually be smaller (and less regular) than the previous paragraph suggests. The dependence on leaf angle of heat and mass transfer will usually be a consequence of differences in radiation absorption rather than in transfer coefficients.

When Chamberlain measured the rate of uptake of lead vapor by real bean leaves growing in a canopy, Sherwood numbers were about 25% greater than predicted from measurements on isolated models. The corresponding value of β (p. 169) is 1.25. This is consistent with measurements of evaporation from *Citrus* leaves in a canopy for which Haseba [1973] found that β increased systematically with leaf area density (area per unit volume of canopy) as Figure 11.3 shows. For arable crops, leaf area density is often of the order of 0.1 so Fig 11.3 implies that β should be between 1.1 and 1.2 in the field, consistent with conclusions for heat. Somewhat larger values would be expected if leaves were fluttering.

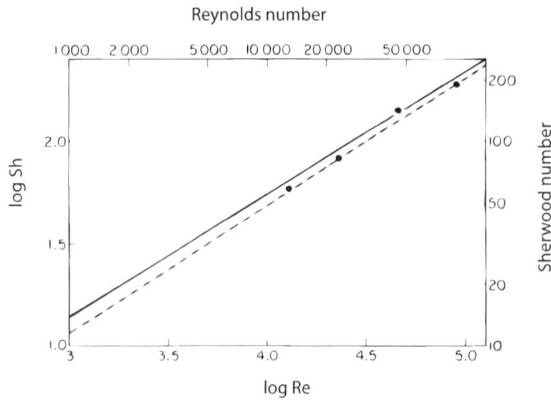

FIGURE 11.4   Relations between Sherwood and Reynolds numbers for a wet cylinder at right angles to the airstream. Continuous line, $Sh = 0.22 Re^{0.6}$; pecked line, $Sh = 0.16 Re^{0.62}$. The points were calculated by Rapp [1970] from measurements by Kerslake on a man covered with sweat.

CYLINDERS

For Reynolds numbers between 10^3 and 5×10^4, the Nusselt number for cylinders can be expressed as $Nu = 0.26 Re^{0.6} Pr^{0.33}$ and the corresponding Sherwood number is $Sh = 0.26 Re^{0.6} Sc^{0.33}$ or, in air, $Sh = 0.22 Re^{0.6}$. Powell's measurements of the evaporation from a wet cylinder in a wind tunnel fit this relation closely and it is shown by the continuous line in Figure 11.4. For the more restricted range of Reynolds number from 4×10^3 to 4×10^4, the relation $Nu = 0.17 Re^{0.62}$ is often used. The corresponding Sherwood number $Sh = 0.16 Re^{0.62}$ is shown by the pecked line in Figure 11.4.

Nobel [1974] measured the resistance to evaporation from wet paper cylinders ($d = 2$ cm) as a function of turbulent intensity and found that turbulence of 10% was enough to decrease boundary layer resistance by 22%, implying that $\beta = 1.22$ in the corresponding expression for Sherwood number. When the intensity of turbulence was increased from 10 to 70%, resistance decreased only slightly further. As with similar experiments on heat transfer, observed values of β must to some extent reflect the size of turbulent eddies in relation to the size of the object and turbulent intensity.

Rapp [1970] established close agreement between measurements of the evaporative loss from nude men covered with sweat and values predicted from the Sherwood number for a cylinder of appropriate diameter. The units of his calculations were transformed to show this agreement in Figure 11.4.

SPHERES

The Nusselt number for a sphere can be expressed in the form $0.34 \, Re^{0.6}$ and the corresponding Sherwood number is $0.34 \, Re^{0.6} \, Le^{0.33}$ or $0.32 \, Re^{0.6}$ for water vapor. Analysis of Powell's [1940] measurements of evaporation from wet spheres gives $Sh = 0.26 \, Re^{0.59}$, about 20% less than the value predicted from heat transfer rates.

The rate of heat transfer from a 6 cm diameter balsa wood sphere was close to the predicted value when the turbulent intensity i was about 1%, but was about 1.1 for $i = 30\%$ and 1.14 for $i = 70\%$ [Nobel, 1975].

11.3 Ventilation

When mass transfer is induced by the ventilation of a system, an equation relating the mass flux to an appropriate potential gradient can be used to define a transfer resistance consistent with the values of diffusion resistance already discussed. Relevant examples are the exchange of carbon dioxide between the air inside and outside a greenhouse, the loss of water vapor by evaporation from the lungs of an animal, and the exchange of gases in open-top field chambers.

If the air in a greenhouse is well stirred so that the volume concentration of CO_2 in the internal air has a uniform value of ϕ_i ($m^3 \, CO_2 \, m^{-3}$ air) when the external concentration is ϕ_e, the rate at which plants in the glasshouse absorb CO_2 from the external atmosphere can be written as

$$Q = \rho_c v N \left(\phi_e - \phi_i \right) g \, h^{-1} \tag{11.8}$$

where v is the volume of air in the house (m^3), N is the number of air changes per hour and ρ_c is the density of CO_2 ($g \, m^{-3}$). Dividing both sides of Equation 9.6 by the floor area A gives a flux of CO_2 per unit floor area

$$F = Q/A = \rho_c v N \left(\phi_e - \phi_i \right) / A \tag{11.9}$$

The resistance to CO_2 diffusion r_c can be defined by writing

$$F = \rho_c \left(\phi_e - \phi_i \right) / r_c$$

and comparison of the two equations gives

$$r_c = A/vN \qquad (11.10)$$

$$= \left(N\bar{h}\right)^{-1}$$

where $\bar{h}$ is the mean height of the house. For example, if N is 10 air changes per hour and $\bar{h} = 3$ m, r_c is $1/30$ h m^{-1} or 1.2 s cm^{-1}, comparable in size with boundary layer and stomatal resistances. Roy et al [2002] reviewed heat and mass transfer in greenhouses considered as well-mixed volumes.

A similar approach is applicable to mass transfer from animals. If V is the volume of air respired by an animal per minute (the *"minute volume"*) and A is the area of skin surface, the loss of water per unit skin area is

$$\mathbf{F} = V\left(\chi_s\left(T_b\right) - \chi\right)/60A \text{ g m}^{-2} \text{ s}^{-1} \qquad (11.11)$$

where $\chi_s\left(T_b\right)$ is the water vapor concentration of air saturated at deep body temperature T_b and χ is the concentration in the environment, both expressed in g m^{-3}. Then the resistance to vapor exchange is

$$r_V = 60A/V$$

For a man at rest $V = 10^{-2}$ m^3 min^{-1} and if $A = 1.7$ m^2, r_V is 10^4 s m^{-1} or 100 s cm^{-1}; i.e., about two orders of magnitude larger than common values of the boundary layer resistance for a sweating nude figure. Even during very rapid respiration when V may reach 10^{-1} m^3 min^{-1}, the diffusion resistance for respiration will exceed the boundary layer resistance by an order of magnitude, and when the skin is covered with sweat, the loss of water by evaporation from the lungs will be much smaller than the cutaneous evaporation rate.

To determine the response of crops to pollutant gases or elevated carbon dioxide concentrations, cylindrical chambers with open tops (Figure 11.5, see color plates) are often used in attempts to alter the quality of air around plants growing in the field without making large changes in other aspects of the microclimate [Heagle, Body and Heck, 1973]. For air quality studies, fans blow air into the base of chambers through filters that absorb pollutants; other chambers have fans without filters to provide "control" treatments. Extra pollutants or CO_2 may also be injected into the ventilation air of some chambers [Unsworth, Lesser and Heagle 1984, Mulholland et al 1998]. However, the concentration of a pollutant within a filtered open-top chamber is never zero because some unfiltered air enters by "incursion" through the open top.Figure 11.6 shows a simple

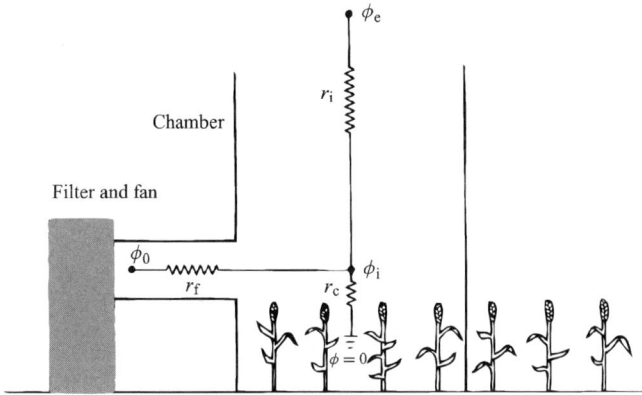

FIGURE 11.6 Schematic diagram of an open-top field chamber and the equivalent resistance analogue.

resistance analogue that can be used to estimate the concentration of the pollutant gas in a chamber in which a filter to absorb pollutants is adjacent to the fan. Exchange of the gas through the open top (incursion) can be regarded as driven by the potential difference between the concentration ϕ_e in the outside air and the concentration ϕ_i in the chamber. The resistance to incursion is given by $r_i = A/vN_i$ where A is the chamber base area, v the volume, and N_i the rate of air change through the open top. Similarly, air flow through the filter is driven by the potential $\phi_o - \phi_i$ and limited by the resistance r_f describing ventilation by the fan (N_f air changes per second); ϕ_o is the gas concentration (usually close to zero) leaving the filter. To complete the general analogue, the pollutant flux in the chamber to plants and soil where ϕ is assumed to be zero is given by $(\phi_i - 0)/r_c$, where r_c is the plant canopy resistance. It follows, from the conservation of flux, that

$$\frac{\phi_o - \phi_i}{r_f} + \frac{\phi_e - \phi_i}{r_i} = \frac{\phi_i - 0}{r_c}$$

which, for the ideal case $\phi_i = 0$, reduces to

$$\phi_i/\phi_e = \left[1 + r_i\left(\frac{1}{r_f} + \frac{1}{r_c}\right)\right]^{-1} \tag{11.12}$$

For a set of chambers designed by Heagle, Body and Heck [1973], Unsworth, Heagle and Heck [1984a,b] estimated the resistances as follows: $r_f = 6$ s m^{-1} (about three air changes per minute); r_i decreased from about 50

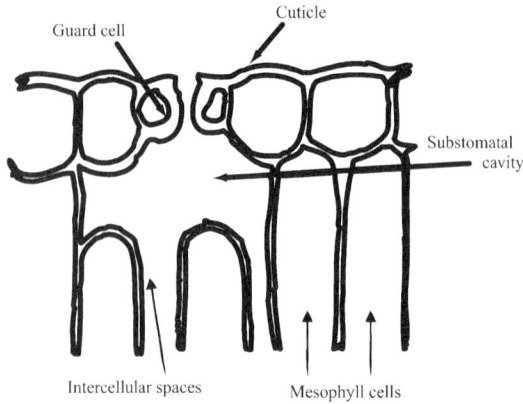

FIGURE 11.7   Schematic diagram of a single stoma (after Jones 1992).

s m^{-1} when the windspeed outside was 1 m s^{-1} to 15 s m^{-1} at a windspeed of 6 m s^{-1}; r_c for ozone uptake by a canopy of soybeans had a minimum value of about 70 s m^{-1}. Equation 11.12 then shows that the pollutant concentration in the chamber would increase from about 0.1 ϕ_e at the lowest windspeed to 0.3 ϕ_e at 6 m s^{-1}. It is unlikely that open-top chambers could be designed to be much more effective than this without substantially altering other features of microclimate such as temperature and humidity.

11.4 Mass Transfer through Pores

When leaves transpire, water evaporates from mesophyll cell walls and escapes to the atmosphere by diffusing into substomatal cavities, through stomatal pores, and finally through the leaf boundary layer into the free atmosphere. During photosynthesis, carbon dioxide molecules follow the same path but in the opposite direction.Figure 11.7 shows a schematic of a single stomatal pore, showing the guard cells that alter the pore aperture, the cuticle that is relatively impervious to water vapor, mesophyll cells, and the substomatal cavity and intercellular spaces. Plants maintain active control over the dimensions of their stomatal pores by changes in the osmotic potential of guard cells, and even though, when fully open, the pores occupy only 0.5 to 5% of the leaf surface area, almost all the water vapor and CO_2 exchanged between leaves and the atmosphere passes through these pores [Jones 1992].

Rigorous treatments of diffusion through pores (e.g., Leuning, 1983) allow for the interaction of diffusing gases and for the difference in (dry) air pressure across pores needed to balance the difference in water vapor pressure. There is a class of problems in which these complications cannot be ignored (e.g., in precise estimates of the intercellular CO_2 concentration), but for many practical purposes the elementary treatment that follows is adequate.

It will therefore be assumed that for a leaf lamina, the resistance of stomatal pores for a particular gas depends only on their geometry, size and spacing whereas the resistance offered by the boundary layer depends on leaf dimensions and windspeed.

As discussed earlier, in plant physiology it is often convenient to express fluxes in units of mol m^{-2} s^{-1} and concentration gradients in mol gas mol^{-1} air. Conductance then has the same units as flux and resistance has reciprocal units. For conversion, conductance in mol m^{-2} s^{-1} must be multiplied by m^3 mol^{-1} (0.0224 at STP) to obtain units of m s^{-1}, and resistance in units of mol^{-1} m^2 s must be multiplied by mol m^{-3} (44.6 at STP) to obtain units of s m^{-1}.

Meidner and Mansfield [1968] tabulated stomatal populations and dimensions for 27 species including crop plants, deciduous trees, and evergreens. The leaves of many species have between 100 and 200 stomata per mm^2 distributed on both the upper and lower epidermis (amphistomatous leaf) or on the lower surface only (hypostomatous leaf). The length of the pore is commonly between 10 and 30 μm and the area occupied by a complete stoma, including the guard cells responsible for opening and shutting the pore, ranges from 25 × 17 μm in *Medicago sativa* to 72 × 42 μm in *Phyllitis scolopendrium*.

Because stomata tend to be smaller in leaves where they are more numerous, the fraction of the leaf surface occupied by pores does not vary much between species. There is much greater variation in the geometry of pores: the stomata of grasses are usually long, narrow, and aligned in rows parallel to the midrib, whereas the elliptical stomata of sugar beet (*Beta vulgaris*) and broad bean (*Vicia faba*) are randomly oriented but uniformly dispersed over the epidermis.

The network of resistances in Figure 11.8 is an electrical analogue for the diffusion of water vapor from between the intercellular spaces of a leaf and the external air. The calculation of boundary layer resistance r_V has already been discussed: values of 30–100 s m^{-1} are expected for small leaves in a light wind. Xerophytes such as succulents and cacti have large minimum stomatal resistances consistent with the need to conserve water.

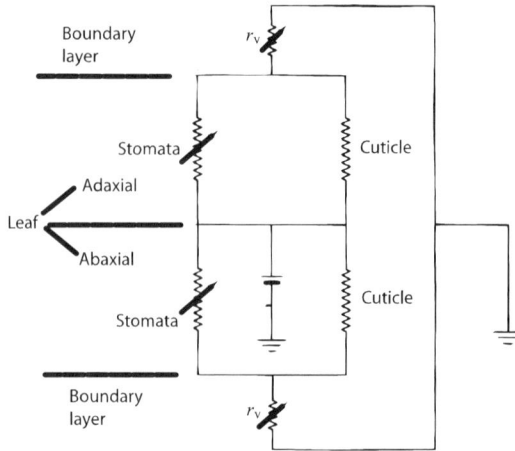

FIGURE 11.8   Equivalent electrical circuit for loss of water vapor from a leaf by diffusion through the stomata and cuticle of the adaxial (upper) and abaxial (lower) epidermis.

Many mesophytes have minimum stomatal resistances in the range 50 to 300 s m^{-1}, with the smallest values associated with cultivated grasses, cereals, and herbaceous crops. Cuticular resistances range from 2000 to 6000 s m^{-1} in mesophytes and from 4000 to 40000 s m^{-1} in xerophytes. In both types of plant, the resistance of the cuticle is usually so much larger than the stomatal resistance that its role in water vapor and CO_2 transfer can generally be ignored. Jones [1992] summarized much of the extensive literature on stomatal resistance and its components.

Nobel [1975] found that the resistance to the loss of water vapor from the fruiting bodies of fungi (Basidlomycetes) fell between 30 and 60 s m^{-1}, but for the fruits of many tree species the range was from 3000 to 700000 s m^{-1}. The resistance generally increased with age as the cuticle became thicker; e.g., from 600 s m^{-1} for a green orange to 150000 s m^{-1} for mature fruit.

Entomologists use resistance networks to describe the diffusion of water vapor from within an insect either through spiracular valves (analogous to stomata) or dermal and cuticular layers (analogous to the epidermis of a leaf). Precise measurements by Beament [1958] of water loss from a cockroach nymph gave a resistance to vapor transfer of about 2×10^3 s m^{-1}, equivalent to the resistance of about 100 condensed monolayers of stearic acid [Gilby, 1980].

Pores in the eggshells of birds have a function similar to the stomata of leaves, allowing the inward diffusion of oxygen to the developing embryo and the outward diffusion of carbon dioxide and water vapor. Tullett [1984] reviewed the structure and function of eggshells. Unfortunately, avian physiologists conventionally express the porosity g of eggshells in units of water loss per egg per day per unit water vapor pressure difference across the shell. The failure to normalize for surface area obscures the fact that the diffusion resistance of eggshells is almost independent of egg size. As an illustration, the relationship between g and egg mass W (g) derived by Ar et al [1974] can be written as

$$g = 37.5 \times 10^{-9} W^{0.78} \text{ g s}^{-1} \text{ kPa}^{-1}$$

For eggs with the same geometry and constant density but different size, surface area A is proportional to $W^{0.66}$ (see p. 259), so the porosity per unit shell area g/A has only a weak dependence on egg mass ($W^{0.12}$). When calculated by the methods used to define leaf resistances, eggshell resistances to water vapor diffusion are typically between about 5×10^4 and 1×10^5 s m^{-1} for eggs that range in mass from about 5–500 g.

RESISTANCE CALCULATIONS

Electrical analogues provide a useful way of visualizing the process of diffusion from the intercellular spaces of a leaf through stomatal pores to the external boundary layer. If the leaf surface was uniformly covered with water, the concentration of water vapor (equivalent to electrical potential) would decrease with distance away from the surface, eventually reaching the concentration of the free air outside the leaf boundary layer. Lines of constant concentration (equipotentials) would be parallel to the surface. Diffusion through the boundary layer (assumed of constant depth) would be one-dimensional, with the diffusive flux (equivalent to electrical current) perpendicular to the equipotential lines as shown in Figure 11.9.

But for leaves, more complex three-dimensional concentration fields are set up near the surface because each stomate is a source of water vapor, whereas the cuticle is almost impermeable to water. In addition, pores have a length l, comparable with their diameter d. Figure 11.10 illustrates diffusion through a single stomatal pore. The broken lines are lines of constant potential representing water vapor concentration. "Current," representing the flux of water vapor, must always flow perpendicular to the potential lines. According to the theory of gaseous diffusion in three dimensions through a circular hole of diameter d, there is a resistance to diffusion associated with each side of the hole given by $r_h = \pi d/8D$

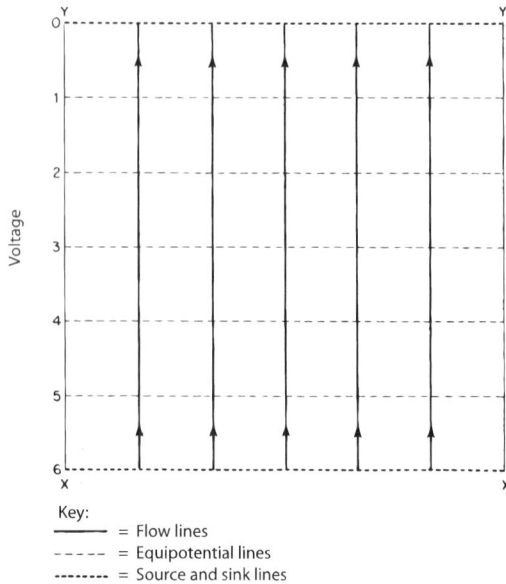

FIGURE 11.9 Two-dimensional electrical analogue of diffusion between parallel planes XX and YY. The continuous lines represent flow of current (mass) and the pecked lines join points of equal potential (concentration).

where D is the diffusion coefficient of the gas. The diffusion resistance of uniform pore of length l is $r_p = l/D$. In principle, the total resistance of a pore r_t is found by adding r_p to the resistance r_h at either end of the pore; i.e.,

$$r_t = r_p + 2r_h \qquad (11.13)$$

For stomata, r_h is often smaller than r_p and is therefore referred to as an "end-correction."

Two points can be made about the applicability of Equation 11.13 for estimating resistances of real pores. First, as the cross section of most stomatal pores is not uniform, the resistance of the pore cannot be calculated accurately without knowing the shape of the cross section and its area A at different distances from the end of the pore. An approximate value of r_t can be obtained from the length l and mean diameter d of a pore with circular cross section or from the major and minor axes of an elliptical pore. Second, although the end-correction for a circular pore is usually assumed to be $2r_h = \pi d/4D$, Figure 11.10 shows that it would be incorrect to apply a conventional end-correction to the inner end of the pore. If the substomatal cavity is assumed to be lined with cell walls from which

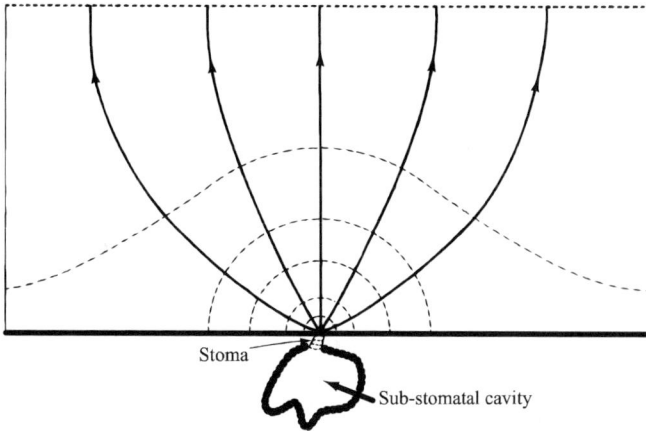

FIGURE 11.10 Electrical analog of diffusion of water vapor from a single stomatal pore. Note the absence of equipotential lines in the sub-stomatal cavity showing that the "end-correction" can be neglected at the inner end of the pore.

water is evaporating (represented by the dotted line), the end-correction resistance of the inner end of the pore is much less than the resistance of the outer end. For many leaves, $r_t = r_p + r_h = (l + \pi d/8)/D$ is probably a better estimate of the resistance of a single pore than $r_t = r_p + 2r_h$.

Finally, the resistance of a multi-pore system can be estimated. As mesophytes usually have about 100 stomata per mm^2 their average separation is about 0.1 mm or 100 μm, an order of magnitude larger than the maximum diameter of the pore. At this spacing, there is little interference between the equipotential shells of individual pores, although the merging of the equipotential lines shown in Figure 11.11 indicates that, for precise calculations, the end-correction resistance for the outer end of each pore could be reduced slightly.

When there are n pores per unit leaf area, the resistance of a set of pores r_s can be readily derived from the resistance of the individual pores r_t. For example, if $\delta\chi$ is the difference of water vapor concentration maintained across a set of circular pores with a mean diameter of d, the transpiration rate can be written either as

$$E = \frac{\delta\chi}{r_s} \text{ or as } \frac{n\pi \left(d^2/4\right)\delta\chi}{r_t}$$

It follows that

$$r_s = \frac{4\left(l + \pi d/8\right)}{\pi nd^2 D}$$

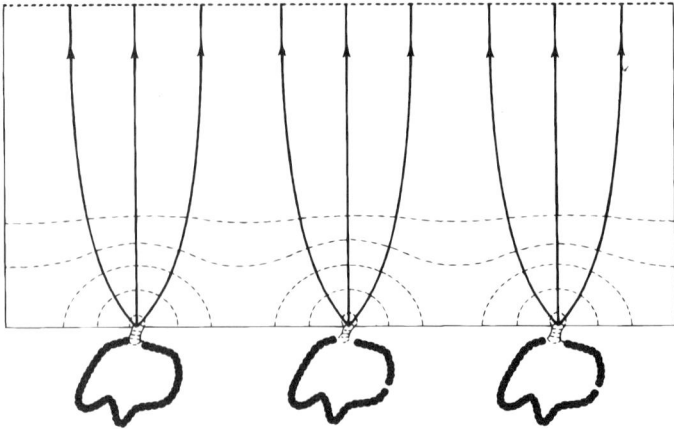

FIGURE 11.11   Electrical analog for a leaf epidermis showing three pores. Note the merging of the equipotential lines which decrease the effective end-correction for the outer end of each pore.

Similar expressions in a variety of units were derived by Penman and Schofield [1951] and Meidner and Mansfield [1968].

Milthorpe and Penman [1967] evaluated the stomatal resistance of wheat leaves with rectangular pores. Refinements in their calculations included: (i) allowing for the stomatal slit getting shorter as the stoma closed; (ii) making the diffusion coefficient a function of the stomatal width to allow for the effect of "slip" at the stomatal walls. This phenomenon is important when the width is comparable with the mean free path of the diffusing molecules. For example, when the width of the throat was 1 μm, the diffusion coefficient for water vapor was 88% of its value in free air; (iii) making the end-correction for the inner end of the pore 1.5 times the correction for the outer end (Figure 11.10 suggests that this factor should have been smaller rather than greater than unity.) Figure 11.12 shows the relation between resistance and slit width (a) from figures tabulated by Milthorpe and Penman for wheat stomata, assumed rectangular, and (b) from figures tabulated by Biscoe [1969] for sugar beet stomata, assumed elliptical.

Figure 11.12 can be used to estimate the effect of stomatal closure on the *total* resistance to diffusion of water vapor or carbon dioxide for a leaf, taking account of the distribution of stomata over the abaxial and adaxial surfaces. On the assumption that wheat has amphistomatous leaves with the same resistance r, on each epidermis, the total resistance for each surface is the sum of two resistances in series, i.e., $r_s + r_V$, where r_V is the

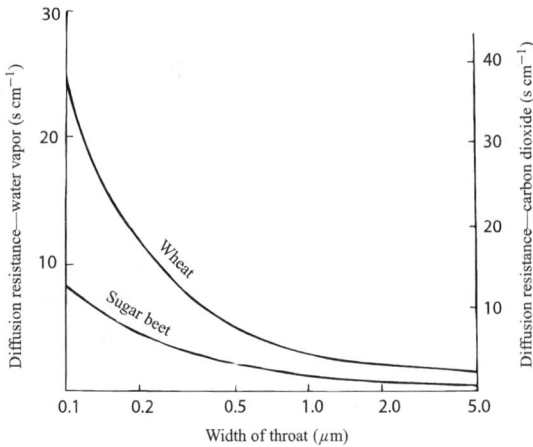

FIGURE 11.12 Diffusion resistance calculated for wheat and sugar beet leaves as a function of throat width. Resistances for water vapor and carbon dioxide are given on left- and right-hand axes respectively.

leaf boundary layer resistance to water vapor diffusion. The resistance of the whole leaf is the sum of the resistances for the two sides in parallel, i.e.,

$$\left(\frac{1}{(r_V + r_s)} + \frac{1}{(r_V + r_s)}\right)^{-1} = (r_V + r_s)/2$$

When r_V is much smaller than r_s, the leaf resistance is approximately $r_s/2$. On the assumption that sugar beet has hypostomatous leaves (i.e., with stomata on only one surface, usually the abaxial surface), the resistances of the two surfaces are $r_V + r_s$ (abaxial) and $r_V + x$ (adaxial), where x is a resistance much larger than r_s representing the resistance of the cuticle. Combining these resistances in parallel gives the total leaf resistance as

$$\left(\frac{1}{(r_V + r_s)} + \frac{1}{(r_V + x)}\right)^{-1} \approx r_V + r_s$$

which is approximately equal to r_s when r_V is much smaller than r_s. (Measurements with model bean leaves by Thom [1968] suggest that the appropriate value of r_V for a hypostomatous leaf may be about 30% smaller than the value for a flat plate of the same size because of increased exchange round the edge of the plate.)

In practice, the two surfaces of an amphistomatous leaf often have different stomatal resistances. Relevant equations have been published in the literature but are too cumbersome to reproduce here. As an additional complication, the stomata on the two surfaces may respond in different ways to levels of irradiance and of the water stress in the mesophyll tissue.

MASS TRANSFER AND PRESSURE

Gale [1972] and others have drawn attention to the fact that the stomatal resistance of a leaf with specified stomatal geometry should be proportional to pressure because the diffusion coefficient of a gas is inversely proportional to pressure. At sea level, the normal range of pressure (95 to 102 kPa) is equivalent to a change in stomatal resistance trivial in comparison with the uncertainty of most measurements, but differences of pressure are much larger between sea level and the tops of mountains. At a height of 3000 m, for example, atmospheric pressure is about 30% less than its value at sea level, implying that stomatal resistance r_s should be smaller by the same factor.

Boundary layer resistance also decreases with decreasing pressure. The boundary layer resistance for water vapor transfer r_V is defined as t/D where the boundary layer thickness t is proportional to $v^{0.5}$ for a flat plate (see Equation 9.1) and therefore to $p^{-0.5}$. But since D is proportional to p^{-1}, r_V is proportional to $p^{0.5}$.

At first sight, the implication of the pressure dependence of r_s and r_V is that evaporation rate E should increase with height, all other factors being equal. This is correct if the gradient for water vapor transfer is defined in terms of a fixed difference in vapor density $\delta\chi$ since $E = \delta\chi/(r_s + r_V)$. However, if the gradient is defined by a difference of specific humidity so that $E = \rho\delta q/(r_s + r_V)$, the presence of density (proportional to pressure) in the numerator makes E proportional to $p^{0.5}$ when r_V is large compared with r_s, and insensitive to pressure in the more common case when r_V is small compared with r_s.

The same argument holds for rates of photosynthesis, which are effectively independent of pressure when the concentration gradient is expressed either as a volume concentration (volume CO_2 per unit volume air) or as a specific mass (g CO_2 per g air).

11.5 Coats and Clothing

Few attempts have been made to determine resistances to water vapor transfer within the coats of animals, but there is increasing use of "breathable" fabrics such as Goretex[®] for outdoor clothing, and textile manufacturers use standard methods (e.g., ISO 11092) to measure the resistances to water vapor transfer in these fabrics.

The physical processes responsible for the diffusion of vapor within hair coats were explored by Cena and Monteith [1975c] who compared rates of transfer through fibreglass and through sections of sheep's fleece, both uncured and cured to remove grease. Only about 2% of the space within each sample was occupied by hair. At 16°C, the resistance of the fiberglass was about 4.3 s cm^{-1} per cm depth, close to the theoretical value for still air (i.e., the reciprocal of the molecular diffusion coefficient D). The resistance per cm (resistivity) of cured and uncured fleece was less than the value for fibreglass, and the difference increased with the depth of the sample, suggesting that the transfer of vapor by diffusion was augmented by the capillary movement of water along hairs. Webster et al [1985] found that the resistivity of pigeon's plumage was about twice the value for still air, presumably because the porosity of the samples was much smaller than that of fleece.

Gatenby et al [1983] measured water vapor concentration within the fleece of a ewe standing in a constant temperature room. The depth of fleece was about 7 cm and the concentration decreased linearly with distance from the skin at about 0.6 g m^{-3} per cm at 5°C ambient temperature, increasing to 1.0 g m^{-3} per cm at 28°C. If molecular diffusion is assumed, corresponding fluxes of latent heat range from about 3.5 to 6.0 W m^{-2}, much smaller than metabolic heat production (see Chapter 12). The real rate of latent heat transfer was probably substantially larger because of free convection associated with temperature gradients (see p. 186 and Cena and Monteith, 1975b).

Breathable fabrics such as Goretex$^{\circledR}$ consist of a thin layer of expanded polytetrafluoroethylene (ePTFE) plastic membrane attached to an outer, and sometimes an inner, layer of fabric. The membrane has a porous structure, similar to that of a leaf, typically with about 10^9 pores per cm^2, of diameters between about 0.1 and 10μm. Thus the pores are 2–3 orders of magnitude smaller than raindrops, which consequently cannot penetrate the membrane, but they are much larger than water vapor molecules, which can diffuse through them. Water vapor evaporated from the skin creates a diffusion gradient across the fabric, and mass transfer by diffusion through the pores reduces the buildup of moisture that occurs with non-breathable waterproof raingear. If the work rate is sufficient to induce sweating, breathable fabrics are seldom able to transport water vapor rapidly enough to avoid some accumulation of sweat inside the raingear, and it becomes important for comfort to wear inner garments made of "wicking" materials such as polypropylene rather than absorbent materials such as cotton so that moisture is more readily lost through the breathable layers. The outermost layer of breathable fabrics is usually given hydrophobic properties so it does not become saturated in rain and cause heat to be lost from the body by conduction and evaporation from the wet exterior.

Free and forced convection also increase vapor transfer in clothing. In a study of transfer in a tropical fatigue uniform cited by Campbell et al [1980], the vapor conductance increased linearly with windspeed from about 0.17 cm s^{-1} in still air to about 5 cm s^{-1} at 6 m s^{-1} (equivalent to a resistance range from 6 to 1.6 s cm^{-1}). The conductance for heat transfer was larger by an amount consistent with radiative transfer (see p. 186).

Both in man and in birds, the resistance of skin to vapor diffusion is of the order of 100 to 200 s cm^{-1} and so is much larger than the resistance of clothing or feathers. Exceptionally, measurements on premature babies lying naked in incubators revealed that skin resistance could be as small as 30 s cm^{-1} and that the associated loss of latent heat could exceed metabolic heat production if the circulating air were not humidified [Wheldon and Rutter, 1982].

11.6 Problems

1. Water vapor is transferred by molecular diffusion down narrow stomatal pores, length 10μm, from the saturated interior of a leaf at $25°$C to the open air where the temperature is also $25°$C and the relative humidity is 60%.

   (i) What are the water vapor concentrations (absolute humidity) in the interior of the leaf and in the open air (assume that the leaf is at sea level)?

   (ii) Calculate (a) the flux density of water vapor F_w (kg m^{-2} s^{-1}) through a single stoma, and (b) the diffusion resistance of the stoma r_p (s m^{-1}).

   (iii) If there are 200 stomata per mm^2 on one surface of the leaf (and no stomata on the other surface), and each pore is circular with a diameter of 5μm, calculate the stomatal diffusion resistance of the leaf (r_l) in s m^{-1}, distinguishing between the stomatal component and the "end correction" component.

2. A breathable fabric used in a ski jacket has 10^9 pores per cm^2, of diameter 4 μm. If air inside the jacket is saturated at a temperature of 30°C, and the external air temperature and relative humidity are $-5°C$ and 30% respectively, calculate the rate at which water vapor diffuses through unit area of the fabric to the exterior, and express this as the latent heat flux in W m^{-2}.

3. A leaf has different populations of stomata on its upper and lower surfaces, and the respective stomatal resistances are 200 and 100 s m^{-1}. The boundary layer resistance to heat transfer of both surfaces in parallel is 40 s m^{-1}. Assuming that both surfaces have the same boundary layer resistance, find the combined stomatal and boundary layer resistance of the leaf.

4. A leaf exposed to air at 20.0°C and 50% relative humidity loses water at a rate of 1.0×10^{-6} g cm^{-2} s^{-1}. Assuming that the leaf and air temperature are the same, calculate the combined stomatal and boundary layer resistance. If the concentration of carbon dioxide in the substomatal cavity is 100 vpm less than in the ambient air, calculate the flux of carbon dioxide (assume that the density of carbon dioxide is 1.87 kg m^{-3}).

5. Calculate the boundary layer resistance for water vapor transfer (in s m^{-1}) at a windspeed of 1.0 m s^{-1} for an isolated flat leaf with characteristic dimension 50.0 mm. State any assumptions that you make. Now assume that the leaf surface is covered with water at 25°C, and the air temperature and relative humidity are 25°C and 60%, respectively. Calculate the evaporation rate of water from the leaf surface (g m^{-2} s^{-1}), and express this as the latent heat flux (W m^{-2}).

6. A flat model leaf, area 100 cm^2, is made of wet green filter paper. When the leaf is suspended in a plant canopy where the air temperature is 25°C, and the relative humidity is 75%, it loses 0.70 g water in 10 minutes. Assuming the leaf is at air temperature, estimate the boundary layer resistance of the leaf for water vapor transfer.

Chapter 12

Mass Transfer (Particles)

Small particles are transferred in the free atmosphere by the same process of turbulent diffusion that is responsible for the mass transfer of
gases. Because particles have inertia, they cannot respond to the most
rapid eddy motions, but this is generally unimportant in the usual scales
of atmospheric turbulence. However, inertia is important close to surfaces
when particles may be thrown against an object if the air stream they are
in changes direction rapidly. A second distinction between particles and
molecules is the importance of gravitational forces, the initial topic in the
discussion of particle transfer.

12.1 Steady Motion

SEDIMENTATION VELOCITY

The gravitational force on a particle is the difference between the weight
of the particle and the weight of the air that it displaces. This is given
by the product of the volume of the particle, gravitational acceleration g,
and the difference between particle density ρ and air density ρ_a. When
spherical particles with radius r and volume $(4\pi r^3/3)$ fall under gravity
they attain a steady *sedimentation velocity* V_s when the gravitational force
balances the drag force; i.e., from Equation 9.11

$$(4/3)\pi r^3 g \left(\rho - \rho_a\right) = (1/2)\,\rho_a V_s^2 c_d \pi r^2 \qquad (12.1)$$

where c_d is the drag coefficient (p. 153). For particles of natural origin,
and for most pollutant particles, ρ is generally much larger than ρ_a, so

Equation 12.1 may be expressed approximately as

$$V_s^2 \simeq 8rg\rho/3\rho_a c_d \qquad (12.2)$$

For particles obeying Stokes Law (i.e., those for which the particle Reynolds number Re_p is less than about 0.1 (p. 153)), Equation 9.12 showed that $c_d = 12v/V_s r$, where v is the kinematic viscosity of air. Substituting in Equation 12.2 gives

$$V_s = 2\rho gr^2/9\rho_a v \qquad (12.3)$$

Equation 12.3 enables V_s to be calculated directly when $Re_p < 0.1$, but for particles with $Re_p > 0.1$, Equation 12.2 should be used together with an estimate of c_d from Figure 9.6 or Equation 9.13 to find V_s by trial and error. For example, to find the sedimentation velocity of a raindrop, radius 100 μm, a *first approximation* can be estimated from Equation 12.3, $V_s = 1.2$ m s^{-1}. The corresponding Reynolds number is 16 (confirming that Equation 12.3 is strictly not applicable), and the drag coefficient (from Equation 9.13) is about 3. The drag force on the drop (Equation 12.1) is

$$(1/2)\,\rho_a V_s^2 c_d \pi r^2 = 82 \times 10^{-9} \text{ N}$$

which does not balance the gravitational force

$$(4/3)\,\pi r^3 \rho g = 42 \times 10^{-9} \text{ N}$$

A smaller value of V_s must therefore be estimated and the drag force recalculated. The value of V_s when drag exactly balances the gravitational force may then be found by interpolation, graphically or otherwise. This method is of general applicability. Figure 12.1 shows a specific case—the variation of V_s with radius for particles of unit density $\rho = 1$ g cm^{-3}.

Biological aerosols such as pollen and spores can be treated as spheres of unit density, so the equations given here are adequate for finding their sedimentation velocities. Fig. 12.1 shows that Stokes Law ($Re_p \lesssim 0.1$) holds to about $r = 30$ μm ($V_s = 0.1$ m s^{-1}), a size range that includes most spores and pollen.

Particles of soil and pollutant materials are seldom spherical, and their density is seldom exactly 1 g cm^{-3} (assumed in Fig. 12.1). Such particles are often characterized by their *Stokes diameter*, which is the diameter of a sphere having the same density and sedimentation velocity. Water drops with radius exceeding about 0.4 mm are appreciably flattened as they fall, and this increases c_d. Figure 12.1 shows observed values of V_s for water drops. When $r > 2$ mm, any further increase in drop weight

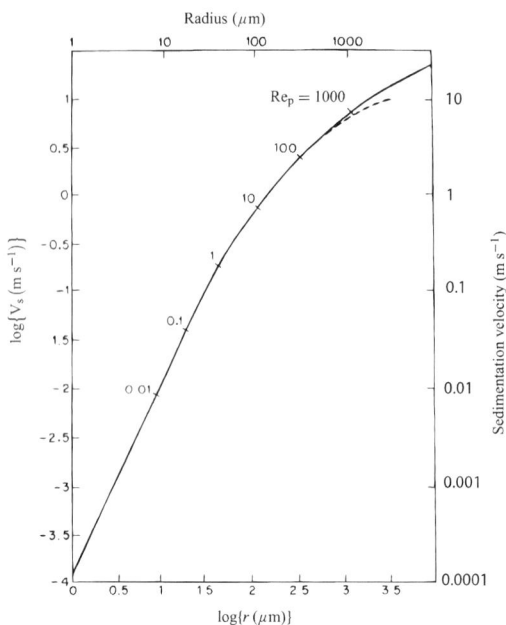

FIGURE 12.1   Dependence of sedimentation velocity on particle radius for spherical particles with density 1 g cm^{-3}. Stokes' Law applies for $Re_p < 0.1$. The pecked line shows the sedimentation velocity of raindrops (from Fuchs, 1964).

is compensated for by an increase in deformation, and hence in c_d, so V_s remains approximately constant. Raindrops much larger than 3 mm radius are seldom observed, as they tend to break up during their fall, but they may be maintained in strong updraughts in storm clouds.

12.2   Non-Steady Motion

If a particle obeying Stokes Law is projected horizontally in still air with velocity V_0 at $t = 0$, its motion is subject to the drag force $6\pi \upsilon \rho r V(t)$ where $V(t)$ is the velocity at time t, and to the gravitational force mg. Resolving the motion into horizontal and vertical components, and writing dx/dt and d^2x/dt^2 for the horizontal velocity and acceleration respectively, the equation of motion for horizontal displacement is

$$m\frac{d^2x}{dt^2} = -6\pi \upsilon \rho r \frac{dx}{dt} \tag{12.4}$$

Similarly, for vertical motion,

$$m\frac{d^2z}{dt^2} = mg - 6\pi \upsilon \rho r \frac{dz}{dt} \tag{12.5}$$

The quantity $m/6\pi \upsilon \rho r$ has the dimension [T], and is called the *relaxation time* τ. Appendix A.6 lists relaxation times of various size particles. For 1μm diameter particles, τ is about 4 μs, and for 20 μm particles it is about 1 ms, so aerosol particles conform to all but very small eddies in turbulent flow.

The equations of motion for a particle may be written

$$\frac{d^2x}{dt^2} = \tau^{-1}\frac{dx}{dt} \tag{12.6}$$

and

$$\frac{d^2z}{dt^2} = g - \tau^{-1}\frac{dz}{dt} \tag{12.7}$$

For horizontal motion, integration of Equation 12.6 gives

$$\frac{dx}{dt} = V_0 \exp\left(-\frac{t}{\tau}\right)$$

and

$$x(t) = \tau V_0 \left[1 - \exp\left(-t/\tau\right)\right]$$

When the horizontal component of velocity is zero, $x = V_0\tau$ and this is called the *stopping distance* l.

Integration of the equation for vertical motion (Equation 12.7) leads to the case considered earlier for sedimentation velocity. The vertical acceleration becomes zero when $dz/dt = g\tau$ and the particle then moves at the sedimentation velocity V_s. Thus the equations of motion lead to

$$l = V_0\tau$$
$$V_s = g\tau$$

The terms *relaxation time*, *stopping distance*, and *sedimentation velocity* have a central role in particle physics.

12.3 Particle Deposition

In the absence of gravitational or other external forces, particles are deposited on objects by two processes: diffusion (Brownian motion, p. 36);

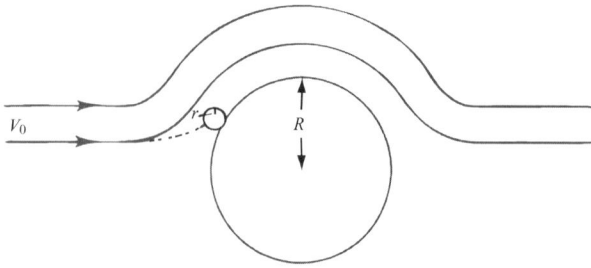

FIGURE 12.2   Impaction of a particle on a cylinder.

and *impaction*. Impaction occurs when particles with significant inertia fail to follow precisely the streamline on which they move initially as they approach an obstacle. Figure 12.2 illustrates the trajectory of a particle near a cylinder.

As the streamline changes direction rapidly, the inertia of the particle does not allow it to turn so sharply. To hit the surface, the particle must penetrate the boundary layer within which the flow velocity decreases to zero. The probability of particle impaction therefore depends on the ratio of the particle stopping distance l to the boundary layer thickness δ. Equation 9.1 (p. 144) shows that boundary layer thickness is proportional to (dimension)$^{0.5}$ and inversely proportional to (velocity)$^{0.5}$; small obstacles and large flow velocities therefore favor impaction.

A good example of the importance of stopping distance and boundary layer thickness in determining impaction can be seen if soap bubbles are released upwind of an isolated tree trunk or other large cylindrical obstruction. The bubbles, with low mass and large drag, have short stopping distances, fail to penetrate the boundary layer, and consequently move around the obstacle. In contrast, golf balls of the same size and traveling towards the obstacle at the same velocity would not slow appreciably in passing through the boundary layer, and would hit the obstacle. (They would also demonstrate dramatically another phenomenon that limits the deposition of dry particles onto surfaces—"*bounce-off*.")

The dimensionless ratio of the stopping distance of a particle to the characteristic dimension of an object (e.g., the radius of a cylinder) is called the *Stokes number*, Stk, which provides a useful way of comparing results of different impaction studies.

To describe the deposition by impaction, two further terms are commonly used. The *efficiency of impaction* c_p for particles on an object is defined as the number of impacts on the object divided by the number of

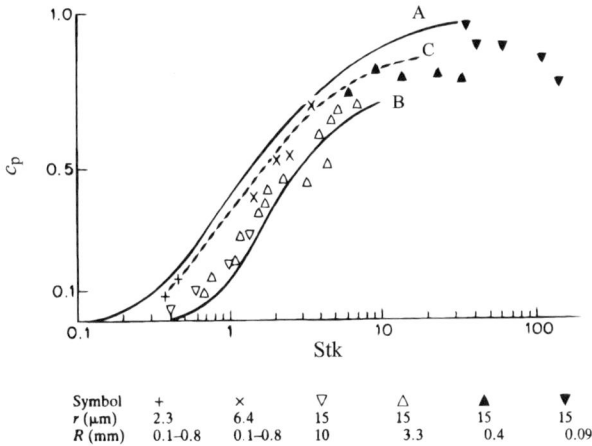

Symbol	+	×	▽	△	▲	▼
r (μm)	2.3	6.4	15	15	15	15
R (mm)	0.1–0.8	0.1–0.8	10	3.3	0.4	0.09

FIGURE 12.3 Efficiency of impaction of particles (radius r) on cylinders (radius R) (from Chamberlain, 1975): A and B are theoretical relationships for flow Reynolds numbers Re > 100 and Re $= 10$, respectively. Line C is fitted to experimental measurements with droplets. The plotted points are measurements with spores impacting on sticky cylinders.

particles that would have passed through the space in the same time if the object had not been there (i.e., the potential number of impacts).

The *deposition velocity* v_d is defined as the number of impacts per unit area per second divided by the number of particles per unit volume in the air stream. If the relevant area of the object is taken as the cross-sectional area exposed to the flow, then

$$c_p = v_d / V$$

where V is the flow velocity.

Figure 12.3 shows the calculated dependence of the efficiency of impaction c_p on Stk for cylinders at two Reynolds numbers: at large Re the streamlines curve more sharply than at small Re, and so c_p is increased. The figure also includes a number of observations with spores, using sticky cylinders, and a curve based on measurements of droplet impaction. Droplets impact with efficiencies close to the theoretical values although they may fragment on impact. The slightly lower values of c_p for spores suggests that some spores bounced off, or were dislodged from the cylinders even though the surfaces were sticky.

A good natural example of the high impaction efficiency of droplets with Stk > 10 occurs when cloud drops, radius ~ 10 μm, impact on pine needles, $R \sim 0.5$ mm, when forests are enveloped in low cloud. If the drops are moving at windspeed of 5 m s^{-1}, $l \simeq 6$ mm and so Stk $\simeq 12$. The impaction efficiency of such drops on larger objects such as rain gauges is much less, and so this source of water to the forest is called *occult precipitation* (occult = hidden) because it is not recorded in normal gauges. Impaction of cloud drops on grass would be less efficient than on needle-leaved trees at the same location because windspeed close to the ground is reduced by momentum transfer (Chapter 9). The capture of mist (Figure 12.4a, see color plates) on spider's threads ($R \sim 0.1$ μm) is even more efficient than the previous example because mist drops ($r = 10$ μm), much larger than the obstacle, are scarcely deflected, and any drops passing within a distance r of a thread are likely to be captured by *interception.* Interception is an important mechanism of deposition for particles with size comparable with or larger than the obstacles they approach.

Dollard and Unsworth [1983] built gauges to capture cloud water using a conical arrangement of fine threads that channelled water into a collector. Using these devices and the aerodynamic method for flux measurement, they showed that cloud water was transferred to grass with an efficiency similar to that for momentum.

Chamberlain [1975] and Chamberlain and Little [1981] reviewed the experimental data for particle deposition on vegetation. Stickiness, or wet-ness of surfaces, seems to be an important factor for the retention of im-pacting dry particles in the size range of spores and pollen (10–30 μm radius) on leaves and stems. For example, Chamberlain [1975] exposed barley straw to ragweed pollen (r about 10 μm) at windspeeds of 1.55 m s^{-1} and found that c_p increased from 0.04 to 0.31 when the straw was made sticky. The presence or absence of a soft layer to absorb particle mo-mentum on impact, thus avoiding "bounce-off," probably explains these results. Bounce-off is most pronounced with large particles, high veloc-ities (large momentum), small obstacles (thin boundary layer), and large coefficients of restitution at the surface (small loss of kinetic energy on impact).

Aylor and Ferrandino [1985] used ragweed pollen (radius about 15 μm, mass 11 ng) and *Lycopodium* spores (radius about 10 μm, mass 4 ng) to study bounce-off from glass rods in a wind tunnel and from wheat stems in the tunnel and in the field. They derived a relative retention factor F defined as the ratio of the catch on non-sticky cylinders to that on sticky cylinders, and related F to the kinetic energy of the particle on impact KE_i. They showed by numerical integration of momentum equations that

KE_i was related to the flow velocity u_0 in the tunnel and to Stokes number by

$$KE_i = 0.5m_pu_0^2 [f\,(Stk)]^2 \qquad (12.8)$$

where m_p is the mass of particle and the function f (Stk) is given by

$$f\,(Stk) = 0.236\,\ln(Stk - 0.06) + 0.684$$

Figure 12.5(a) shows the variation of F with KE_i for *Lycopodium* and ragweed particles impacting on glass rods with diameters 3, 5, and 10 mm. For both types of particles, F was near unity for $KE_i < 10^{-12}$ J then decreased by about 2 orders of magnitude as KE_i increased to 10^{-11} J. The corresponding particle speed at the threshold for bounce-off was about 40 cm s^{-1} for *Lycopodium* and 70 cm s^{-1} for ragweed; the threshold kinetic energy for bounce-off was independent of particle diameter.

Relative retention on wheat stems in the wind tunnel (Figure 12.5b, see color plates, filled symbols) did not decrease as steeply with KE_i as for glass rods, but the threshold for bounce-off appeared similar to that for the rods. The greater scatter is probably a consequence of variability in surface structure of the stems. Bounce-off occurs when the kinetic energy of a particle after impact exceeds the potential energy of attraction at the surface. Surface structure is likely to influence both the energy absorption on impact and the potential energy of attraction, and so variation in bounce-off is likely with biologically variable surfaces.

In the field (Figure 12.5(b), open symbols), the average value for F for a given KE_i was smaller than in the tunnel. At low values of KE_i this was probably because turbulent variation of windspeed about the mean caused KE_i to range from sub-critical (constant F, no bounce-off) to well above super-critical (F decreasing rapidly with KE_i), and time spent at super-critical KE_i lowered the average value of F. At higher values of KE_i this explanation is unlikely to apply, and the lower retention observed may then have been the result of strong gusts of wind removing particles that had previously deposited.

In contrast to the factors influencing capture of particles larger than about 10 μm, the capture of small particles ($r < 10\,\mu$m) by leaves is relatively uninfluenced by surface wetness or stickiness, but is enhanced by the presence of hairs or surface irregularities that probably act as efficient micro-impaction sites. Table 12.1 shows the relative deposition of particles of a range of sizes to various surfaces exposed in a stand of artificial grass in a wind tunnel operating at about 4.5 m s^{-1}. The sticky surface was most efficient at capturing the spores and pollen but was not effective for the smaller particles.

TABLE 12.1   Deposition of particles on segments of real leaves, and on filter paper, as a proportion of deposition on sticky artificial leaves made of PVC (from Chamberlain, 1975).

Particle	Diameter (μm)	Relative deposition Grass	Plantain	Clover	Filter paper	Sticky pvc
Lycopodium spore	32	0.45	0.26	0.18	0.70	1.00
Ragweed pollen	19	0.15	0.11	0.23	0.68	1.00
Polystyrene	5	1.74	1.82	3.25	1.98	1.00
Tricresylphosphate	1	1.70	2.60	5.50	6.40	1.00
Aitken nuclei	0.08	1.06	1.70	0.86	1.54	1.00

The relative importance of deposition by the mechanisms of impaction, interception, sedimentation, and diffusion can be assessed from Table 12.2. Displacement by Brownian motion dominates the movement of submicron size particles, whereas impaction and sedimentation increase rapidly with particle size above about 1 μm.

As an illustration of the combined influence of all mechanisms on deposition, Figure 12.6 shows how the deposition velocity of particles to an artificial sward of short grass in a wind tunnel depended on particle size. The steep decrease in v_d as r decreased below 5 μm was caused by decreasing impaction efficiency. When r was less than 0.1 μm, Brownian motion began to be effective and v_d increased. The figure also shows sedimentation velocity, demonstrating that sedimentation accounted for a large proportion of the deposition only when particles exceeded 10 μm radius.

Knowledge of the physical factors influencing deposition of particles on vegetation can be combined with models of atmospheric diffusion to estimate fractions of pollen depositing at various distances downwind of a "source" crop [Chamberlain 1975; Di-Giovanni and Beckett 1990]. This

TABLE 12.2   Characteristic distances for particle transport.

Particle Radius r (μm)	0.01	0.1	1	10
Stopping distance (μm) given initial velocity of 1 m s^{-1}	14×10^{-3}	0.23	13	1230
Distance (μm) travelled in 1 second by virtue of:				
Terminal velocity	14×10^{-2}	2.2	128	1200
Brownian diffusion	160	21	5	1.5

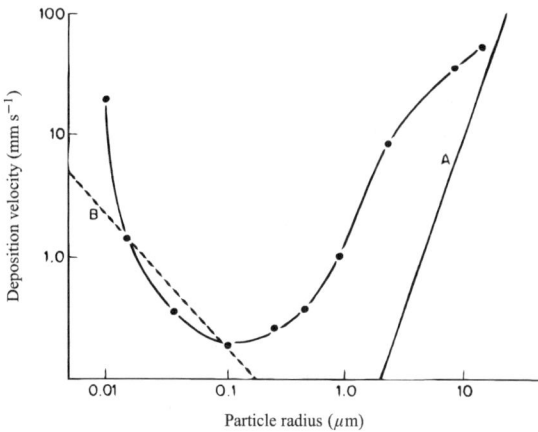

FIGURE 12.6 The variation with particle radius of the deposition velocity to short grass exposed at a windspeed of about 2.5 m s^{-1} (from Little and Whiffen, 1977). Line A is the sedimentation velocity of particles with density 1 g cm^{-3}; line B was calculated from Brownian diffusion theory.

topic has particular relevance to estimating the probability of gene flow from fields of genetically modified (GM) crops to other crops or the natural environment (Walklate et al 2003). In field measurements of pollen dispersion and deposition from small plots, the airborne concentration of pollen decreases rapidly with distance from the source because turbulent dispersion in three dimensions expands the plume and deposition occurs. For maize pollen, which is spherical with diameter about 90 μm (sedimentation velocity 0.2–0.3 m s^{-1}), airborne concentrations decreased by a factor of 2–3 between 3 and 10 m downwind of a source plot [Raynor et al 1972; Jarosz et al 2003]. Jarosz et al [2003] found that the deposition rate at 20 m downwind of the source was about 20% of the rate at 1 m, and Raynor et al [1972] reported an even sharper decline; differences are probably related to differences in wind speed and turbulence. Jarosz et al estimated that about 99% of the pollen emitted was deposited within 30 m of the plot. Values of the deposition velocity of maize pollen in the field measured close to the plot were 2–3 times the sedimentation velocity of the grains. This may be because turbulence downwind of a roughness change enhanced particle deposition [Reynolds 2000].

Considered together, the deposition processes of sedimentation, diffusion and impaction are least efficient for particles of about 0.1–0.2 μm radius, and it is significant that man-made aerosols in this size range are found widely distributed in the earth's atmosphere. In particular, soluble sulphate particles in this size range, formed by the oxidation of sulphur dioxide, can be carried extremely long distances in the atmosphere, and tend to persist until they encounter conditions of high humidity in which they can grow by condensation into larger droplets that are more effectively deposited or captured by falling rain. The relationship between the diameter D_0 of a dry deliquescent particle and its diameter D_S at a water vapor *saturation ratio S* (defined as % relative humidity/100) is

$$D_S/D_0 = (1 - S)^{-\gamma} \qquad (12.9)$$

where γ is a hygroscopic growth parameter that depends on particle chemical composition. Values of γ are around 0.2 for aged European aerosol, but are larger for marine aerosol. Equation 12.9 demonstrates that European aerosol particles would approximately double their dry size when the relative humidity reached 97%.

Sea salt (sodium chloride, NaCl) particles resulting from wave action and bubbles in oceans are another important example of soluble aerosols, used here to illustrate principles that apply to any soluble particle. Salt dissolved in water lowers the equilibrium vapor pressure over a water surface, which allows sea salt particles to grow by condensation at a smaller saturation ratio than pure water (p. 23). The affinity for water also allows stable droplets to exist in saturated or unsaturated environments. For

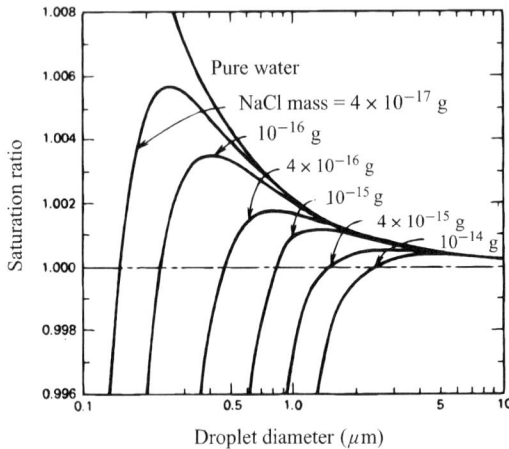

FIGURE 12.7 Variation of droplet diameter with saturation ratio for pure water and for droplets containing the indicated mass of sodium chloride at 20°C (from Hinds 1999).

a droplet formed from a soluble nucleus, two competing physical effects determine the saturation ratio at which equilibrium is achieved. First, for a droplet growing from a dry salt particle of a specific mass, the salt concentration decreases as the particle grows by condensation. Thus, as the droplet size increases, the equilibrium vapor pressure over its surface increases towards that of pure water. Second, as the droplet size increases, the influence of curvature, responsible for an *increase* of equilibrium vapor pressure above that of a plane surface (p. 23), decreases.

Figure 12.7 illustrates the result of these competing processes for salt particles of various initial mass, indicating the unique relation existing between humidity, particle mass, and the equilibrium droplet size. If a droplet is displaced to an environment corresponding to the area above its specific curve, it must grow by condensation until it reaches the curve; if it is displaced below the curve, it must evaporate to reach equilibrium. Figure 12.8 illustrates how a relatively large sea-salt particle of dry mass 10^{-14} g responds to increases and decreases in humidity over a wider range. As humidity increases, the particle (0.21 μm dry equivalent diameter) undergoes a rapid transition to a droplet (0.38 μm diameter) at 76% relative humidity, then continues to grow to about 1.0 μm at 100% relative humidity. If humidity then decreases, the droplet does not recrystallize

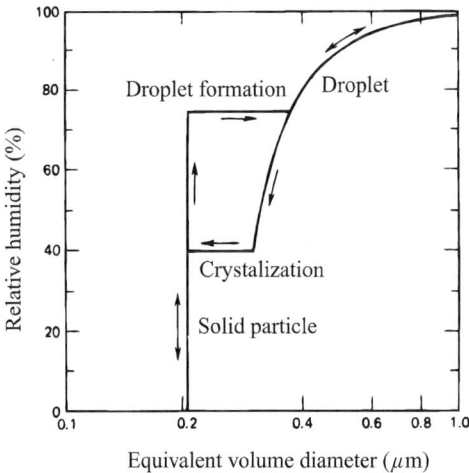

FIGURE 12.8 The dependence of particle/droplet diameter on relative humidity for a sodium chloride particle of dry mass 10^{-14} g, showing the transition from a particle to a droplet as humidity increases above 76% and the recrystallization as decreasing humidity reaches 40% (from Hinds 1999).

until the relative humidity is about 40%. The hysteresis effect depends on the chemical composition and can be used to identify aerosol types. Since aerosol interactions with radiation are size dependent (Chapter 5), aerosol scattering and absorption properties alter as humidity increases, accounting for the haziness of humid air masses. Hinds [1999] includes a good discussion of these topics.

Growth of soluble aerosol particles is very rapid as humidity changes, so that most particle sizes are essentially in equilibrium with the local relative humidity. Since humidity varies considerably with height in the boundary layer close to wet or transpiring surfaces, changes in deliquescent particle size need to be taken into account when deducing aerosol fluxes from vertical transport of particles. Vong [2004] used the eddy covariance method (p. 305) to measure turbulent fluxes of aerosols to grassland. He measured aerosol concentration and diameter at ambient humidity with a fast optical counting instrument. Since humidity increased toward the grass surface, deliquescent particles transported upward in eddies were larger than those of the same dry diameter moving downward, so the gradient in humidity caused the measured aerosol flux to be apparently upward (i.e., from the grass to the atmosphere). After corrections were applied for particle growth, Vong deduced that the "true" particle flux was downward, with a deposition velocity of about 0.3 cm s^{-1} for 0.52μm diameter particles in neutral stability.

PARTICLE DEPOSITION IN THE LUNGS

The figures in Table 12.2 are also relevant to particle inhalation and deposition in the respiratory system. Hazards of particle inhalation depend on the chemical properties of the particles and the site of deposition. Understanding how and where particles deposit is also important for the effective design of inhalers used to dispense aerosol medication for asthma and for assessing risks of exposure to airborne pathogens. The design of the respiratory system in animals provides some defense against hazards of aerosol inhalation, and is discussed more fully by Hinds [1999].

Airway Characteristics

The human respiratory system can be considered as three regions (Figure 12.9).

The *head airways* include the nose, mouth, pharynx, and larynx. In this region, inhaled soluble particles grow by condensation as the air is warmed and humidified. The *tracheobronchial airways* consist of the path from the trachea to the terminal bronchioles, resembling a tree with increasingly

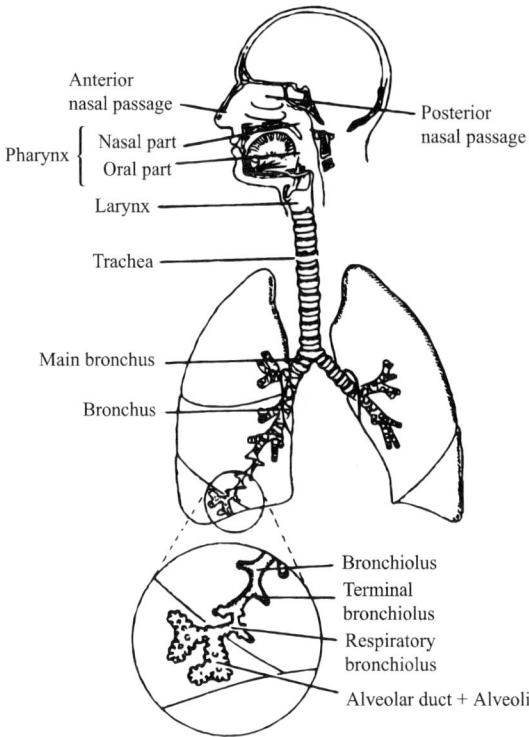

FIGURE 12.9 The human respiratory system (from Hinds, 1999).

finer branches. The *pulmonary* or *alveolar* region is the part of the lung system beyond the terminal bronchioles where gas exchange of oxygen and carbon dioxide takes place in the alveoli, small sacs of the order of 0.1–1 mm diameter. The area of the gas exchange system in adult humans is about 75 m^2, about half the size of a tennis court, allowing very efficient exchange. During inhalation at a typical rate of 1 l s^{-1}, the velocity of air increases from about 4 to 5 m s^{-1} as it passes through the head airways region to the bronchi, then it decreases rapidly as the lung system branches into many small airways that increase the cross-sectional area of the flow path by a factor of about 250. Consequently, the residence time for air increases from about 4 ms in the bronchi to about 600 ms in the alveoli. About 0.5 l of tidal air is taken in with each breath. About 2.4 l of reserve air in the lungs is not exchanged with each breath, but there is some mixing of inhaled air with this reserve air, increasing with exercise.

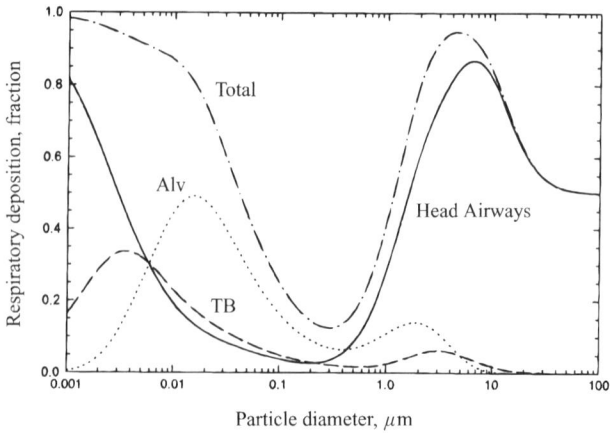

FIGURE 12.10   Predicted total and regional deposition of aerosol for light exercise (nose breathing) (from Hinds, 1999).

Deposition Mechanisms

Particles deposit in the various regions of the respiratory system principally by impaction, sedimentation, and (Brownian) diffusion. During inhalation, air is forced to change direction several times as it flows from the mouth or nose through the branching airways. Inertial impaction is most effective when the air speed and particle size are large, giving large stopping distances (Table 12.2). Consequently, impaction is particularly effective for depositing large particles (3–10 μm diameter) that are passing close to the airway walls in the head and tracheobronchial regions. Sedimentation is more important in the smaller tracheobronchial airways and alveolar regions where flow velocities are slow and duct dimensions are small. In these conditions, larger residence times allow particles 1–3 μm diameter to settle to surfaces, but smaller particles settle too slowly to deposit effectively by this mechanism (Table 12.2). Brownian diffusion of particles between 0.01 and 0.1 μm diameter is particularly effective for deposition in the alveoli where the ratio (root mean square displacement/ airway diameter) is large.

Figure 12.10 illustrates the predicted deposition sites for particles of different sizes based on data for males and females undertaking light exercise and inhaling a typical mixed aerosol through the nose. The outcome of the various deposition mechanisms is that 80–95% of 5–10 μm particles are trapped by impaction in the head airways. Particles 2–5 μm are deposited mainly in the tracheobronchial airways by sedimentation and impaction. As a result of this size-selective deposition, particles larger than 10 μm do not reach the alveolar region, and numbers of 2–10 μm particles are greatly depleted. About 10–20% of particles 0.1–1 μm deposit in the alveolar region by Brownian diffusion and by mixing of the inhaled (tidal) air with the reserve air, followed by sedimentation.

One further mechanism, interception, is important for deposition of fibrous particles such as silica and asbestos that are large in one dimension but have small aerodynamic diameters, so behave like fine aerosols (i.e., they have small stopping distances and sedimentation velocities). These materials can effectively negotiate the pathways to the small airways, where they have a high probability of being intercepted on the walls.

Once they are deposited, particles remain in the lungs for varying times depending on their composition, location, and the type of clearance mechanism. The surfaces of the head airways and tracheobronchial airways are covered with a layer of mucus that is propelled upward by cilia to the pharynx where it is swallowed. This transports particles captured in these regions out of the lungs in a few hours. But alveolar regions do not have mucus or cilia, which would interfere with gas exchange, so particles remain for months to years. This emphasizes the hazards of inhalation of fine particles, radionuclides, or pathogens.

12.4 Problems

1. Find the sedimentation velocities of (i) a pollen grain, diameter 10 μm, density 0.8 g cm^{-3}, and (ii) a hailstone, diameter 6 mm, density 0.5 g cm^{-3}. (Hint: You will have to estimate the Reynolds number and decide whether to use the trial and error method described in the text.)

2. Find the relaxation times and stopping distances of spores of 10 μm radius and aerosol particles of 0.5 μm radius in a gas moving at 2 m s^{-1}. Hence confirm that the spores would be deposited rapidly to the walls of a bronchus 4 mm diameter while the aerosol would penetrate effectively along the tube.

3. Particles with radius 0.1 μm and 10 μm and density 1 g cm^{-3} are approaching a cylindrical branch, diameter 1 cm, at a velocity of 0.5 m s^{-1}. Determine the stopping distance and Stokes number for each particle size and the boundary layer thickness around the branch. What do you conclude about the likelihood of the particles being deposited?

4. Calculate the fraction of an aerosol of 10 μm diameter particles that would be washed out of the atmosphere in one hour by rainfall of 1 mm h^{-1},assuming that the raindrops are all the same radius, namely (i) 100 μm, and (ii) 1000 μm. (Hints: Use Figure 12.1 or the method in Problem 1 to calculate the sedimentation velocity v of the raindrops. Assume that the stopping distance of a 10 μm diameter particle impacting on a droplet moving at velocity v is $S_0 = 3 \times 10^4 v$ where S_0 is measured in m and v is in m s^{-1}. The efficiency of impaction C_p of the aerosol on the droplet may be assumed similar to that for impaction on cylinders of the same diameter. Then the rainfall rate enables you to calculate how often a raindrop passes through a fixed point in the atmosphere, and if this happens n times per second, the *washout coefficient* of the particles by the rain drops is nC_p.)

Chapter 13

Steady State Heat Balance (i) Water Surfaces, Soil, and Vegetation

The heat budgets of plants and animals will now be examined in the light of the principles and processes considered in previous chapters. The First Law of Thermodynamics states that when a balance sheet is drawn up for the flow of heat in any physical or biological system, income and expenditure must be exactly equal. In micrometeorology, radiation and metabolism are the main sources of income; radiation, convection, and evaporation are methods of expenditure.

For any component of a system, physical or biological, a balance between the income and expenditure of heat is achieved by adjustments of temperature. If, for example, the income of radiant heat received by a leaf began to decrease because the sun was obscured by cloud, leaf temperature would fall, reducing expenditure on convection and evaporation. If the leaf had no mass and therefore no heat capacity, the drop in expenditure would exactly balance the drop in income, second by second. For a real leaf with a finite heat capacity, the drop in temperature would lag behind the drop in radiation and so would the drop in expenditure, but the First Law of Thermodynamics would still be satisfied because the income from radiant heat would be supplemented by the heat given up by the leaf as it cooled. This chapter is concerned with the heat balance of relatively simple systems in which (a) temperature is constant so that changes in heat storage are zero; and (b) metabolic heat is a negligible term in the heat budget. The heat budget of warm-blooded animals, controlled by metabolism, is considered in the next chapter, and examples of diurnal changes of heat storage are considered in Chapter 15.

13.1 Heat Balance Equation

The heat balance of any organism can be expressed by an equation with the form

$$\bar{R}_n + \bar{M} = \bar{C} + \lambda\bar{E} + \bar{G} \tag{13.1}$$

The individual terms are

$\bar{R}_n$ = net gain of heat from radiation

$\bar{M}$ = net gain of heat from metabolism

$\bar{C}$ = loss of sensible heat by convection

$\lambda\bar{E}$ = loss of latent heat by evaporation

$\bar{G}$ = loss of heat by conduction to environment

The over-bars in Equation 13.1 indicate that each term is an average heat flux per unit surface area. (In the rest of this chapter, they are implied but not printed.) In this context, it is convenient to define surface area as the area from which heat is lost by convection, although this is not necessarily identical to the area from which heat is gained or lost by radiation. The conduction term $\bar{G}$ is included for completeness but is negligible for plants and has rarely been measured for animals. An equation similar to 13.1 applies to bare soil surfaces or water bodies but without the term $\bar{M}$; in these cases, storage of heat by conduction is often significant.

The grouping of terms in the heat balance equation is dictated by the arbitrary sign convention that fluxes directed away from a surface are positive. (When temperature decreases with distance z from a surface so that $\partial T/\partial z < 0$, the outward flux of heat $C \propto -\partial T/\partial z$ is a positive quantity, see p. 31.) The sensible and latent heat fluxes C and λE are therefore taken as *positive* when they represent losses of heat from the surface to the atmosphere, and as *negative* when they represent gains. On the left-hand side of the equation, R and M are positive when they represent gains and negative when they represent losses of heat. When both sides of a heat balance equation are positive, the equation is a statement of how the total supply of heat available from sources is divided between individual sinks. When both sides are negative, the equation shows how the total demand for heat from sinks is divided between available sources.

The sections that follow deal with the size and manipulation of individual terms in the heat balance Equation 13.1, with some fundamental physical implications of the equation, and with a few biological applications.

CONVECTION AND LONG-WAVE RADIATION

When the surface of an organism loses heat by convection, the rate of loss per unit area is determined by the scale of the system and by its geometry, by windspeed, and by temperature gradients (Chapter 10). Convection is usually accompanied by an exchange of long-wave radiation between the organism and its environment at a rate that depends on geometry and on differences of radiative temperature but is *independent* of scale. The significance of scale can be demonstrated by comparing convective and radiative losses from an object such as a cylinder with diameter d and uniform surface temperature T_0 exposed in a wind tunnel whose internal walls are kept at the temperature T of the air flowing through the tunnel with velocity u. When Re exceeds 10^3, the resistance to heat transfer by convection increases with d according to the relation $r_H = d/(\kappa Nu) \propto d^{0.4} u^{-0.6}$ (see Appendix A.5). In contrast, the corresponding resistance to heat transfer by radiation r_R (p. 43) is independent of d. Figure 13.1 compares r_H and r_R for cylinders of different diameters at windspeeds of 1 and 10 m s^{-1} chosen to represent outdoor conditions. Corresponding rates of heat loss are shown on the right-hand axis for an arbitrary surface temperature excess $(T_0 - T)$ of 1 K.

Because the dependence of r_H on scale and windspeed is similar for planes, cylinders, and spheres provided the appropriate dimension is used to calculate Nusselt numbers, a number of generalizations may be based on Figure 13.1. For organisms on the scale of a small insect or leaf ($0.1 < d < 1$ cm), r_H is much smaller than r_R, implying that convection is a much more effective mechanism of heat transfer than long-wave radiation. The organism is tightly coupled to air temperature but not to the radiative temperature of the environment. For organisms on the scale of a farm animal or a man ($10 < d < 100$ cm) r_H and r_R are of comparable importance at low windspeeds. For very large mammals ($d > 100$ cm), r_H can exceed r_R at low windspeeds, and in this state the surface temperature will be coupled more closely to the radiative temperature of the environment than to the air temperature. These predictions are consistent with measurements on locusts and on piglets, for example, and they emphasize the importance of wall temperature as distinct from air temperature in determining the thermal balance of large farm animals in buildings with little

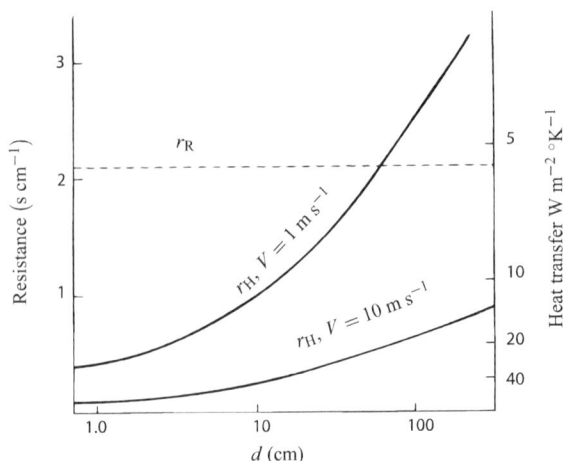

FIGURE 13.1   Dependence of resistance to convective heat transfer r_H and resistance to radiative heat exchange r_R as a function of body size represented by the characteristic dimension of a cylinder d. The cylinder is assumed exposed in a wind tunnel where air and wall radiative temperature are identical. The right-hand axis shows the rate of heat transfer assuming a difference of 1 K between cylinder surface temperature and air/wall radiative temperature.

ventilation. They also demonstrate that warming the air in a cold room would not induce a feeling of comfort for people occupying the room if the radiative temperature of the walls remained low.

When an organism with an emissivity of unity and a surface temperature of T_0 exchanges heat (a) by convection to air at temperature T and (b) by radiation to an environment with a mean radiative temperature equal to air temperature, the net rate at which heat is gained or lost is

$$\rho c_p \left\{ (T - T_0)/r_H + ((T - T_0)/r_R) \right\} = \rho c_p \left\{ (T - T_0)/r_{HR} \right\} \quad (13.2)$$

where $r_{HR} = \left(r_H^{-1} + r_R^{-1} \right)^{-1}$ is a combined resistance for convection (p. 33) and long-wave radiation (p. 43) formed by grouping the component resistances in parallel because the fluxes are in parallel.

13.2   Heat Balance of Thermometers

DRY-BULB

As an introduction to the more relevant physics of the wet-bulb thermometer, it is worth considering the implications of the general heat balance equation for a dry-bulb thermometer (i.e. a thermometer with a dry sensing element). For measurements in the open, it is essential to avoid heating a thermometer by direct exposure to sunlight, so screening is employed. In several common designs used for precise measurements of air temperature, a thermometer with a cylindrical bulb is housed in a tube through which air is drawn rapidly, and for the sake of the following discussion we assume the tube completely surrounds the bulb. If the tube itself is exposed to sunshine, its temperature, T_s, may be somewhat above the temperature of the air (T) and of the thermometer (T_t).

The net long-wave radiation received by the thermometer from the housing (assuming emissivity $\varepsilon = 1$) is

$$\mathbf{R}_n = \sigma \left(T_s^4 - T_t^4 \right) = \rho c_p \left(T_s - T_t \right) / r_R \qquad (13.3)$$

using the definition of r_R on p. 43. The loss of heat by convection from the bulb to the air is

$$\mathbf{C} = \rho c_p \left(T_t - T \right) / r_H \qquad (13.4)$$

In equilibrium, the heat balance Equation 13.1 reduces to $\mathbf{R}_n = \mathbf{C}$ and rearrangement of terms in Equation 13.4 gives

$$T_t = \frac{r_H T_s + r_R T}{r_R + r_H} \qquad (13.5)$$

implying that the temperature recorded by the thermometer (the apparent temperature) is a weighted mean between the true temperature of the air and the temperature of the thermometer housing. There are two main ways in which the difference between apparent and true air temperature can be minimized:

(i) by making r_H much smaller than r_R, either by adequate ventilation or by choosing a thermometer with very small diameter (see Figure 13.1). The numerator in Equation 13.5 then tends to $r_R T$ and the denominator to r_R.

(ii) by making T_s very close to T; e.g., by using a reflective metal or white painted screen, by introducing insulation between outer and inner surfaces or by increasing ventilation on both sides of the screen.

In the standard Assmann psychrometer, the screen is a double-walled cylinder, nickel-plated on the outer surface and aspirated at about 3 m s^{-1}. Since the diameter of the mercury-in-glass thermometer it contains is about 3 mm, Figure 13.1 illustrates that it is effectively decoupled from its radiative environment.

WET-BULB

The concept of a "wet-bulb temperature" is central to the environmental physics of systems in which latent heat is a major heat-balance component, and it has two distinct connotations: the *thermodynamic wet-bulb temperature*, which is a theoretical abstraction; and the temperature of a thermometer with its sensing element covered with a wet sleeve, which, at best, is a close approximation to the thermodynamic wet-bulb temperature.

A value for the thermodynamic wet-bulb temperature can be derived by considering the behavior of a sample of air enclosed with a quantity of pure water in a container with perfectly insulating walls. This is an *adiabatic system* within which the sum of sensible and latent heat remains constant. The initial state of the air can be specified by its temperature T, vapor pressure e, and total pressure p. Provided e is smaller than $e_s(T)$ (the saturated vapor pressure at T), water will evaporate and both e and p will increase. Because the system is adiabatic, the increase of latent heat represented by the increase in water vapor concentration must be balanced by a decrease in the amount of sensible heat, which is realized by cooling the air. The process of humidifying and cooling continues until the air becomes saturated at a temperature T', which, by definition, is the thermodynamic wet-bulb temperature. The corresponding saturated vapor pressure is $e_s(T')$.

To relate T' and $e_s(T')$ to the initial state of the air, the initial water vapor concentration is approximately $\rho \varepsilon e / p$ when p is much larger than e (see p. 15). When the vapor pressure rises from e to $e_s(T')$, the total change in latent heat content per unit volume is $\lambda \rho \varepsilon \left[e_s(T') - e \right] / p$. The corresponding amount of heat supplied by cooling unit volume of air from T to T' is $\rho c_p (T - T')$. (A small change in the heat content of water vapor is included in more rigorous treatments but is usually unimportant in micrometeorological problems.) Equating latent and sensible heat

$$\lambda \rho \varepsilon \left[e_s(T') - e \right] / p = \rho c_p \left(T - T' \right) \qquad (13.6)$$

and rearranging terms gives

$$e = e_s(T') - (c_p p / \lambda \varepsilon)(T - T') \qquad (13.7)$$

The group of terms $(c_p p / \lambda \varepsilon)$ is often called the "*psychrometer constant*" for reasons explained shortly, but it is neither constant (because p is atmospheric pressure and λ changes somewhat with temperature) nor exact (because of the approximations made). The psychrometer constant is often assigned the symbol γ and, at a standard pressure of 101.3 kPa, has a value of about 66 Pa K^{-1} at 0°C increasing to 67 Pa K^{-1} at 20°C. Thus

$$e = e_s(T') - \gamma (T - T') \qquad (13.7a)$$

Another useful quantity that has the same dimensions as γ is the change of saturation vapor pressure with temperature or $\partial e_s(T)/\partial T$, usually given the symbol Δ (or s). This quantity can be used to obtain a simple (but approximate) relation between the saturation vapor pressure deficit $D = e_s(T) - e$ and the wet-bulb depression $B = T - T'$. When the saturation vapor pressure at wet-bulb temperature T' is written as

$$e_s(T') \approx e_s(T) - \Delta \left(T - T'\right) \qquad (13.8)$$

where Δ is evaluated at a mean temperature of $(T + T')/2$, the psychrometer Equation 13.7 becomes

$$e \approx e_s(T) - (\Delta + \gamma)\left(T - T'\right) \qquad (13.9)$$

or

$$D \approx (\Delta + \gamma)\, B \qquad (13.10)$$

Equation 13.7 can be represented graphically by plotting $e_s(T)$ against T (Figure 13.2).The curve QYP represents the relation between saturation vapor pressure and temperature and the point X represents the state of any sample of air in terms of e and T. Suppose the wet bulb temperature of the air is T' and that the point Y represents the state of air saturated at this temperature. The equation of the straight line XY joining the points (T, e), $(T', e_s(T'))$ is

$$e - e_s\left(T'\right) = \text{slope} \times \left(T - T'\right) \qquad (13.11)$$

Comparison of Equations 13.7 and 13.11 shows that the slope of XY is $-\gamma$. The thermodynamic wet bulb temperature of any sample of air can therefore be obtained graphically by drawing a line with slope $-\gamma$ through the appropriate coordinates T and e to intercept the saturation curve at a point whose abscissa is T'.

If a sample of air in the state given by X was cooled towards the state represented by the point Y, the path XY shows how temperature and vapor pressure would change in adiabatic evaporation; i.e., with the total heat content of the system constant. Similarly, starting from Y and moving to X, the path YX shows how T and e would change if water vapor were condensed adiabatically from air that was initially saturated. As condensation proceeded, the temperature of the air would rise until all the vapor

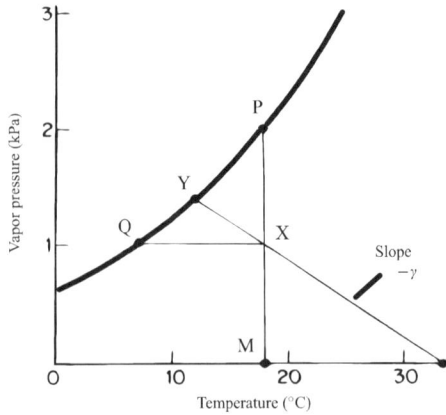

FIGURE 13.2   The relation between dry-bulb temperature, wet-bulb temperature, equivalent temperature, vapor pressure and dew point. The point X represents air at 18°C and 1 kPa vapor pressure. The line YXZ with a slope of $-\gamma$ gives the wet bulb temperature from Y (12°C) and the equivalent temperature from Z (33.3°C). The line QX gives the dew point temperature from Q (7.1°C). The line XP gives the saturation vapor pressure from P (2.1 kPa).

had condensed. This state is represented by the point Z at which $e = 0$. The corresponding temperature T_e, is called the "*equivalent temperature*" of the air. As Z has coordinates $(T_e, 0)$, the equation of the line ZX can be written in the form

$$T_e = T + e/\gamma \qquad (13.12)$$

Alternatively, the equation of YZ can be written

$$T_e = T' + e_s\left(T'\right)/\gamma \qquad (13.13)$$

showing that the equivalent and wet-bulb temperature are uniquely related. Both T' and T_e remain unchanged when water is evaporated or condensed adiabatically within a sample of air.

Moving from theoretical principles to a real wet-bulb thermometer, it is necessary to account for the finite rate at which heat is lost by evaporation and gained by convection and radiation.

Suppose that a thermometer bulb covered with a wet sleeve has a temperature T_w when it is exposed to air at temperature T and surrounded by a screen at air temperature. The rate at which heat is gained by convection and radiation is

$$\mathbf{C} + \mathbf{R}_n = \rho c_p\left(T - T_w\right)/r_{HR} \qquad (13.14)$$

The rate at which latent heat is lost may be found using Equation 3.12a to give

$$\lambda E = \lambda \left\{ \chi_s(T_w) - \chi \right\} / r_V \quad (13.15)$$

Using the relations in Equations 2.32 and on p. 234, Equation 13.15 may be written

$$\begin{aligned} \lambda E &= (\lambda \rho \varepsilon / p) \left\{ e_s(T_w) - e \right\} / r_V \\ &= \rho c_p \left\{ e_s(T_w) - e \right\} / \gamma r_V \end{aligned} \quad (13.16)$$

In equilibrium, $\lambda E = R_n + C$ from which

$$e = e_s(T_w) - \gamma \left(r_V / r_{HR} \right) (T - T_w) \quad (13.17)$$

It is often convenient to regard $(\gamma r_V / r_{HR})$ (or simply $(\gamma r_V / r_H)$ in some circumstances explained later) as a *modified psychrometer constant*, written γ^*. That is

$$e = e_s(T_w) - \gamma^* (T - T_w) \quad (13.18)$$

Comparing Equations 13.7a and 13.17, it is clear that the measured wet-bulb temperature will not be identical to the thermodynamic wet-bulb temperature unless $r_V = r_{HR}$. Because $r_V = (\kappa / D)^{0.67} r_H$, which may be written $r_V = 0.93 r_H$ (p. 192), this condition implies that

$$0.93 r_H = (r_H^{-1} + r_R^{-1})^{-1} \quad (13.19)$$

from which $r_H = 0.075 r_R$. At 20°C, $r_R = 2.1$ s cm^{-1}, so the thermodynamic and measured wet-bulb temperatures would be identical when $r_H = 0.17$ s cm^{-1}. A wet-bulb thermometer would therefore record a temperature above or below the thermodynamic wet-bulb temperature depending on whether r_H was greater or less than this value.

Because both r_V and r_H are functions of windspeed and r_R is not, γ^* decreases with increasing windspeed and, when r_V is much less than r_R, tends to a constant value independent of windspeed, viz.

$$\gamma^* = \gamma \left(r_V / r_H \right) = 0.93 \gamma \quad (13.20)$$

In the Assmann psychrometer, regarded as a standard for measuring vapor pressure in the field, the resistances corresponding to the instrument specification already given are $r_V = 0.149$ s cm^{-1}, $r_H = 0.156$ s cm^{-1} giving $\gamma^* = 63$ Pa K^{-1}. A much more detailed discussion of psychrometry leading to a standard value of $\gamma^* = 62$ Pa K^{-1} for this instrument was given by Wylie [1979]. With this and similar instruments, the error involved in using γ instead of γ^* is often negligible in micrometeorological work.

A further source of psychrometer error not considered here is the conduction of heat along the stem of the thermometer, which can be minimized by using a long sleeve and/or a thermometer with very small diameter.

Heat Balance of Surfaces

WET SURFACE

The physics of the wet-bulb thermometer is the key to solving a wide range of problems concerned with the exchange of sensible and latent heat between wet surfaces and their environment. In this context, "wet" can mean covered with pure water or with a salt solution.

Consider first a surface of pure water over which air is moving. In the free atmosphere, temperature and vapor pressure are T and e and corresponding potentials at the water surface are T_0 and $e_s(T_0)$. Resistances for heat and vapor transfer between the surface and the point where T and e are measured are r_H and r_V, respectively.

Using the standard convention that fluxes away from surfaces are inherently positive, the surface will *gain* heat by convection at a rate given by

$$\mathbf{C} = \rho c_p (T - T_0) / r_H \tag{13.21}$$

and will *lose* latent heat at a rate given by Equation 13.16 replacing T_w by T_0 and γr_V by $\gamma^* r_H$ to give

$$\lambda \mathbf{E} = \rho c_p \{e_s(T_0) - e\} / \gamma^* r_H \tag{13.22}$$

Adiabatic Systems

To start with the simplest heat balance, the system will be treated as adiabatic so that

$$\lambda \mathbf{E} + \mathbf{C} = 0 \tag{13.23}$$

If $e_s(T_0)$ were simply a linear function of T, it would now be possible to eliminate T_0 from Equations 13.21 to 13.23 and to evaluate $\lambda \mathbf{E}$ and $\mathbf{C}$ as functions of the temperature and vapor pressure of the airstream and of the two resistances. To achieve this objective, it is legitimate to *assume* a linear relation between e_s and T over a narrow range of, say, 10 K. The saturation vapor pressure at T_0 may then be related to the corresponding vapor pressure at air temperature by using Equation 13.8 in the form

$$e_s(T_0) \approx e_s(T) - \Delta(T - T_0) \tag{13.24}$$

where Δ must be evaluated at T because T_0 is unknown at this stage of the analysis.

Substituting this (approximate) value of $e_s (T_0)$ in Equation 13.22 and eliminating T_0 using Equations 13.21 and 13.23 gives

$$\lambda E = -C = \frac{\rho c_p \{e_s (T) - e\} r_H^{-1}}{\Delta + \gamma^*} \tag{13.25}$$

Provided the process is adiabatic, C is the rate at which heat is released when air is cooled to the wet-bulb temperature T_0 (Figure 13.1). More precisely, it is the heat released by unit volume of air per unit area of surface and per unit time. (Unit volume per unit area and time has dimensions of $m \, s^{-1}$, which is the reciprocal of the resistance unit used to specify the rate of exchange in these equations.)

Non-Adiabatic Systems

Suppose now that the process of heat exchange is *not* adiabatic. If the surface receives additional heat by radiation at a rate R_n, the heat balance equation becomes

$$\lambda E + C = R_n \tag{13.26}$$

The solution of Equations 13.21, 13.22, and 13.26 is, using Equation 13.24,

$$\lambda E = \frac{\rho c_p \{e_s (T) - e\} r_H^{-1}}{\Delta + \gamma^*} + \frac{\Delta R_n}{\Delta + \gamma^*} \tag{13.27}$$

Comparing Equations 13.22 and 13.27 reveals that 13.27 still contains the adiabatic term unchanged, but cooling a parcel of air to wet-bulb temperature is now only part of the story. The surface receives additional energy from radiation and may therefore be warmer than the wet-bulb temperature. When a parcel of air stays saturated as it is warmed by an amount δT, the sensible heat content of the air increases in proportion to δT, and Equations 13.22 and 13.24 imply that the latent heat content increases in proportion to $(\Delta / \gamma^*) \delta T$. It follows that the fraction of R_n allocated to sensible heat will be $R_n / (1 + \Delta / \gamma^*)$, and the complementary fraction allocated to latent heat will be $(\Delta / \gamma^*) R_n / (1 + \Delta / \gamma^*)$. The second term in Equation 13.27 can therefore be identified as the *diabatic* component of latent heat loss associated with the additional supply of heat from radiation (Figure 13.3).

To recap, to estimate the rate at which a wet surface exchanges sensible and latent heat with the air above it, it is necessary to know the temperature of the surface. When this is not given, it can be eliminated from

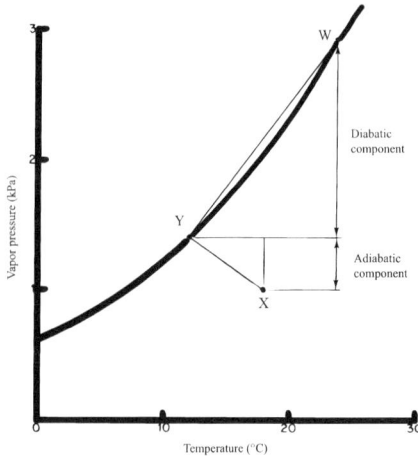

FIGURE 13.3   Basic geometry of the Penman equation (from Monteith, 1981a). A parcel of air at X is cooled adiabatically to Y (cf. Figure 13.2) and heated diabatically to W. Corresponding quantities of latent heat are given in Equation 13.27.

the equations describing the system by assuming that the saturation vapor pressure is a linear function of temperature—a valid approximation when the temperature range is small. (Alternatively, a solution could be found by iteration [McArthur 1990].) Howard L. Penman [1948] was the first to demonstrate this procedure and his formula is often written by combining the terms in Equation 13.27 to give the *Penman equation* in the form

$$\lambda E = \frac{\Delta R_n + \rho c_p \{e_s(T) - e\} r_H^{-1}}{\Delta + \gamma^*} \tag{13.28}$$

(See also (13.27a) for a related form of this equation, p. 242.)

The complementary expression for sensible heat loss, found by putting $C = R_n - \lambda E$, and substituting from Equation 13.28 is

$$C = \frac{\gamma^* R_n - \rho c_p \{e_s(T) - e\} r_H^{-1}}{\Delta + \gamma^*} \tag{13.29}$$

To find an expression for surface temperature, Equations 13.21 and 13.30 are combined to give

$$T = T_0 + \frac{(\gamma^* r_H / \rho c_p) R_n}{\Delta + \gamma^*} - \frac{\{e_s(T) - e\}}{\Delta + \gamma^*} \tag{13.30}$$

$$\text{(diabatic)} \qquad\qquad \text{(adiabatic)}$$

HOWARD LATIMER PENMAN

Howard Penman is known worldwide as the author of an equation for estimating the rate at which a stand of vegetation is likely to lose water by evaporation, given relevant conditions of weather and water supply. He was born in 1909 in northeast England, studied physics and mathematics at Armstrong College (University of Durham), and taught for some time at a boy's school before returning to Armstrong College to earn a PhD.

In 1937, Penman joined the staff of the physics department at Rothamsted Experimental Station and became departmental head in 1954. He became particularly interested in the rate at which crops lose water by evaporation, initially through analyzing data from drainage and rain gauges that had been operating at Rothamsted since 1870. Later he established sites for measurement of evaporation both on clay soil at Rothamsted and on contrasting sandy soil near Woburn. Toward the end of the second World War he applied his developing theory in a crash program of research to estimate the trafficability of soil in Europe for troops following the D-day landings. His classic paper [Penman 1948] "Natural evaporation from open water, bare soil, and grass" showed how partitioning of available energy was related to windspeed and saturation deficit, and included the unique analysis that enabled surface temperature to be eliminated from the relevant equations. The paper has probably been quoted more frequently than any other in the literature of agricultural meteorology and hydrology.

Penman also wrote seminal papers on diffusion of gases in pores. He continued to publish until 1976, and a full list of his 104 papers and reports was published in the Biographical Memoirs of the Royal Society, 32, 379–404, 1986.

This equation implies that the temperature of a wet surface will be warmer or cooler than that of the air passing over it depending on whether the diabatic term proportional to R_n is greater or less than the adiabatic term proportional to $\{e_s(T) - e\}$.

Isothermal Net Radiation

A minor defect in the original Penman equation was that R_n was a specified quantity although its exact value is always a (weak) function of surface temperature. This difficulty can be overcome by the expedient of replacing R_n by the *isothermal net radiation* R_{ni}, the radiation the surface would receive if it were at air temperature (p. 82). The fact that the surface is not usually at air temperature may be taken into account by using the resistance for convective and radiative heat loss in parallel, r_{HR}, in place of

r_H (p. 232). Then

$$\lambda E = \frac{\Delta \mathbf{R}_{ni} + \rho c_p \{e_s(T) - e\} r_{HR^{-1}}}{\Delta + \gamma^*} \tag{13.27a}$$

where $\gamma^* = \gamma r_V / r_{HR}$. McNaughton and Jarvis [1995] provide examples of more complex analysis using net isothermal radiation.

OPEN WATER SURFACES

To calculate the evaporation from an open water surface for periods of a week or more, Penman estimated net radiation, saturation deficit, temperature and windspeed from climatological data, used an empirical function of windspeed to estimate r_H, and assumed $r_V = r_H$. He also made the assumption, implicit in the derivation given here, that heat storage in the water was negligible compared with the value of $\mathbf{R}_n$. For relatively shallow tanks of water of the type used by Penman, this assumption is valid when the averaging period is several days. The greater the depth of water, the greater must be the averaging period for heat storage to be safely neglected, and for very deep lakes the period would have to be at least a year. As depth and heat storage increase, the month of maximum evaporation moves later and later in the year until it is out of phase with the annual radiation cycle. In the ocean, the large heat capacity of water results in much of the heat gained in the summer months being stored in the surface layers (to about 100 m) and returned to the atmosphere in winter. Because heat storage is so effective, sea surface temperature changes much less rapidly than that of the land surface.

DEPENDENCE OF EVAPORATION RATE ON WEATHER

To conclude this section, the dependence of evaporation rate on weather is briefly considered. It is clear from inspection of Equation 13.28 that the rate of evaporation from water increases linearly with the absorption of net radiation and with the value of the saturation deficit $\{e_s(T) - e\}$. It also increases with windspeed because r_H decreases with increasing wind. Because λE is a function of Δ, which increases with temperature, evaporation rate depends on temperature. Differentiation of Equation 13.28 with respect to T (with saturation deficit kept constant) shows that the fractional change of λE is related to the fractional change of T by

$$\frac{1}{\lambda E} \frac{\partial(\lambda E)}{\partial T} = \left(\frac{\Delta}{\Delta + \gamma^*}\right) \left\{\frac{\mathbf{R}_n}{\lambda E} - 1\right\} \frac{1}{\Delta} \frac{\partial \Delta}{\partial T} \tag{13.31}$$

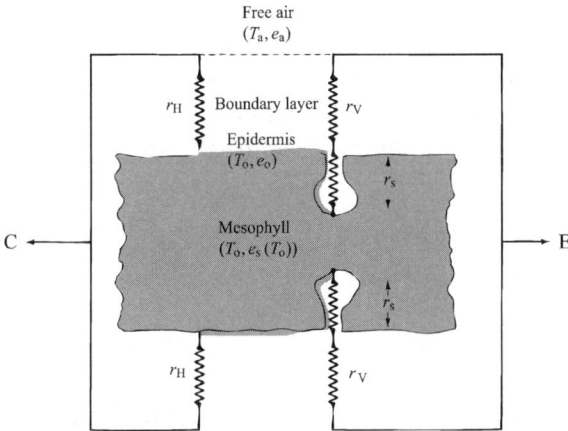

FIGURE 13.4 Electrical analog for transpiration from a leaf and leaf heat balance (see also Figure 11.8).

The algebraic sign of the right-hand side of this equation depends on whether $R_n/\lambda E$ is greater or less than unity. The equation thus shows that λE will increase or decrease with temperature depending on whether its initial value is less than or greater than R_n; i.e., on whether the surface is cooler or warmer than the air (since if λE exceeds R_n, the sensible heat flux C must be negative (ignoring heat storage); i.e., the surface must be cooler than the air, and when λE is less than R_n the surface must be warmer than the air). However, because $\Delta/(\Delta + \gamma^*)$ is less than 1, because $R_n/\lambda E$ is often close to unity, and because $(\partial\Delta/\partial T)/\Delta$ is only about 0.003 K^{-1} at 20°C, the temperature dependence of λE is negligible.

Finally, Equation 13.30 implies that increasing windspeed (decreasing r_H) will always decrease surface temperature in a system where γ^* is effectively independent of windspeed.

LEAF HEAT BALANCE

The equations for the sensible and latent heat exchange of a leaf are formally identical to those presented for a wet surface provided the resistances for heat and water vapor transfer are specified in an appropriate way. This approach was started by Penman, and extended by Monteith [1965]. The resulting equation is often termed the *Penman–Monteith equation*.Figure 13.4 shows an equivalent circuit in which the resistance for heat transfer is r_H for each side of the leaf (i.e., $r_H/2$ for the two sides in parallel). The resistance to vapor transfer for each side of the leaf is the sum of a boundary layer resistance r_V and a stomatal resistance r_s. The ratio γ^*/γ therefore assumes values of $(r_V + r_s) \div (r_H/2)$ for a hypostomatous leaf (stomata on one side only) and $(r_V + r_s)/2 \div r_H/2$ for an amphistomatous leaf with the same stomatal resistance on both surfaces. In general

$$\gamma^* = n\gamma \, (r_V + r_s)\,/r_H$$
$$\approx n\gamma \, (1 + r_s/r_H) \tag{13.32}$$

where $n = 1$ (amphistomatous leaf) or $n = 2$ (hypostomatous leaf), and $r_V \approx r_H$.

To apply Equation 13.28 to a leaf, it is necessary to assume that the mean temperature of the epidermis, which is the site for sensible heat exchange, is the same as the mean temperature of wet cell walls, which are the site of latent heat exchange. Most leaves are so thin that this assumption is fully justified. It is also safe to assume that the metabolic term in Equation 13.1 is negligible compared with $\mathbf{R_n}$. During the day when a net amount of energy is stored by photosynthesis, $\mathbf{M}/\mathbf{R_n}$ is never more than a few percent, and it is even less at night when heat is generated by respiration.

Despite the fact that Equation 13.28 can be applied to a leaf and to a wet surface, the two systems differ significantly in the following ways.

(i) *Dependence of transpiration rate on radiation and saturation deficit.*

For a wet surface, evaporation rate is finite even when *either* net radiation *or* saturation deficit is zero, and it increases linearly with either of these variables when the other is held constant. For leaves in their natural environment, stomatal aperture (and consequently r_s) depends strongly on solar radiation [Jones 1992] and, in the absence of light, stomata are usually closed so that transpiration is effectively zero. (Evaporation may occur very slowly through a waxy cuticle.) Moreover, substantial evidence both from the field and from work in controlled environments reveals that many plants close their stomata as saturation deficit increases, presumably as a mechanism for conserving water. When this response is evinced, transpiration rate will not increase in proportion to saturation deficit and may even reach a maximum value beyond which it decreases as the air gets drier still [Monteith 1995]. The response comes about through changes in the turgor of the guard cells that open and close the stomatal pore, and an attractive theory proposes that the mechanism exists to avoid the

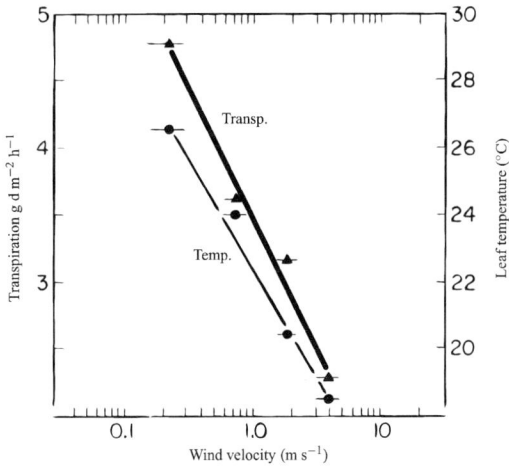

FIGURE 13.5 The change of transpiration rate and leaf temperature with windspeed for a *Xanthium* leaf exposed to radiation of 700 W m^{-2} at an air temperature of 15°C and 95% relative humidity (from Mellor et al 1964).

water potential gradient between the soil and the foliage becoming sufficiently negative to cause breakage (*cavitation*) of the continuous column formed by water moving from roots to foliage [Sperry 1995, Williams et al 2001, Tuzet et al 2003].

(ii) *Dependence of transpiration and temperature on windspeed.*

With increasing windspeed, the rate of evaporation from a wet surface always increases and surface temperature decreases (Equation 13.30). For a leaf, it can be shown by differentiating Equation 13.27a with respect to r_{HR} that λE is independent of r_{HR} (and hence of windspeed) when $\lambda E/C = \Delta/(n\gamma)$. (Note that Equation 13.27a is used in preference to 13.28 for this analysis to avoid the complication that R_n is a function of surface temperature and therefore of r_H). When $\lambda E/C$ exceeds this critical value, an increase of windspeed increases the latent heat loss at the expense of the sensible loss in such a way that $\lambda E + C$ stays constant. This behavior is expected intuitively because the rate of evaporation from a free water surface always increases with windspeed. When $\lambda E/C$ is smaller than the critical value, however, an increase of windspeed increases C at the expense of λE: evaporation decreases as windspeed increases. This behavior has been demonstrated in the laboratory (Figure 13.5) and inferred from measurements in the field. One of the practical implications is that providing shelter does not necessarily benefit plants by saving water.

The dependence of leaf temperature on r_H, and by implication on windspeed, is shown in Figure 13.6 for an arbitrary set of weather variables corresponding to bright sunshine in a temperate climate. When stomata are shut (resistance assumed infinite), the temperature excess of the leaf surface increases somewhat more slowly than r_H (i.e., the response is concave to the resistance axis) because allowance was made for the decrease of $\mathbf{R_n}$ with increasing surface temperature. When stomata are partly closed so that r_s is very large (500 s m^{-1} is a representative figure for partly closed stomata), curvature becomes much greater because, provided $\lambda E/C < \Delta/\gamma$, sensible heat loss decreases (and latent heat loss increases) as windspeed declines. When stomata are fully open ($r_s = 50$ s m^{-1}), the curvature becomes convex to the resistance axis (for small resistances at least) because in this regime $\lambda E/C > \Delta/\gamma$ and sensible heat loss increases with decreasing windspeed.

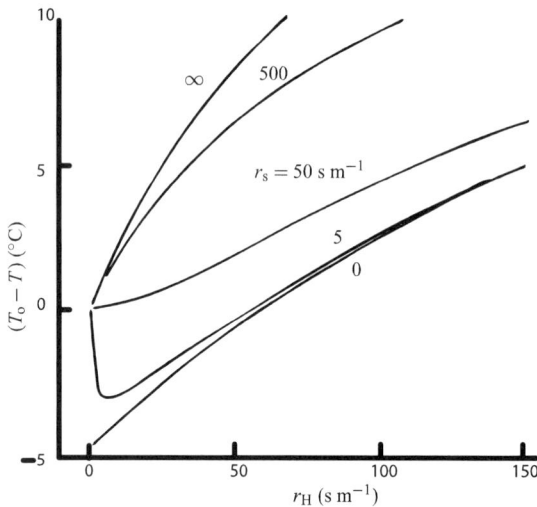

FIGURE 13.6 Predicted difference between surface and air temperature for an amphistomatous leaf in sunlight with specified boundary layer and stomatal resistances (for both laminae in parallel). Assumed microclimate: $\mathbf{R_{ni}} = 300$ W m^{-2} , $T = 20°C$, saturation deficit 1 kPa (from Monteith, 1981b).

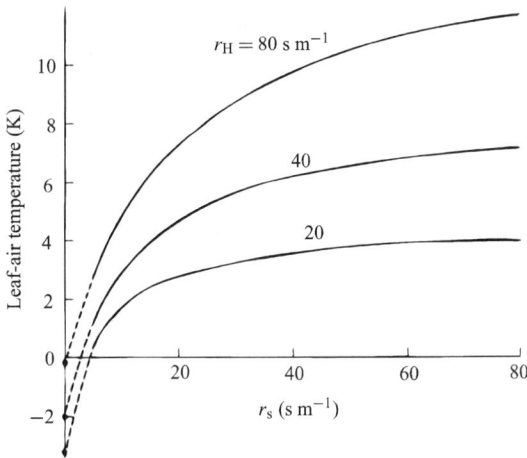

FIGURE 13.7 Excess of leaf over air temperature as function of stomatal and aerodynamic resistances when $R_{ni} = 300$ W m^{-2}, $D = 1$ kPa, $T = 20°$C (see also Figure 13.6).

During rain, some falling drops strike the foliage and are retained on leaf surfaces. Typically, 1–2 mm of water (per unit ground area) of this *intercepted rainfall* may be retained on the canopy of crops and young forests, but old growth forests may intercept two to three times this amount because they typically have mosses and lichens that intercept additional rainfall (Pypker et al, 2006). The rate of evaporation of intercepted rainfall is faster than the transpiration rate because it is not limited by the stomatal resistance. The rate may be particularly high for needle-leaved vegetation exposed to large windspeeds because values of r_H are small in such situations (see Equation 13.28). In regions where rain showers are frequent, the total loss of water by evaporation from coniferous forests can be substantially increased by the direct evaporation of intercepted water [Calder, 1977; Shuttleworth, 1988]. The line marked $r_s = 0$ in Figure 13.6 corresponds to a leaf with water covering its surface and therefore behaving like a wet-bulb exposed to radiation. In very strong wind, the limiting value of $T - T_0$ is the wet-bulb depression of the air, $T - T_w$.

When a dry leaf is given the unrealistically small value $r_s = 5$ s m^{-1}, the dependence of surface temperature on r_H is almost the same as for a water-covered leaf when r_H exceeds 50 s m^{-1} but $T_0 - T$ has a minimum at a very small value of r_H and approaches zero as r_H approaches zero. By writing Equation 13.30 in terms of $\mathbf{R}_{ni}$ and r_{HR} and evaluating $\partial T/\partial r_{HR}$, it can be shown that a minimum temperature is reached when

$$\frac{r_{HR}^2}{r_s}\left\{\left(1 + \frac{r_s}{r_{HR}}\right) + \frac{\Delta}{\gamma}\right\} = \frac{\rho c_p D}{\gamma\,\mathbf{R}_n} \qquad (13.33)$$

The weather assumed for Figure 13.6 gives a value of about 30 s m^{-1} for the right-hand side of Equation 13.33. Because stomatal resistances are very rarely less, it must be very unusual for a leaf to get warmer as windspeed increases, but a wet-bulb thermometer could behave like this if it were over-ventilated so that part of the wick became slightly dry and presented a small additional resistance to vapor transfer.

(iii) *Relation between leaf and air temperature.*

Equation 13.30 can be used to explain why the excess of leaf temperature over air temperature in bright sunshine is often reported to be large in cold climates and small (or even negative) in hot climates. Two features of the equation are implicated. First, because Δ increases with temperature, $T_0 - T$, if positive, will tend to increase with decreasing temperature. Strong radiative heating at low temperatures may be important for the survival of arctic species growing in a very short summer. Second, saturation deficit is commonly much larger in hot than in cold climates so the negative term in Equation 13.30 is larger. It follows that a leaf of a tropical plant adapted to keep stomata open at air temperatures in the range 30 to 40°C should be able to maintain a tissue temperature close to air temperature provided that water is available to its roots.

(iv) *Transpiration rate and stomatal resistance.*

Increasing the stomatal resistance of a leaf decreases the rate of evaporation and increases the sensible heat loss when $\mathbf{R}_n$ is constant. Stomatal closure therefore increases the temperature of leaf tissue as shown in Figure 13.7. When air temperature is in the range 20 to 30°C, many species have a minimum stomatal resistance of 100–200 s m^{-1}, so in bright sunshine and in a breeze, small leaves are expected to be only 1–2°C hotter than the surrounding air. Greater excess temperatures are observed on very large leaves in a light wind because r_H is large. From the same set of calculations, it can be shown that the relative humidity of air in contact with the epidermis will usually be similar

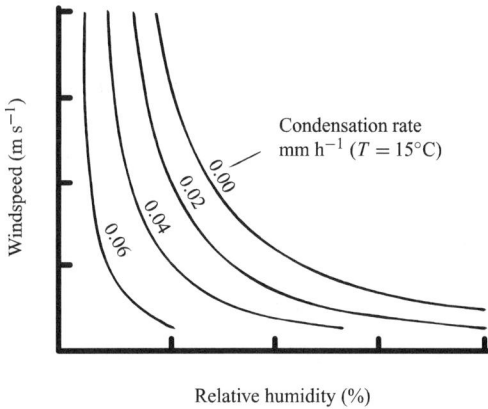

FIGURE 13.8 The rate of condensation on a horizontal plane exposed to a cloudless sky at night when the air temperature is 15°C, as a function of windspeed and relative humidity (from Monteith, 1981b).

to the relative humidity of the ambient air, and this feature of leaf microclimate may have important implications for the activity of fungi which need high relative humidity to reproduce and grow.

DEW

When $\mathbf{R}_n$ is negative at night, condensation will occur on a leaf when the numerator of Equation 13.28 is negative; i.e., when $-\Delta\mathbf{R}_n$ exceeds $\rho c_p \{e_s(T) - e\}/r_H$. The rate of dew formation can be calculated from the formula putting $\gamma^* = \gamma (r_V/r_H)$. When the air is saturated, the predicted maximum rate of dew formation on clear nights is about 0.06 to 0.07 mm per hour but may be much less in unsaturated air (Figure 13.8). These estimates are consistent with the maximum quantities of dew observed on leaves and on artificial surfaces—about 0.2 to 0.4 mm per night depending on site and circumstances [Monteith 1957]. As these quantities are an order of magnitude smaller than potential evaporation rates, dew rarely makes a significant contribution to the water balance of vegetation even in arid climates.

Because $\mathbf{R}_n$ is negative at night, the temperatures of leaves are always less than air temperature. Figure 13.9 shows the dependence of $T_0 - T$ on r_H on the same scale as Figure 13.6. For a dry leaf, the temperature difference is about -5 K in light winds in the absence of dew but is less than -2 K when dew forms from air with a relative humidity more than 90%.

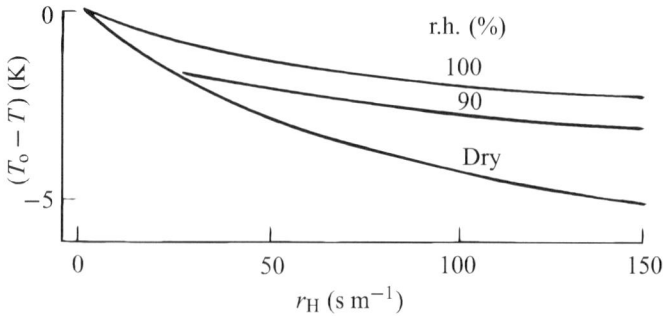

FIGURE 13.9   Predicted difference between surface and air temperature for a leaf with specified boundary layer resistance in the dark ($r_s = \infty$). Assumed microclimate: $R_n = -100$ W m^{-2} ; $T = 10°$C. Dew formation occurs when r.h. is 100 or 90%. The "dry" curve is appropriate when the r.h. is too low to allow condensation (from Monteith, 1981b).

When minimum temperature is critical, dew may therefore be important for the thermal regime of cold- or frost-sensitive plants and for the water supply of plants in very dry areas. When leaf temperatures falls below the frost-point of the air, hoar frost, rather than dew, is deposited (Figure 13.10, see color plates).

13.4 Developments from the Penman Equation

All the equations derived here for an isolated leaf can be applied to a uniform stand of vegetation provided the resistances of the system are described in an appropriate way [Monteith 1965, 1981]. The Penman–Monteith equation used in this mode is sometimes called a *big leaf model* because it specifies the stand-atmosphere exchange in terms of an *aerodynamic resistance* r_a analogous to the boundary layer resistance r_H around a single leaf (Chapter 9), and a *canopy resistance* r_c that corresponds to the stomatal resistance r_s of a leaf. The complications that arise from this procedure are described in Chapter 17. This chapter concludes by considering ways in which the Penman and Penman–Monteith equations can be extended to apply to surfaces whose wetness is specified either in terms of a fixed relative humidity or a fixed wet-bulb depression. The concept of equilibrium evaporation arising from the latter type of specification is also discussed.

SPECIFIED SURFACE HUMIDITY

When a chemical compound is dissolved in water, the free energy of the water molecules is reduced and there is a corresponding reduction in the vapor pressure of air in equilibrium with the water surface. Equation 2.42 gives the unique relation between the free energy of a solution (which depends on the molar concentration of the compound and on the degree of dissociation of its ions) and the relative humidity h of air in equilibrium with the solution. In deriving the heat balance equation for the surface of a solution, the saturated vapor pressure term $e_s(T)$ has to be replaced wherever it occurs by $he_s(T)$, so the Penman equation becomes

$$\lambda E = \frac{\Delta' \mathbf{R}_n + \rho c_p \{he_s(T) - e\} / r_H}{\Delta' + \gamma^*} \quad (13.34)$$

where $\Delta' = h\Delta$. This form of the equation was used by Calder and Neal [1984] to estimate annual evaporation and surface temperature for the Dead Sea, assuming $h = 0.75$.

Equation 13.34 is valid for any surface at which the free energy of water can be treated as a constant. In principle, it could be applied to bare soil, making h a function of the water content at the surface, but this is not a practical method of estimating evaporation from soil because the water content changes very rapidly with depth below the surface. Another possible application is the drying of hay or straw where h may depend on the porosity of the material and on the free energy of cell contents [Bristow and Campbell 1986].

SPECIFIED SURFACE WET-BULB DEPRESSION

Slatyer and McIlroy [1961] derived a form of the Penman equation in which the adiabatic term was obtained by assuming that a parcel of air at an initial wet-bulb depression of B made contact with a surface and was cooled (adiabatically) until it reached equilibrium with the surface in terms of its temperature and vapor pressure. It then had a smaller wet-bulb depression B_0. When the rate of heat and vapor exchange was specified by the resistance $r_H = r_V$, the equation became

$$\lambda E = \frac{\Delta \mathbf{R}_n}{\Delta + \gamma} + \rho c_p (B - B_0) / r_H \quad (13.35)$$

As the saturation vapor pressure deficit is given by $D \approx B(\Delta + \gamma)$ (p. 235), Equation 13.35 is equivalent to

$$\lambda E \approx \frac{\Delta \mathbf{R}_n + \rho c_p (D - D_0) / r_H}{\Delta + \gamma} \quad (13.36)$$

where D_0 is the saturation deficit of air in equilibrium with the surface.

EQUILIBRIUM EVAPORATION

The Slatyer–McIlroy equation has not been widely adopted because the term B_0 (or D_0) depends not only on the surface resistance to vapor transfer but also on prevailing weather. However, Equation 13.36 draws attention to the fact that the adiabatic (second) term in Penman's Equation 13.28 represents a lack of equilibrium between the state of the atmosphere at a reference height as given by D and the corresponding state of air in equilibrium with the surface as given by D_0. Priestley and Taylor []1972] suggested that air moving over an extensive area of uniform surface wetness (but not necessarily water) should come into equilibrium with the surface when $D = D_0$, giving the *equilibrium evaporation rate* as

$$\lambda E_q = \frac{\Delta R_n}{\Delta + \gamma} \tag{13.37}$$

However, the measurements which they reviewed convinced them that, on average, the observed latent heat of evaporation λE_o from water or from well-watered short vegetation exceeded λE_q by a factor α of about 1.26; i.e.,

$$\lambda E_o = \alpha \frac{\Delta R_n}{\Delta + \gamma} \tag{13.38}$$

Equation 13.38 is known as the *Priestley–Taylor equation*. Subsequent analysis of measurements over a wider range of surfaces has shown that α varies over a large range. In attempts to examine the properties of the atmosphere responsible for the fact that the loss of latent heat from extensive areas of well-watered vegetation may be both greater and less than the equilibrium rate of evaporation (Equation 13.37), De Bruin [1983], McNaughton and Spriggs [1986] and others have explored the behavior of the *Convective Boundary Layer* (CBL), the layer of the atmosphere about 1 km deep within which temperature and humidity change diurnally in response to the inputs of sensible and latent heat at the surface. For the case when the CBL is capped by an inversion (common on fine summer days), rising parcels of warm air penetrate a small way into the inversion layer, and, as they sink, they mix dry, warm air from above the inversion into the CBL by *entrainment*. The CBL therefore increases in depth during daylight hours as a consequence of the input of heat both from above and from below (when the ground is warm relative to the air above it). McNaughton pointed out that the saturation deficit D is the atmospheric potential that controls the non-equilibrium flux of heat and water vapor

(Equation 13.37). The vertical gradient of D can be written as

$$\partial D/\partial z = \partial \{e_s(T) - e\}/\partial z$$
$$\approx \Delta \partial T/\partial z - \partial e/\partial z$$
$$\propto \Delta \mathbf{C} - \gamma \lambda \mathbf{E} \tag{13.39}$$

The process of equilibrium between a relatively dry air mass and a moist surface can therefore be described in terms of a decrease in the *atmospheric* value of saturation deficit D complemented by an increase in the *surface* value D_0 until the two quantities are the same. When the gradient of D vanishes, $\Delta \mathbf{C} = \gamma \lambda \mathbf{E}$ which is equivalent to Equation 13.37 when $\mathbf{C} = \gamma \mathbf{R}_n/(\Delta + \gamma)$. When moist air moves over dry ground, equilibration proceeds in the opposite direction.

The preceding discussion emphasizes that plants and their atmospheric environment are mutually dependent: the state of the atmosphere is altered by fluxes of latent and sensible heat from vegetation; and plants respond to changes in air temperature and humidity in ways that alter heat fluxes. Monteith [1995] proposed a simple scheme for how vegetation and the CBL interact during the day. The response of vegetation to a pre-

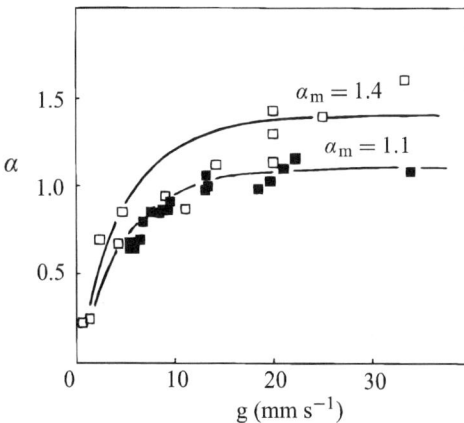

FIGURE 13.11 Increase in the Priestley–Taylor coefficient α with canopy conductance g $(= 1/r_c)$. Filled squares are from De Bruin [1983]; open points are from Monteith [1965]. The curves are given by the empirical equation $\alpha = \alpha_m(1 - \exp(-g/g_c))$, with the scaling conductance $g_c = 5$ mm s^{-1} (g_c probably depends on conditions at the top of the CBL). Curves with the scaling factor $\alpha_m = 1.1$ or 1.4 bracket the observed data (from Monteith 1995).

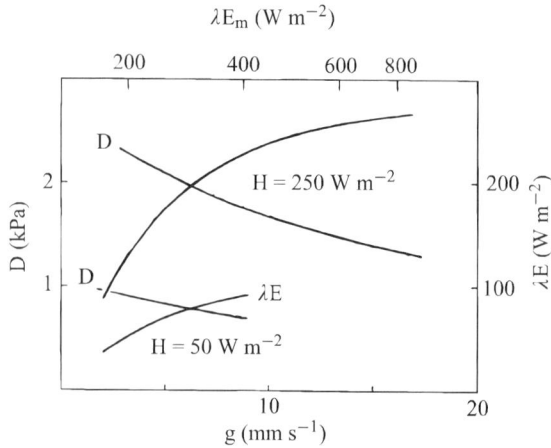

FIGURE 13.12   Predicted dependence of latent heat flux λE and atmospheric saturation vapour pressure deficit D on canopy conductance $g\ (=\ 1/r_c)$ for two levels of net
heat input $\mathbf{H}$ (net radiation and stored heat). The upper horizontal axis indicated the corresponding maximum rates at latent heat flux for the case $\mathbf{H} = 250$ W m^{-2}.

scribed atmospheric state (net radiation, temperature, saturation deficit)
was described by the Penman–Monteith equation (Equation 13.29 used
with aerodynamic and canopy resistances, see pp. 250 and 335–344) em-
ploying an empirical expression that described how canopy conductance g
(the reciprocal of canopy resistance) is often observed to decrease as transpiration rate increases. A complementary relation was found between g
and the Priestley–Taylor coefficient α that fits observed values for a diverse range of vegetation. Figure 13.11 illustrates that measured values
of α tend to an upper limit between 1 and 1.5 when canopy conductance
is large, consistent with the value of α estimated by Priestley and Taylor
[1972]. Combining the two relations allows λE and D to be expressed as
functions of canopy conductance and net radiation (Figure 13.12). This
simple method is likely to be valid only over periods of several hours and
when the land surface is sufficiently uniform and extensive for the CBL to
reach equilibrium with the underlying surface, but the scheme illustrates
the principles of how vegetation and the atmosphere interact so that plant
water supply and the atmospheric demand for evaporation are balanced
at regional scales. Huntingford and Monteith [1998] developed the simple approach further, and McNaughton [1989] discussed more complex
models for linking the convective boundary layer and vegetated surfaces.

COUPLING

McNaughton and Jarvis (1983) extended the concept of an equilibrium evaporation rate by re-writing the Penman–Monteith equation in the form

$$E = \Omega E_q + (1 - \Omega) E_i \qquad (13.40)$$

where E_q is the equilibrium evaporation rate (Equation 13.37), E_i is termed the *imposed evaporation rate* (see below), and the *decoupling coefficient* Ω is defined by

$$\Omega = (\Delta + \gamma) / (\Delta + \gamma^*) \qquad (13.41)$$

Comparing Equations 13.40 and 13.27 reveals that ΩE_q is the rate of evaporation that would occur if the heat budget of a surface were dominated by the diabatic (radiative) term. This condition tends to be satisfied when a large leaf or a stand of short well-watered vegetation is exposed to bright sunshine, humid air, and a light wind. The evaporation rate is then effectively independent of the saturation deficit of the ambient air, and the surface may be described as "decoupled" from the prevailing weather. The term "decoupled" must not be taken too literally, however, because E_q depends on the absorption of radiation and is slightly dependent on temperature (through Δ).

The complementary quantity $(1 - \Omega)E_i$ is the rate of evaporation "imposed" by the environment when the leaf or surface is "fully coupled" to the prevailing weather, i.e. when the adiabatic term $\rho c_p D / r_H$ in Equation 13.27 is much larger than the diabatic term, a condition satisfied when r_H (or its canopy-scale equivalent r_a) is very small (small leaf or rough canopy, strong wind) and R_n is also small. In these conditions, the saturation deficit at the leaf surface (or in a canopy) is effectively equal to the value (D) at a reference height in the free atmosphere so that Equation 13.27 reduces to

$$\lambda E_i \approx \rho c_p D / (\gamma r_s) \qquad (13.42)$$

For vegetation freely supplied with water, the value of the decoupling coefficient Ω for leaves depends mainly on windspeed and leaf size, and ranges from about 0.1 for pine needles to 0.8 for rhubarb leaves. For canopies, Ω depends mainly on surface roughness and wind speed, varying from about 0.1 for needle-leaved forests (strong coupling) to 0.8 to 0.9 for short crops (decoupled) [Jarvis and McNaughton, 1986].

Substituting Equation 13.32 into Equation 13.28, differentiating, and writing $g_s = 1/r_s$, it may be shown, with some rearranging of terms, that the relative change in evaporation for a prescribed change in conductance is given by

$$dE/E = (1 - \Omega)dg_s/g_s \qquad (13.43)$$

Thus, for leaves that are well-coupled to the atmosphere (small Ω), stomata exert strong control over water loss, but transpiration by weakly coupled leaves is poorly controlled by stomata and depends primarily on radiant energy availability (including its influence on leaf temperature and stomatal aperture). A similar conclusion applies to the sensitivity of *canopy* evaporation to canopy conductance.

When vegetation starts to run short of water, Ω decreases because γ^{*} increases, implying better coupling between vegetation and air. In this case, however, the rate of water loss is more and more determined not by the state of the atmosphere but by the rate at which roots can extract water. During any period when the rate of extraction is approximately constant, plants in a stand have to adjust the canopy resistance r_c, so it is approximately proportional to the imposed saturation deficit. "Coupling" then implies that it is stomatal resistance rather than transpiration rate that responds to the state of the atmosphere.

13.5 Problems

1. A cylindrical thermometer element, diameter 3mm is enclosed in a radiation shield that excludes all solar radiation but is $5.0°C$ warmer than the true air temperature, which is $20.0°C$. If the thermometer is to record a temperature within $0.1°C$ of true air temperature, at what windspeed must the element be ventilated? (Assume longwave emissivities of 1.0.)

2. A wet-bulb thermometer in a radiation shield has a cylindrical element 4 mm diameter. Estimate the radiative resistance assuming that the temperatures of the wet bulb and the shield are about $10°C$. Hence, plot a graph to show how the difference between measured and thermodynamic wet-bulb temperatures would vary with ventilation speed over the range 0.5 to 5 m s^{-1}.

3. A leaf has a boundary layer resistance for heat transfer (both surfaces in parallel) of 40 s m^{-1}, and a combined stomatal and boundary layer resistance of 110 s m^{-1}. Taking air temperature as $22°C$ and vapor pressure as 1.0 kPa, set up a spreadsheet to calculate the rates of sensible heat transfer $\mathbf{C}$, latent heat transfer $\lambda\mathbf{E}$, and the sum $\mathbf{C} + \lambda\mathbf{E}$ as a function of leaf temperature over the range $20–26°C$. Hence find graphically or otherwise the leaf temperature when the net radiation absorption by the leaf is 300 W m^{-2}.

4. For the data in question 3, use the Penman–Monteith equation to determine the latent heat flux from the leaf and hence determine the sensible heat flux and leaf temperature.

5. Show, using the Penman–Monteith equation, that the necessary condition for dew-fall is $-\Delta \mathbf{R}_n \succ \rho c_p \{e_s(T) - e\} r_H^{-1}$, and that the maximum rate of dew-fall is $\mathbf{E}_{max} = \Delta \mathbf{R}_n / \lambda (\Delta + \gamma^*)$. On a cloudless night, the net radiation flux density of an isolated leaf at air temperature $T(\mathrm{K})$ is given by the empirical expression $\mathbf{R}_n = (0.206e^{0.5} - 0.47)\sigma T^4$, where e (kPa) is the vapor pressure. Neglecting other sources of heat, plot a graph to show how the maximum rate of dew-fall on the leaf varies with air temperature on the range 0–25°C, and explain your results.

6. A marathon runner, treated as a cylinder with diameter 33 cm moving at 19 km h^{-1} relative to the surrounding air, has a net radiation load of 300 W m^{-2}. The air temperature and vapor pressure are 30°C and 2.40 kPa, respectively. Assuming the runner's skin is covered with sweat that is a saturated salt solution for which the relative humidity of air in contact with the solution is 75%, estimate the rate of latent heat loss. If all the salt could be washed off when the runner was sprayed with water, what would be the new rate of latent heat loss?

7. On the Oregon coast, Sitka spruce is frequently exposed to wind-driven sea spray. In the extreme case where water on needle surfaces is saturated with sea salt (equilibrium relative humidity 75%), by how much would the needle decoupling coefficient be changed relative to needles with pure water on their surfaces?

   Assume a windspeed of 5 m s^{-1} and needle dimension of 1 mm.

Chapter 14

Steady State Heat Balance (ii) Animals

The heat balance of a leaf or stand of vegetation is determined mainly by its environment, and over periods of a few hours the temperature of tissue cannot be controlled except to a limited extent by the movement of leaves with respect to the sun's rays. In contrast, warm-blooded animals exposed to a changing environment are able to control deep body temperature within narrow limits by adjusting the rate at which heat is produced by metabolism or dissipated by evaporation. They are therefore classed as *homeotherms* (from the Greek *homoios*—similar, and *therme*—heat). Cold-blooded animals or *poikilotherms* (*poikilos*-various) form a class with heat balance principles intermediate between vegetation and homeotherms. They have relatively low metabolic rates and no thermostat, but tend to avoid extremes of heat and cold by seeking shade or shelter. Lizards, tortoises, and bees, for example, appear to prefer environments where exposure to solar radiation raises body temperature to between 30 and 35°C, close to the range in which homeotherms operate.

The heat balance equation of any animal can be written in the general form

$$\mathbf{M} + \mathbf{R}_n = \mathbf{C} + \lambda\mathbf{E} + \mathbf{G} + \mathbf{S} \tag{14.1}$$

where each term refers to the gain or loss of heat per unit of body surface area. $\mathbf{M}$ is the rate of heat production by metabolism, $\mathbf{G}$ is conduction to the substrate on which the animal is standing or lying, and $\mathbf{S}$ is the rate of heat storage in the animal. The latent heat loss $\lambda\mathbf{E}$ is the sum of components representing losses from the respiratory system λE_r and from the skin when sweat evaporates, λE_s. The terms R_n and C have the same significance as previously.

Methods of calculating the radiative exchange of animals were considered in Chapter 8. The size and significance of the remaining terms will now be considered separately as an introduction to the "thermo-neutral diagram," which shows how the terms are related for homeotherms.

14.1 Heat Balance Components

METABOLISM (M)

Standard measurements of basal metabolic rate are made when a subject has been deprived of food and is resting in an environment where metabolic rate is independent of external temperature. The basal metabolism of an animal M_b (expressed here in watts and not watts per unit surface area as elsewhere) can be related to body mass W by

$$M_b = BW^n \qquad (14.2)$$

where B is a constant, implying that $\ln M_b$ is a linear function of $\ln$ W (Figure 14.1).

According to many sets of measurements by animal physiologists, $n = 0.75$ and Kleiber [1965] suggested that, for *intra*-specific work, a value of $70\,\text{kcal day}^{-1}$ per $\text{kg}^{0.75}$ (3.4 W per $\text{kg}^{0.75}$) should be adopted for the term B. Hemmingsen [1960] found that $B = 1.8$ W per $\text{kg}^{0.75}$ was a better value for inter-specific comparisons over a range of mass from 0.01 to 10 kg. His review also showed that the metabolism of a poikilotherm kept at a body temperature of $20°C$ is about 5% of the value for a homeotherm of the same mass with a deep body temperature of $39°C$. Hibernating mammals metabolize energy at about the same rate as poikilotherms of the same mass.

Referring metabolic rates to a power of body mass is awkward in studies of heat transfer where fluxes are referred to unit area. Conveniently, Prothero [1984] found that n was closer to $2/3$ than to $3/4$ when bias produced by circadian rhythms was removed from metabolic measurements. For any set of objects with identical geometry but different in size, surface area is proportional to the $2/3$ power of mass, assuming constant mass per unit volume. It follows that if basal metabolic rates are proportional to the $2/3$ power of mass, they should be proportional to surface area, at least when animals of like geometry are compared. This proportionality has

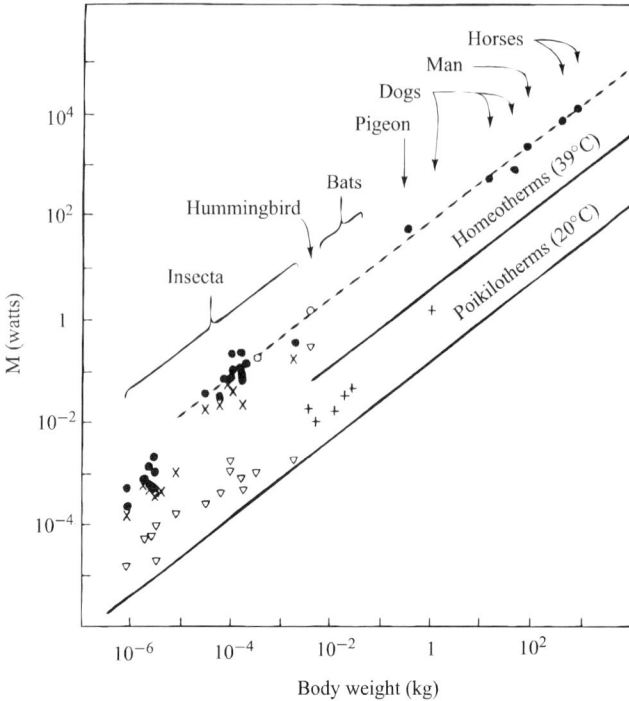

FIGURE 14.1   Relation between basal metabolic rate of homeotherms (upper continuous line), maximum metabolic rate for sustained work by homeotherms (upper dashed line) and basal rate for poikilotherms at 20°C (lower continuous line) (from Hemmingsen, 1960).

been demonstrated for mammals ranging in size from a mouse (0.02 kg) to an elephant (1400 kg). The average basal metabolic rate for a mammal per unit of body surface area is about 50 W m^{-2}. This is much smaller than the flux of short-wave radiation absorbed by a dark-coated animal in bright sunshine (say 300 W m^{-2}), but is comparable with the energy that might be absorbed under shade. On a cloudless night, the basal metabolic rate is comparable to the net loss of heat by long-wave radiation from a surface close to air temperature.

For walking or climbing, the efficiency with which man and domestic animals use additional metabolic energy is about 30%. To work against gravity at a rate of 20 W m^{-2}, for example, metabolism must increase by about 60 W m^{-2}. The power expended in walking increases with velocity and, in man, reaches 700 W at 2 m s^{-1}.

For rapid forms of locomotion such as running or flying, the work done against wind resistance is proportional to the wind force times the distance traveled. When the drag coefficient of a moving body is independent of velocity, the drag force should be proportional to V^2 (p. 145), and as the distance traveled is proportional to V, the rate of energy dissipation should increase with V^3. However, studies on birds by Tucker [1969] revealed a range of velocities (e.g., 7 to 12 m s^{-1} for gulls) in which the rate of energy production was almost independent of V, implying a remarkable decrease of drag coefficient with increasing velocity.

LATENT HEAT (λE)

In the absence of sweating or panting, the loss of latent heat from an animal is usually a small fraction of metabolic heat production and takes place both from the lungs during breathing (*respiratory evaporation*) and from the skin as a result of the diffusion of water vapor, sometimes politely referred to as "insensible perspiration." For a man inhaling completely dry air, the vapor pressure difference between inhaled and exhaled air is about 5.2 kPa and the heat used for respiratory evaporation λE_r is the latent heat equivalent of about 0.8 mg of water vapor per ml of absorbed oxygen [Burton and Edholm, 1955]. With round figures of 2.4 J mg^{-1} for the latent heat of vaporization of water and 21 J ml^{-1} oxygen for the heat of oxidation, $\lambda E_r/M$ is about 10%. When air with a vapor pressure of 1.2 kPa is breathed, the difference of vapor pressure decreases to 4 kPa and $\lambda E_r/M$ is about 8%.

In the absence of sweating, the latent heat loss from human skin (λE_s) is roughly twice the respiratory loss, implying a total evaporative loss of the order of 25 to 30% of **M** depending on vapor pressure. When sweating, however, a human can produce about 1.5 kg of fluid per hour, equivalent to 600 W m^{-2} if the environment allows sweat to evaporate as fast as it is produced. More commonly, the rate of evaporation is restricted to a value determined by resistances and vapor pressure differences. Excess sweat then drips off the body or soaks into hair and clothing.

Sheep lack glands of the type that allow man to sweat profusely, but species such as cattle can lose substantial amounts of water by sweating. Sheep and dogs compensate for their inability to sweat by panting in hot environments. For cattle exposed to heat stress, the respiratory system can account for 30% of total evaporative heat loss, the remaining 70% coming from the evaporation of sweat from the skin surface and from wetted hair. The maximum evaporative heat loss from ruminants is only a little larger than metabolic heat production, whereas a sweating man can lose far more heat by evaporation than he produces metabolically if he is exposed to strong sunlight.

Interspecific differences in $\lambda E/M$ and in mechanisms of evaporation may play an important part in adaptation to dry environments. Relatively small values of $\lambda E/M$ have been reported for a number of desert rodents which appear to conserve water evaporated from the lungs by condensation in the nasal passages where the temperature is about 25°C [Schmidt-Nielsen, 1965]. The respiratory system operates as a form of counter-current heat exchanger. In contrast, measurements of total evaporation from a number of reptiles yield figures ranging from 4 to 9 mg of water per ml of oxygen. These figures imply that nearly all the heat generated by metabolism was dissipated by the evaporation of water, lost mainly through the cuticle.

For different animal species, the amount of body water available for evaporation depends on body volume, whereas the maximum rate of cutaneous evaporation depends on surface area. At one extreme, insects have such a large surface:volume ratio that evaporative cooling is an impossible luxury. Man and larger animals can use water for limited periods to dissipate heat during stress and, in general, the larger the animal the longer it can survive without an external water supply.

CONVECTION (C)

The left side of Figure 14.2 demonstrates how resistance to convective heat transfer between an organism and its environment decreases as body size decreases, a significant relation in animal ecology (see also Figure 13.1). Schmidt-Nielsen (1965) regarded both convective and radiative exchanges as proportional to surface area and concluded that, in the desert, "the small animal with its relatively larger relative surface area is in a much less favorable position for maintaining a tolerably low body temperature." It is true that small animals are unable to keep cool by evaporating body water but when convection is the dominant mechanism of heat loss they can lose heat more rapidly (per unit surface area) than large animals exposed to the same windspeed. In this context, the main disadvantage of smallness is microclimatic: because windspeed increases with height above the ground, small animals moving close to the surface are exposed to slower windspeeds than larger mammals. As r_{H} is approximately proportional to $(d/V)^{0.5}$, an animal with $d = 5$ cm exposed to a wind of 0.1 m s^{-1} will be coupled to air temperature in the same way as a much larger animal with $d = 50$ cm exposed to wind at 1 m s^{-1}. Insects and birds in flight and tree-climbing animals escape this limitation.

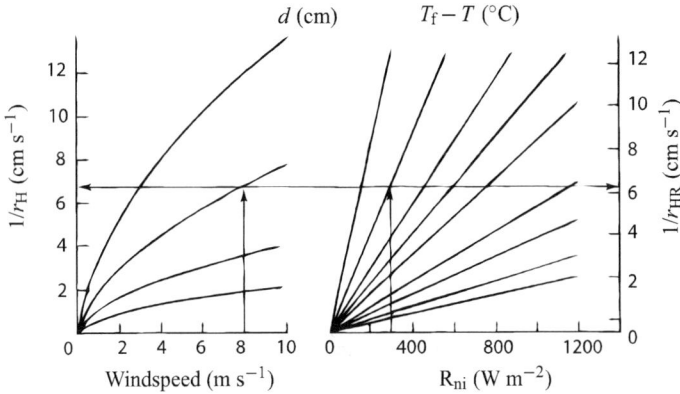

FIGURE 14.2 Diagram for estimating thermal radiation increment when windspeed, body size, and net radiation are known. For example, at 8 m s^{-1} an animal with $d = 2$ cm has $r_H^{-1} = 6.7$ cm s^{-1} and $r_{HR}^{-1} = 6.2$ cm s^{-1}. When $R_{ni} = 300$ W m^{-2}, $T_f - T$ is $40°$C.

CONDUCTION (G)

Few attempts have been made to measure the conduction of heat from an animal to the surface on which it is lying. Mount (1968) measured the heat lost by young pigs to different types of floor material and found that the rate of conduction was strongly affected by posture and by the temperature difference between the body core and the substrate. When the temperature of the floor (and air) was low, the animals assumed a tense posture and supported their trunks off the floor, but as the temperature was raised, they relaxed and stretched to increase contact with the floor. Figure 14.3 shows the heat loss per unit area, recalculated from Mount's data, for newborn pigs in a relaxed posture. As the heat flow was approximately proportional to the temperature difference, the thermal resistance for each type of floor can be calculated from the slope of the lines. The resistances are about 8, 17, and 58 s cm^{-1} for concrete, wood and polystyrene respectively. As the corresponding resistances for convective and radiative transfer are usually around 1 to 2 s cm^{-1} (Figure 13.1), it follows that heat losses by conduction to the floor of an animal house are likely to be significant only when it is made of a relatively good heat conductor such as concrete. Conduction will be negligible when the floor is wood or concrete covered with a thick layer of straw.

Gatenby [1977] measured the conduction of heat beneath a sheep with a fleece length of about 2 cm lying on grass in the open. When the deep soil temperature was $10°$C, downward fluxes of about 160 W per m^2 of

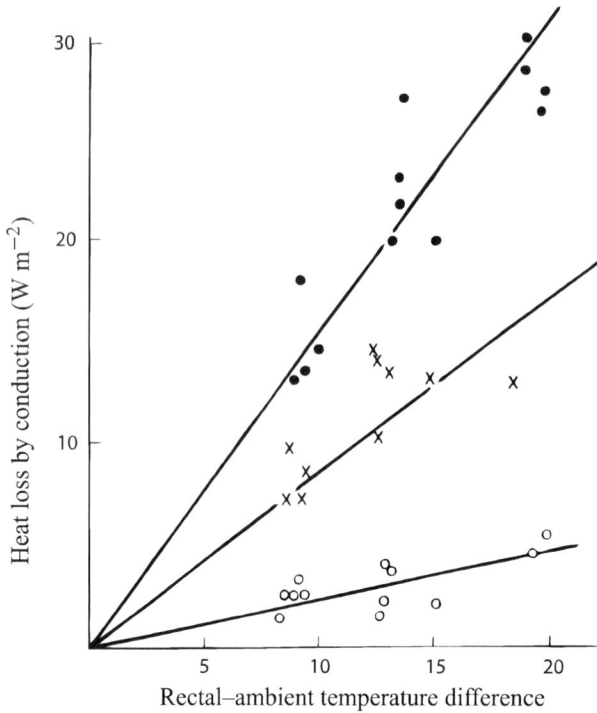

FIGURE 14.3   Measurements of heat lost by conduction from a pig to different types of floor covering expressed as watts per square meter of total body area (after Mount, 1967). ●–concrete; X—wood; ◯—polystyrene.

contact surface were measured when the sheep lay down, equivalent to about 40 W per m² of body surface and therefore comparable with the loss by convection. Energy requirements for free-ranging animals therefore depend to some extent on the time spent lying, on soil temperature, and on the thermal properties of the ground.

STORAGE (S)

As discussed in more detail here, most homeotherms have mechanisms that maintain their core body temperature almost constant over a wide range of environmental conditions. On the other hand, the body temperatures of poikilotherms vary in response to external heat inputs. All animals alter their behavior to avoid extreme conditions, e.g., seeking shade, sheltering from wind, or adopting a nocturnal lifestyle. In desert envi-

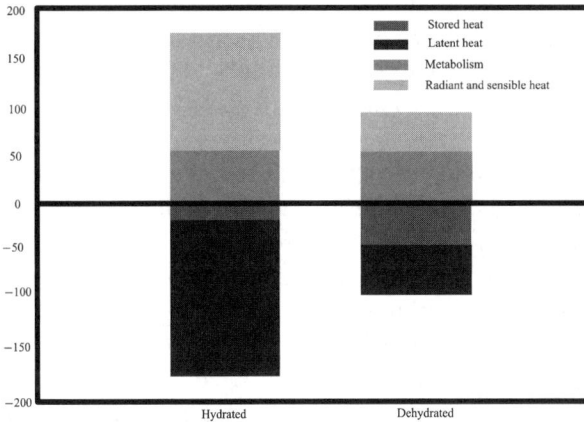

FIGURE 14.4 Heat balance components of a camel in a desert environment when hydrated and dehydrated, expressed as mean rates over the 10h during which external heat loads were largest (based on data from Schmidt-Neilsen et al 1957).

ronments where daytime air temperatures may be much larger than body temperature and minimizing evaporative loss is important, it has been reported that some homeotherms allow their body temperature to increase during the day, thus reducing the need for evaporative cooling by storing heat and by reducing sensible heat gain (i.e., reducing the air-skin temperature gradient) and increasing longwave radiation loss. The stored heat is dissipated at night by sensible heat loss. Schmidt-Nielsen et al [1957] studied the body temperature of dromedary camels kept in full desert sun. When the camels were allowed to drink water at the beginning of each day, body temperature varied by about 2.1°C between day and night, but when the camels were deprived of water their body temperatures during the day were up to 6.5°C greater than at night. Such a range would be life-threatening for many mammals. For a camel weighing 260 kg, assuming a body specific heat of 3.6 kJ kg^{-1} K, the daytime heat storage rate **S** averaged 15 W m^{-2} for the hydrated and 42 W m^{-2} for the dehydrated animals. Schmidt Neilsen also measured evaporation rate **E** from the skin (ignoring the small respiratory evaporation rate), and estimated metabolic heat production **M**. Using the heat balance equation, 14.1, he interpreted the residual **E + S − M** as the heat gained from net radiation and sensible heat.Figure 14.4 illustrates the heat balances over the hottest 10 hours of the day for the hydrated and dehydrated camels. When water was available, the animals balanced their energy gains principally by evaporation of sweat; when dehydrated, the sweating rate was much decreased, and heat storage was similar in magnitude to the loss of heat by evaporation. Some of the large reduction in the radiant and sensible heat gain is probably due to postural changes the camels adopted to reduce radiation interception when short of water.

Although heat storage is not viable as a long-term component of the heat balance of most homeotherms, it is important during short bursts of activity. For example, Taylor and Lyman [1972] found that running antelopes generated heat at about 40 times their resting metabolic rate, and the increase of their body temperature of up to 6°C over 5–15 minutes (depending on running speed) accounted for 80–90% of this heat. For humans, body temperature T_b also increases during exercise in direct proportion to the metabolic rate $\mathbf{M}$ until the limits for thermoregulation are reached. A convenient approximation [Kerslake 1972] is

$$T_b = 36.5 + 4.3 \times 10^3 \, \mathbf{M}$$

where T_b is in Celsius when $\mathbf{M}$ is in W m^{-2}.

14.2 The Thermo-Neutral Diagram

The fundamental relation between the metabolic heat production of a homeotherm and the temperature of its environment is often represented by a thermo-neutral diagram (Figure 14.5),for which measurements of metabolism can be obtained in one of two ways. In "direct" calorimetry, the subject is placed in a *calorimeter*, an enclosure usually with wall temperature equal to air temperature to simplify estimates of radiative transfer, and the flow of heat through the walls is measured with transducers. For "indirect" calorimetry, measurements of oxygen consumption are widely used to determine heat loss from the subject. Ways of defining the effective temperature of more complex environments are considered later.

To simplify interpretation of the thermo-neutral diagram without concealing major details, conduction of heat to the ground will be neglected and the animal (or man) will be assumed covered by a uniform layer of hair (or clothing). It is convenient to work with isothermal net radiation $\mathbf{R}_{ni}$ (p. 242) because this flux is independent of surface temperature, and to introduce a combined resistance for convection and long-wave radiative exchange r_{HR} (p. 232).

The heat balance equation for the surface of a coat whose mean temperature is T_0 then becomes

$$\mathbf{M} + \mathbf{R}_{ni} - \lambda \mathbf{E}_r - \lambda \mathbf{E}_s = \rho c_p \left(T_0 - T \right) / r_{HR} \qquad (14.3)$$

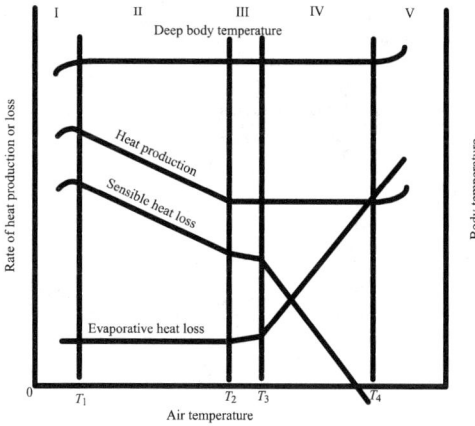

FIGURE 14.5 Diagramatic representation of relations between heat production, evaporative and non-evaporative heat loss and deep-body temperature in a homeothermic animal (from Mount, 1979).

where T is air temperature.

If the resistance of the coat is r_c, if it is impermeable to radiation of all wavelengths, and if the evaporation of sweat is confined to the surface of the skin with mean temperature T_s, the flux of sensible heat through the coat is

$$M - (\lambda E_r - \lambda E_s) = \rho c_p (T_s - T_0)/r_c \qquad (14.4)$$

where r_c is the resistance of the coat. The flux of heat through the skin surface is

$$M - \lambda E_r = \rho c_p (T_b - T_s)/r_d \qquad (14.5)$$

where r_d is the resistance of the body tissue and T_b is the body temperature.

Eliminating T_0 and T_s from Equations 14.3 to 14.5 gives the relation between heat balance components as

$$M + (r_{HR}/r_t)\, R_{ni} = \rho c_p (T_b - T)/r_t + \lambda E_r + \lambda E_s (r_{HR} + r_c)/r_t \qquad (14.6)$$

where $r_t = r_{HR} + r_c + r_b$. This equation can be used to explore the physiologically significant air temperatures shown in Figure 14.5 if it is written in the form

$$T = T_b - \{r_t M + r_{HR} R_{ni} - r_t \lambda E_r - (r_{HR} + r_c)\lambda E_s\}/\rho c_p \qquad (14.7)$$

Equations 14.6 and 14.7 will now be used to examine five discrete regimes incorporated in the thermo-neutral diagram.

Table 14.1 Cold limit and lower critical temperature for three species (from Mount, 1979), with corresponding values of $r_t\mathbf{M}$ (see text).

	Cold limit $T_1(°C)$	Critical temperature $T_2(°C)$	$r_t\mathbf{M}_{max}$ (kJ m^{-3})	$r_t\mathbf{M}_{min}$ (kJ m^{-3})
Man, newborn	27	33	15	6
mature	14	28	35	14
Pig, newborn	0	34	56	5
mature	−50	10	131	41
Sheep, newborn	−100	30	206	11
mature	−200	20	355	86

I $T < T_1$

At an air temperature of T_1, sometimes referred to as the "*cold limit*," the body produces heat at a maximum rate $\mathbf{M}_{max}$. Assuming that the sum of the two latent heat terms in Equation (14.7) is $0.2\mathbf{M}_{max}$ (see p. 261), the value of T_1 is given by

$$T_1 \approx T_b - \{0.8 r_t\mathbf{M}_{max} + r_{HR}\mathbf{R}_{ni}\}\, \rho c_p \qquad (14.8)$$

Table 14.1 contains values of measurements of T_1 summarized by Mount [1979] for newborn and mature individuals of three species. Comparisons emphasize the value of hair in establishing the size of r_t and therefore of the product $r_t\mathbf{M}_{max}$. Estimates of this quantity given in the Table were derived by assuming that $T_b = 37°C$ for all species, that $\mathbf{R}_{ni}$ was zero in the experimental conditions, and that $\rho c_p = 1200$ J m^{-3} K^{-1}.

When air temperature falls below the cold limit, heat is dissipated faster than it is made available by metabolism and radiation, so that the steady-state heat balance represented by Equation 14.7 cannot be achieved, and body temperature must then decrease with time (see p. 229). This depresses the metabolic rate so that temperature falls still further, leading to death by hypothermia if the process is not reversed.

II $T_1 < T < T_2$

In this regime, sensible heat loss decreases linearly with increasing air temperature, inducing an identical decrease in metabolic rate provided the latent heat component of the heat balance remains constant. Differentiation of Equation 14.6 then gives

$$\frac{\partial \mathbf{M}}{\partial T} = \frac{-\rho c_p}{r_t} \qquad (14.9)$$

If $r_t = 600$ s m^{-1}, for example, $\partial \mathbf{M}/\partial T$ is -2 W m^{-2} K^{-1}.

Equation 14.9 is valid between the cold limit T_1 and a temperature T_2 at which the metabolic rate reaches a minimum value $\mathbf{M}_{min}$, usually somewhat larger than the basal metabolic rate (p. 259) on account of physical exertion and/or the digestion of food. Repeating the assumptions used to find an approximation for T_1 gives

$$T_2 \approx T_b - \{0.8\, r_t \mathbf{M}_{min} + r_{HR}\mathbf{R}_{ni}\}\, / \rho c_p \qquad (14.10)$$

Observed values of T_2 and estimates of $r_t \mathbf{M}_{min}$ are in Table 14.1.

Livestock exposed to a temperature below T_2 need more feed to achieve the same liveweight gain or production of milk, eggs, etc. than a comparable animal kept in the thermo-neutral zone. The temperature T_2 is therefore referred to as the (*lower*) *critical temperature*.

III $T_2 < T < T_3$

This is the zone of "least thermoregulatory effort" for all animals and is also identified as the comfort zone for man. As temperature rises above T_2, metabolic rate remains constant, so the heat balance equation can be satisfied only if (a) the total resistance to sensible heat loss r_t decreases, and/or (b) the latent heat loss λE increases. In many animals, including man, a decrease of r_t is a consequence of blood vessels near the skin surface dilating so that blood circulates faster, effectively decreasing the tissue resistance r_d (see p. 183). This process is sometimes accompanied by a small increase of sweat rate.

IV $T_3 < T < T_4$

In this regime, $\mathbf{M}$ and r_t are constant and the maintenance of thermal equilibrium as temperature rises requires that the decrease of sensible heat loss should be balanced by an identical increase of latent heat loss. This increase is made possible by more rapid production and evaporation of sweat and/or by faster breathing and other mechanisms which increase respiratory evaporation (p. 261). The temperature range T_2 to T_4, in which heat production is at its minimum value and is independent of air temperature, is usually called the *thermo-neutral zone*.

V $T > T_4$

As temperature rises further, loss of latent heat cannot increase indefinitely. The rate of evaporation of sweat is limited by the aerodynamic resistance of the body and therefore by windspeed, and the rate of evaporation from the respiratory system is likewise limited by the rate of panting and the volume of air inhaled and exhaled. Both rates of evaporation depend on the vapor pressure of the atmosphere. When λE reaches a maximum (or even at a somewhat lower temperature as Figure 14.4 suggests), body temperature starts to rise, a primary symptom of hyperthermia. Differentiating Equation 14.7 with respect to T when λE is constant gives

$$1 = \frac{\partial T_b}{\partial T} - \frac{\partial M}{\partial T_b} \frac{\partial T_b}{\partial T} \frac{r_t}{\rho c_p}$$

or

$$\frac{\partial T_b}{\partial T} = \left\{ 1 - r_t \left(\partial M / \partial T_b \right) / \rho c_p \right\}^{-1} \qquad (14.11)$$

The requirement that $\partial T_b / \partial T$ should be positive implies that

$$\frac{\partial M}{\partial T_b} < \rho c_p / r_t \qquad (14.12)$$

In man, the increase of metabolic rate with temperature is about 7% per K, so if $M = 150$ W m^{-2}, r_t should not exceed 170 s m^{-1} ($\rho c_p = 1200$ J m^{-3} K^{-1}), implying good ventilation and a minimum of clothing, familiar requirements for avoiding extreme discomfort in tropical climates. When Equation 14.12 is not satisfied, body temperature and metabolic rate tend to rise uncontrollably with consequences that can be fatal.

14.3   Specification of the Environment

The thermo-neutral diagram has been used mainly to summarize observations of metabolic rate in livestock or in man exposed to the relatively simple environment of a calorimeter. In the real world, metabolic rate depends on several microclimatic factors in addition to air temperature, notably radiation, windspeed, and vapor pressure. Attempts have therefore been made to replace the simple measure of actual air temperature used in the previous discussion with an effective temperature that incorporates the major elements of microclimate. Two examples are now given, the first dealing with the radiative component of the heat balance equation.

RADIATION INCREMENT

When the radiant flux is expressed in terms of net isothermal radiation, the heat balance equation may be written

$$R_{ni} + (M - \lambda E) = \rho c_p (T_0 - T) / r_{HR} \qquad (14.13)$$

where λE is here assumed a relatively small term, independent of T. Suppose that the thermal effect of radiation on an organism can be handled by substituting for air temperature an *effective temperature* T_f such that

$$(M - \lambda E) = \rho c_p (T_0 - T_f) / r_{HR} \qquad (14.14)$$

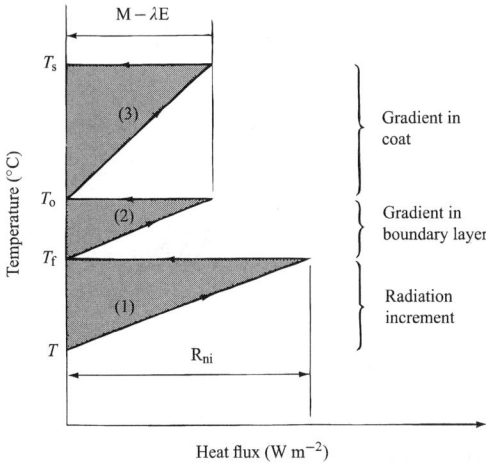

FIGURE 14.6 Cold limit and lower critical temperature for three species (from Mount, 1979) with corresponding values of $r_t\mathbf{M}$ (see text).

Then eliminating $\mathbf{M} - \lambda\mathbf{E}$ from Equations 14.13 and 14.14 gives

$$T_f = T + \left\{ \mathbf{R}_{ni}r_{HR}/\rho c_p \right\} \tag{14.15}$$

where the term in curly brackets is similar to the radiation increment used by Burton and Edholm [1955], Mahoney and King [1977], and many others.

Figure 14.2 provides a graphical method of estimating the temperature increment $T_f - T$ when windspeed and net radiation are known. Given the characteristic dimension of an organism d and the velocity of the surrounding air V, the corresponding value of r_H can be read from the left-hand vertical axis and r_{HR} derived from Equation 13.2 can be read from the right-hand axis. From the right-hand section of the figure, the coordinates of r_{HR} and $\mathbf{R}_{ni}$ define a unique value of $T_f - T$. An example is shown for $V = 8$ m s^{-1}, $d = 2$ cm, which gives $1/r_H = 6.8$ cm s^{-1} from the left-hand axis and $1/r_{HR} = 6.3$ cm s^{-1} from the right-hand axis. At $\mathbf{R}_{ni} = 300$ W m^{-2}, $T_f - T = 4$ K.

In a system where the resistances are fixed, the relation between temperature gradients and heat fluxes can be displayed by plotting temperature against flux as in Figure 14.6. By definition, a resistance is proportional to a temperature difference divided by a flux, and is therefore represented by the slope of a line in the figure. From a start at the bottom left-hand corner, T_f is determined by drawing a line (1) with slope $r_{HR}/\rho c_p$ to intercept the line $x = \mathbf{R}_{ni}$ at $y = T_f$. The equation of this line is

$$T_f - T = r_{HR}\mathbf{R}_{ni}/\rho c_p \qquad (14.16)$$

The temperature of the surface T_0 is now found by drawing a second line (2) with the same slope as (1) to intersect $x = \mathbf{M} - \lambda\mathbf{E}$ at $y = T_0$. The equation of this line is

$$T_0 - T_f = r_{HR}(\mathbf{M} - \lambda\mathbf{E})/\rho c_p \qquad (14.17)$$

Finally, for an animal covered with a layer of hair, a mean skin temperature T_s can be determined if the mean coat resistance r_c is known. Provided that evaporation is confined to the surface of the skin and the respiratory system, the increase of temperature through the coat is represented by the line (3) whose equation is

$$T_s - T_0 = r_c(\mathbf{M} - \lambda\mathbf{E})/\rho c_p \qquad (14.18)$$

This form of analysis can be used to solve two types of problem. When the environment of an animal is prescribed in terms of windspeed, temperature and net radiation, it is possible to use a thermo-neutral diagram to explore the physiological states in which the animal can survive. Conversely, when physiological conditions are specified, a corresponding range of environmental conditions can be established, forming an ecological "niche" or "climate space" [Gates, 1980]. Examples now given for a locust, a sheep, and a man are based on case studies reported in the literature.

14.4   Case Studies

LOCUST

In a monograph describing the behavior of the Red Locust (*Nomadacris septemfasciata*), Rainey, Waloff, and Burnett [1957] derived radiation budgets and heat balances for an insect of average size basking on the ground or flying. Both states are represented in Figure 14.7. At the bottom of the diagram, the basking locust is exposed to an air temperature of $20°C$ but the effective temperature of the environment is about $25°C$ ($\mathbf{R}_{ni} = 150$ W m^{-2}, $r_{HR} = 0.38$ s cm^{-1}). Because $\mathbf{M} - \lambda\mathbf{E}$ is trivial, the surface temperature is only a fraction of a degree above T_f. The flying locust is exposed to a larger radiant flux density (230 W m^{-2}), but with a relative windspeed of 5 m s^{-1} the value of r_{HR} is 0.33 s cm^{-1}, somewhat smaller than for the basking locust. The increment $T_f - T$ is therefore only slightly

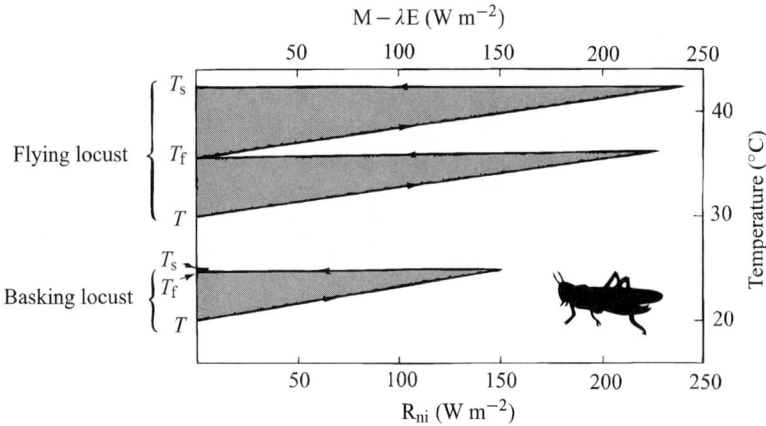

FIGURE 14.7 Temperature/heat-flux diagram for locust basking (lower section of graph) and flying (upper section).

larger than for basking. The difference between body surface and effective temperature is relatively large for the flying locust because the rate of metabolism is much larger than the basal rate; from measurements of oxygen consumption in the laboratory, M is estimated to be 270 W m^{-2}. The graph predicts that a flying locust should be about 12°C hotter than the surrounding air, consistent with maximum temperature excesses measured by sticking hypodermic thermocouple needles into locusts captured in the field.

14.5 Sheep

Figure 14.8 shows the heat balance of sheep with fleeces 1, 4, and 8 cm long exposed to

(i) An air temperature of -10°C and a net radiative *loss* of $R_{ni} = -50$ W m^{-2}.

(ii) An air temperature of 40°C and a radiative *gain* of 160 W m^{-2}.

In both cases, the animal is assumed to behave like a cylinder with a diameter of 50 cm exposed to a wind of 2 m s^{-1}, a special case of the conditions analysed by Priestley [1957].

For the cold state, T_f is found by drawing a line with slope $r_{HR}/\rho c_p$ to give $T - T_f = 3$°C, at $R_{ni} = -50$ W m^{-2}. From $T_f = -13$°C, three lines

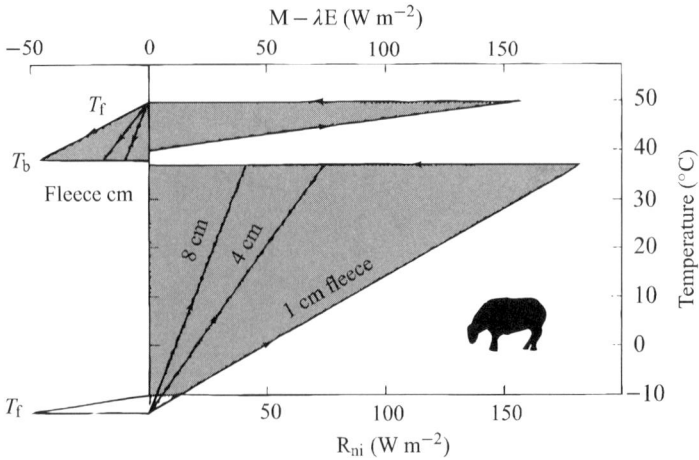

FIGURE 14.8   Temperature/heat-flux diagram for sheep with fleece lengths of 1, 4, and 8 cm exposed to air temperatures of $-10°C$ and net radiation of -50 W m^{-2} (lower section); and $40°C$ with net radiation 160 W m^{-2} (upper section).

were drawn corresponding to the total resistance $(r_{HR} + r_c)$ for the three values of fleece length taking r_c as 1.5 s cm^{-1} per cm length from Table 10.2. For the heat balance equation in the form

$$T_b - T_f = (M - \lambda E)(r_{HR} + r_c)/\rho c_p \qquad (14.19)$$

and if $T_b = 37°C$, the lines intercept $T = 37°C$ at three values of $M - \lambda E$; 42, 74, and 182 W m^{-2}. As the average daily metabolism of healthy, well-fed sheep is expected to be about 60 or 70 W m^{-2}, the graph implies that a fleece at least 4 cm long is needed to withstand effective temperatures between -10 and $-15°C$.

For the hot state, the top section of the graph shows that T_f is $50°C$ in the conditions chosen.

Thermal equilibrium cannot be achieved when T_f exceeds T_b unless $M - \lambda E$ is negative; i.e., unless more heat is lost by evaporation than is generated by metabolism. Assuming $T_b = 38°C$ in this case, $M - \lambda E$ would need to be -46 W m^{-2} if the fleece length is 1 cm, decreasing to -10 W m^{-2} for an 8 cm fleece; i.e., there would be a flow of sensible heat into the sheep from its environment that would decrease as insulation increased. Studies on sheep in controlled environments show that λE can reach 90 W m^{-2} under extreme heat stress; but even during a period of minimum activity, M is unlikely to be less than 60 W m^{-2}. A figure of

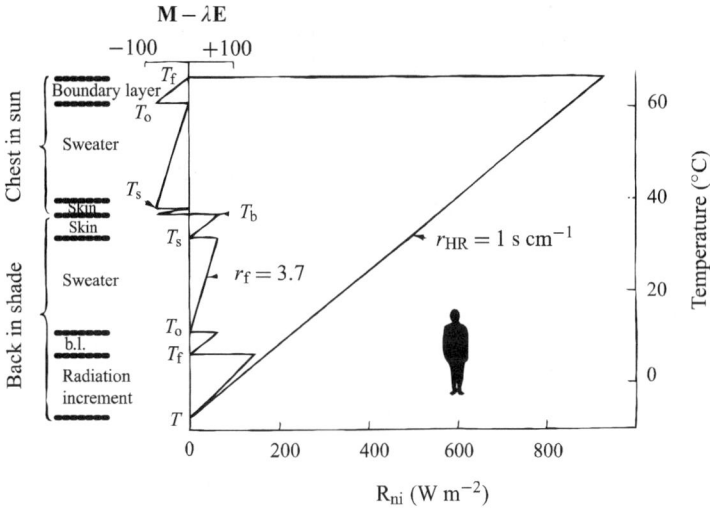

FIGURE 14.9 Temperature/heat-flux diagram for a man wearing a black sweater exposed to arctic sunlight and air temperature of $-7°C$. The lower part of the diagram shows the equivalent temperature and sweater surface temperature on his back (in shade), and the upper part shows the same temperature on his chest (in full sun).

-30 W m^{-2} can therefore be taken as a lower limit for $M - \lambda E$. The diagram implies that a minimum fleece length of about 2 cm would be needed to withstand the conditions chosen to represent heat stress.

MAN

The radiation and heat balance of men working in Antarctica were studied by Chrenko and Pugh [1961], and Figure 14.9 is based on their analysis for a man wearing a black sweater standing facing the sun. The air temperature was only $-7.5°C$, but because the sun was $22°$ above the horizon, the radiative load on vertical surfaces facing the sun was exceptionally large. As the wind was light, r_{HR} was relatively large, about 1 s cm^{-1}. The top left-hand side of the diagram, referring to the sunlit chest, was constructed from measured temperatures and from the heat flow through the clothing measured with a set of heat flow transducers.

The radiation increment was $74°C$, the surface of the sweater was at $61°C$, and the skin was at $38°C$. The conduction of heat into the body was -75 W m^{-2} and, assuming the resistance of the skin over the chest was

0.2 s cm^{-1}, the deep body temperature would be $37°$C. The same deep body temperature can be reached starting from the bottom left-hand side of the diagram representing the temperature gradients on the man's back. As the net radiation on the back was only 145 W m^{-2}, T_f was only $6.5°$C (cf. $74°$C on the chest); the outer surface of the sweater was at $12°$C and the skin was at $32°$C. There was an outward flow of heat through the skin of 60 W m^{-2} and, assuming $T_b = 37°$C, the resistance of the skin is about 1 s cm^{-1}, consistent with the value for vaso-constricted tissue in Table 10.2.

These measurements demonstrate a flow of solar energy through the trunk from chest to back and it would be necessary to integrate this heat flow over the whole body to determine $\overline{M} - \overline{\lambda E}$. Such an exercise is obviously impractical. The heat balance of an animal in a natural environment can be established when $\overline{M} - \overline{\lambda E}$ is measured directly or is estimated from relevant laboratory studies but the converse operation of determining $\overline{M} - \overline{\lambda E}$ when other terms in the heat balance are known is confined to men or animals in a calorimeter or controlled environment chamber.

APPARENT EQUIVALENT TEMPERATURE

A second type of environmental index is needed when the loss of latent heat from an organism is predominantly by the evaporation of sweat from the skin. When the loss of sensible heat occurs from the same surface, temperature and vapor pressure can be combined in a single variable that may be called the *apparent equivalent temperature*. To derive this quantity, the heat balance equation is written,

$$\mathbf{R_{ni}} + \mathbf{M} = \frac{\rho c_p (T_0 - T)}{r_{HR}} + \frac{\rho c_p (e_0 - e)}{\gamma r_v} \qquad (14.20)$$

where T_0 and e_0 are mean values for the skin surface and T and e refer to the air.

If γ is replaced by $\gamma^* = \gamma (r_v/r_{HR})$, Equation 14.20 can be written

$$\mathbf{R_{ni}} + \mathbf{M} = \rho c_p \left(T_{eo}^* - T_e^* \right) / r_{HR} \qquad (14.21)$$

where T_e^* is the apparent equivalent temperature of ambient air given by $T + e/\gamma^*$ and therefore equal to the equivalent temperature derived on p. 13 when $r_v = r_{HR}$. By analogy with the radiation increment, e/γ^* may be regarded as a humidity increment. The mean value of the apparent equivalent temperature at the surface is T_{eo}^*.

In principle, the apparent equivalent temperature should be used in a thermo-neutral diagram in place of conventional temperature when the metabolic rate of an animal is measured in an environment where vapor pressure changes and temperature. Extending this process a stage further, the apparent equivalent temperature of the environment can be modified to take account of the radiation increment by writing

$$T_{eR}^* = T_e^* + \mathbf{R}_{ni} r_{HR}/\rho c_p \qquad (14.22)$$

The quantity T_{eR}^* is an index of the thermal environment that allows the heat balance equation to be reduced to the form

$$(1 - x)\,\mathbf{M} = \rho c_p \left(T_{eo}^* - T_{eR}^*\right)/r_{HR} \qquad (14.23)$$

where x is the fraction of metabolic heat dissipated by respiration. A thermo-neutral diagram using T_{eR}^* as a thermal index rather than T or T_e^* would be valid for changes in radiant heat load and in vapor pressure and temperature.

This type of formulation provides a relatively straightforward way of investigating the heat balance of a naked animal whose skin is either dry or entirely wetted by sweat. The intermediate case of partial wetting is more difficult to handle because the value of T_e^* depends on fractional wetness which is not known *a priori*. A complete solution of the relevant equations for an animal leads to a complex expression within which the structure of the original Penman equation can be identified [McArthur, 1987].

SWEATING MAN

To illustrate how apparent equivalent temperatures can be used, the heat balance of a sweating man will be analysed graphically. The net heat **H** to be dissipated will be taken as the sum of the metabolic heat load **M** and the isothermal net radiation $\mathbf{R}_{ni}$, which can be calculated on the assumption that a body intercepts radiation like a cylinder (p. 104). The resistance to vapor transfer r_v is $d/(D\mathrm{Sh})$ where the characteristic dimension d is often taken as 34 cm for a standard man, and the resistances to heat transfer are $r_H\ (= d/(\kappa\mathrm{Nu})$ and r_R in parallel. For forced convection in a windspeed of about 2 m s^{-1}, r_H is 1 s cm^{-1} and with $r_H = 2.1$ s cm^{-1}, $r_{HR} = 0.68$ s cm^{-1}. The corresponding value of r_v is 0.9 s cm^{-1} so $\gamma^* = \gamma r_V/r_{HR} = 87$ Pa K^{-1}. These values will be taken as standard in the following discussion.

The rate at which sweat evaporates from a man in a given environment cannot be determined without knowing how fast it is produced. The maximum rate of sweating a normal man can sustain for several hours is about 1 kg h^{-1}, and if his surface area is 1.8 m^2 the equivalent rate of evaporative

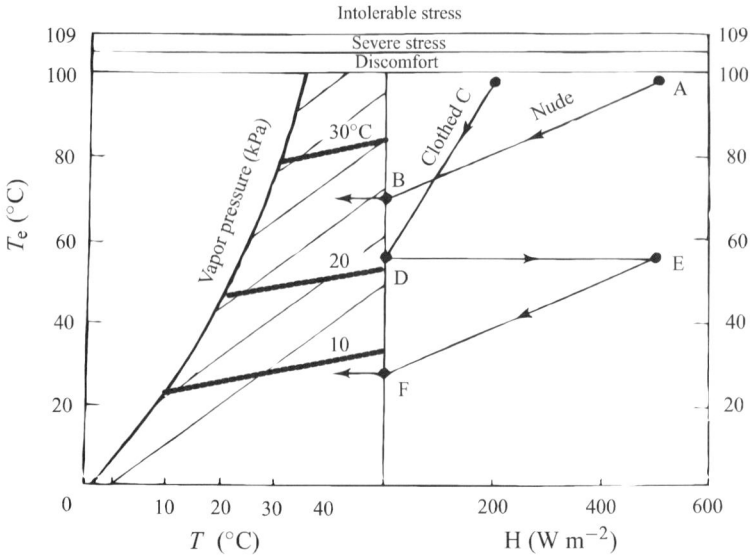

FIGURE 14.10 Apparent equivalent temperature and heat-flux diagram for a clothed and a nude man (right-hand section) and the relation between apparent equivalent temperature, vapor pressure, and air temperature (left-hand section). The nude man with a total heat load of 500 W m^{-2} can avoid discomfort if the apparent equivalent temperature of the environment is less than the value at B (70°C) but the clothed man with a heat load of only 200 W m^{-2} must stay in an environment with an apparent equivalent temperature less than 28°C (point F). The dashed lines are isotherms of wet bulb temperature.

heat loss is 375 W m^{-2}. The rate of sweating is determined partly by skin temperature and partly by metabolic rate. When a subject reports that a particular combination of clothing and environment is "comfortable" and there is no visible evidence of sweat, his mean skin temperature is usually about 32 to 33°C.

Laboratory experiments show that many subjects regard the environment as very uncomfortable when skin temperature rises above about 35°C (95°F) and as intolerable when the skin temperature reaches about 37°C (98°F). When γ^* is 87 Pa K^{-1} the corresponding equivalent temperatures for a wet surface are 100°C and 109°C. Figure 14.10 shows these limits on a diagram of the type already used to present "dry" heat balances, but the equivalent temperature T_e^* now replaces T as the ordinate. The left-hand side of the diagram shows how T_e^* is related to the temperature and vapor pressure of the ambient air. Three lines of constant wet bulb temperature

T' (10, 20, 30°C) are plotted as a reminder that T' is closely related to T_e^* when $\gamma \approx \gamma^*$.

The right-hand side of the diagram shows gradients of equivalent temperature for a nude and for a clothed subject, both assumed to be sweating freely when M is 200 W m^{-2} (light work) and $\mathbf{R}_{ni}$ is 300 W m^{-2} (bright sunshine). The mean value of T_{e0}^* for the skin of the nude man is assumed to be 98°C. The equivalent temperature of the air T_e^* must therefore satisfy Equation 14.21 where $\rho c_p/r_{HR} = 17.7$ W m^{-2} K^{-1}. Thus, when $\mathbf{H} = \mathbf{R}_{ni} + \mathbf{M} = 5000$ W m^{-2}, $T_{e0}^* - T_e^*$ is 28.5 K, T_e^* is 69.5°C and Equation 14.21 is represented by the line AB. Reference to the left side of the figure shows the range of air temperature and humidity for which T_e^* is 70°C; e.g., 35°C, 3 kPa. The increase of $\mathbf{R}_{ni} + \mathbf{M}$ that would make the same environment seem either "severe" or "intolerable" can be derived from the figure by extending the line BA upwards. Conversely if $\mathbf{R}_{ni} + \mathbf{M}$ is constant, the effect of increasing T_e^* by raising T or e is found by displacing AB upward without changing its slope.

The effect of clothing can be demonstrated on the same diagram if γ^* is assumed to have the same value between the skin and the surface of the clothing as it has in the free atmosphere and if all the radiation from the environment is assumed to be intercepted at the surface of the clothing. Then the flux of heat from the wet skin (at an equivalent temperature T_{ec}^*) to the surface of the clothing (at T_{e0}) is given by

$$\mathbf{M} = \rho c_p \left(T_{e0}^* - T_{ec}^* \right) / r_c \tag{14.24}$$

The diffusion resistance of the clothing r_c is taken as 2.5 s cm^{-1} (equivalent to about 1 clo, p. 181) so the factor $(\rho c_p/r_c)$ is 4.8 W m^{-2} K^{-1}. The line CD drawn with this slope shows that when T_{e0}^* is 98°C, T_{ec}^* is 56°C. To represent the gradient of T_e^* from the surface of the clothing to the ambient air, the line EF is drawn parallel to AB. The presence of clothing decreases the equilibrium value of T_e^* from 70°C to 27°C and, because CD is much steeper than AB, a severe or intolerable heat stress would be imposed by a relatively small increase of metabolic rate.

The atmospheric conditions for thermal equilibrium can be represented to a good approximation by a wet bulb temperature of 25°C for the nude subject and 10°C for the clothed subject. The wet bulb temperature has frequently been used as an index of environmental temperature in human studies. Figure 14.10 confirms that it is a good index for a fully wetted skin provided $r_{HR} \approx r_V$, but it is inappropriate when the skin temperature is below the limit for rapid sweating.

HEAT BALANCE OF LONG DISTANCE RUNNERS

The thermo-neutral diagram (Figure 14.5) demonstrates that when an athlete is exercising in hot environments for long periods, almost all of the heat liberated by metabolic processes must be balanced by evaporation of sweat if the body temperature is to remain within safe limits. Neilsen [1996] applied these principles to explore the limits the environment could impose on endurance competitions such as marathon running and bicycle racing. She based her calculations on a hypothetical world-class marathon runner, weight 67 kg, surface area 1.85 m^2, running at a mean speed of 5.4 m s^{-1} (i.e., completing a marathon race in 2 hours 10 minutes). The metabolic heat production associated with running is about 4 kJ (kg body weight)$^{-1}$ km^{-1}, so the runner would generate about 780 W m^{-2} of heat. The range of air temperatures and humidities in which the runner could achieve equilibrium could be investigated using apparent equivalent temperature and a heat flux diagram as in the previous example. For simplicity we analyze the extreme case Neilsen considered, of a runner in an environment where air temperature and skin temperature are 35°C. In this situation, the entire heat load (metabolic heat plus net radiation) must be dissipated by evaporation of sweat if the body temperature is to remain stable. Equation 14.21 therefore becomes

$$\mathbf{M} + \mathbf{R}_{ni} = \lambda \mathbf{E} = \rho c_p \left(T_{eo}^* - T_e^* \right) / r_{HR} \qquad (14.25)$$

Treating the runner as a cylinder, diameter 34 cm, and assuming a relative windspeed of 5.4 m s^{-1}, the boundary layer resistance to heat transfer r_H is 49 s m^{-1}, and to vapor transfer is 46 s m^{-1} (p. 192). The resistance to radiative transfer is 170 s m^{-1}, giving $r_{HR} = 38$ s m^{-1}. The modified psychrometer constant $\gamma^* = \gamma \left(r_V / r_{HR} \right) = 81$ Pa K^{-1}. Assuming that air next to the skin is saturated at 35°C, T_{eo}^* is 104°C. Equation 14.25 then gives $T_e^* = 79$°C when $\mathbf{R}_{ni} = 0$ W m^{-2}, and $T_e^* = 70$°C when $\mathbf{R}_{ni} = 300$ W m^{-2}. When the air temperature is 35°C, these apparent equivalent temperatures cannot be reached if the relative humidity exceeds critical values.

Figure 14.11 demonstrates that these critical values are about 70% when $\mathbf{R}_{ni} = 0$, and 60% when $\mathbf{R}_{ni} = 300$ W m^{-2}. The corresponding values of $\mathbf{E}$ are about 2.1 l h^{-1} and 3.0 l h^{-1}. At greater humidities, this work rate would cause life-threatening hyperthermia as the excess heat was stored in the body. Maximum sweat rates of acclimated, trained athletes can reach about 3 l h^{-1}, but are eventually limited by the inability of the gut to absorb water at more than about 1 l h^{-1}. These calculations demonstrate that it is impossible for even the best-trained athlete to

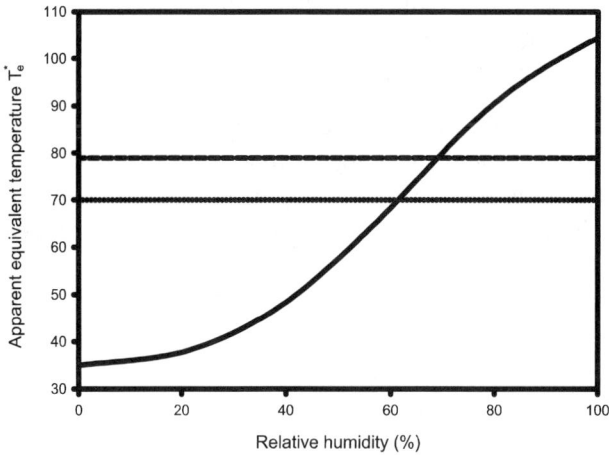

FIGURE 14.11 Variation of apparent equivalent temperature T_e^* with relative humidity for a marathon runner when air temperature is $35°C$ (see text). The dashed lines are the solutions to the heat balance equation when $M + R_{ni} = 780$ (– – –) and 1080 W m^{-2} ($\cdots$) respectively.

achieve a world class marathon time in a hot, humid climate, and they are are a reminder of the extreme risks that less fit individuals may face when exercising in hot weather.

14.6 Problems

1. The metabolic heat production **M** of a cow standing in strong sunshine at air temperature $22.0°C$ is 140 W m^{-2}. The average solar irradiance of the animal is 300 W m^{-2} and the effective radiative temperature of the surroundings is equal to air temperature. The mean temperature of the skin and coat surface are 34.0 and $31.6°C$ respectively, and the rates of evaporative loss from the respiratory system and the skin surface are 4.5×10^{-3} g m^{-2} s^{-1} and 40×10^{-3} g m^{-2} s^{-1} respectively. Calculate the thermal resistances of (i) the boundary layer around the cow and (ii) the coat. If shading reduces the average solar irradiance of the animal by two thirds, and other environmental variables remain unchanged, calculate the corresponding reduction in the rate of evaporation from the skin necessary to maintain a steady heat balance if **M** remains constant.

(Assume that changes in respiratory evaporation can be neglected and that the absorption of solar radiation occurs close to the outer surface of the coat, which has a reflection coefficient of 0.40)

2. Pigs wallow in wet mud to increase body cooling by evaporation. Calculate the maximum air temperature at which a pig could maintain a state of thermal equilibrium (i) if the skin was completely dry, and (ii) if the skin was completely covered in wet mud. Assume the following:

Minimum metabolic heat production rate	60 W m^{-2}
Respiratory heat loss	10 W m^{-2}
Radiative heat load from the environment	240 W m^{-2}
Skin temperature	$33°C$
Vapor pressure of air	1.0 kPa
Combined resistance to radiative and convective heat transfer	80 s m^{-1}
Mean resistance to conduction through mud layer	8 s m^{-1}

Transient Heat Balance

In the last two chapters, it was possible to treat temperature as constant in time because, at every point in the systems considered, the input of heat energy was assumed to balance the output exactly. This state of equilibrium is rare in natural environments except in the depths of caves and in deep water. In the habitats of most plants and animals, air temperature has a marked diurnal variation, more or less in phase with radiation, with superimposed short-term fluctuations associated with changes in cloudiness and with turbulence. In this chapter, we consider the equations that describe how a system responds thermally to changes of external temperature, initially making the simplifying assumption that the system itself is isothermal and contains no source of heat. More complex cases are treated in detail by Gates [1980].

Examples are described for the three types of temperature change illustrated in Figure 15.1: an instantaneous or "step" change; a "ramp" change at steady rate; and an harmonic oscillation.

15.1 Time Constant

If we take first the simplest case of a system in which the latent heat exchanges are negligible, the steady state heat budget is simply

$$\mathbf{R}_{ni} = \mathbf{C} = \rho c_p \left(T_0 - T\right) / r_{HR} \tag{15.1}$$

where $\mathbf{R}_{ni}$ and $\mathbf{C}$ are fluxes per unit area, T_0 is mean surface temperature, T is air temperature, and r_{HR} is the corresponding resistance for the loss

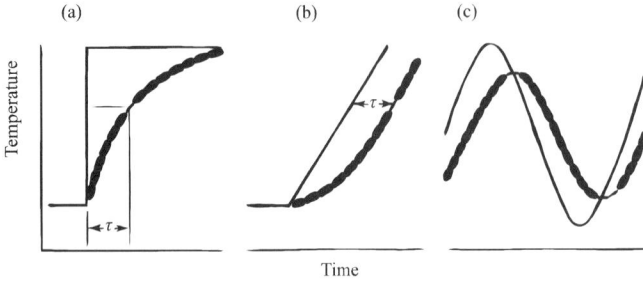

FIGURE 15.1   Change of surface temperature (dashed line) in response to change of environmental temperature (full line). (a) Step change (Equation 15.10); τ is the time for a fractional change of $1 - e^{-1}$ or 0.63. (b) Ramp change (Equation 15.16); τ is the constant time lag established after the term $\exp(-t/\tau)$ becomes negligible. (c) Harmonic oscillation for the case $\phi = \pi/4$ (Equation 15.20) (from Monteith, 1981 b).

of sensible heat and long-wave radiation. Introducing an effective temperature T_f defined by Equation 14.15, i.e.,

$$T_f = T + \left\{ \mathbf{R}_{ni} r_{HR}/\rho c_p \right\} \tag{15.2}$$

the heat budget equation can be reduced to

$$T_0 = T_f \tag{15.3}$$

A similar equation can be written for a system in which the rate of evaporation is determined by a resistance r_v so that the heat budget can be expressed as

$$\mathbf{R}_{ni} = \mathbf{C} + \lambda \mathbf{E} = \rho c_p \left(T_{e0}^* - T_e^* \right)/r_{HR} \tag{15.4}$$

where T_e^* is an apparent equivalent temperature (see p. 276). Equation 15.3 can now be re-written as

$$T_{e0}^* = T_{eR}^* \tag{15.5}$$

where

$$T_{eR}^* = T_f + e/\gamma^* \tag{15.6}$$

For simplicity, however, we shall return to Equation 15.3 and postulate that, as a consequence of an increase either in $\mathbf{R}_{ni}$ or in T, T_f increases to a new value T_f'. If the system has a finite heat capacity, T_0 will not instantaneously increase to T_f' but will approach it at a rate depending on the physical properties of the system. If the heat capacity per unit *area* of the system is C, heat will be stored at a rate $C\partial T_0/\partial T$ so that the heat budget equation becomes

$$\mathbf{R}_{ni} = \mathbf{C} + C\partial T_0/\partial T \qquad (15.7)$$

Substituting $\mathbf{C}$ from Equation 15.1 and $\mathbf{R}_{ni}$ from 15.2 in Equation 15.7 leads to

$$\partial T_0/\partial T = \left(T'_f - T_0\right)/\tau \qquad (15.8)$$

where τ, known as the "*time constant*" of the system because it has dimensions of time, is given by

$$\tau = C r_{HR}/\left(\rho c_p\right) \qquad (15.9)$$

In vegetation, the heat capacity per unit *volume* ranges from about 1 MJ m^{-3} K^{-1} in the heartwood of red pine to 2 to 3 MJ m^{-3} K^{-1} for organs such as leaves and fruits consisting mainly of water (which has a heat capacity of 4.2 MJ m^{-3} K^{-1}). Corresponding time constants are of the order of a few seconds for small leaves, a few minutes for large leaves, and several hours for the trunks of trees [Monteith, 1981b]. Values reported for animals range from about 9 min for a cockroach [Buatois and Croze, 1978], 0.5 h for a shrew, 2 h for a large Cardinal bird, and 330 h for a sheep [Gates, 1980].

15.2 General Cases

STEP CHANGE

When the effective temperature changes instantaneously from T_f to T'_f, the boundary conditions for the solution of Equation 15.8 are

$$T_0 = T_f \quad t = 0$$
$$T_0 = T'_f \quad t = \infty$$

and the solution is

$$T_0 = T'_f - \left(T'_f - T_f\right) \exp\left(-t/\tau\right) \qquad (15.10)$$

Examples

Leaves

Linacre [1972] measured the mean temperature of vine leaves that he covered with vaseline to stop transpiration and then suddenly shaded or unshaded to produce a step change of $\mathbf{R}_{ni}$ and therefore of $\mathbf{C}$ (Figure 15.2).

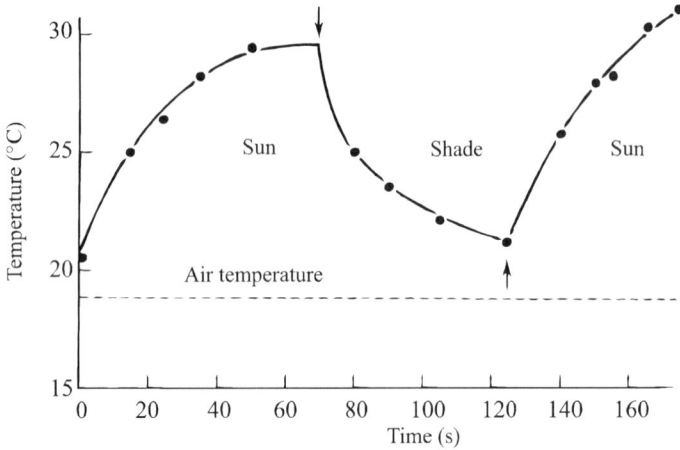

FIGURE 15.2   Heating and cooling of a pepper leaf in sun and shade. (Redrawn by Gates [1980] from an example by Ansari and Loomis).

In one set of measurements the mean value of τ was 20 s and the value of $C/\rho c_p$ was 720 (J m^{-2} K^{-1})/1200(J m^{-3} K^{-1}) = 0.6 m. The corresponding resistance was

$$r_{HR} = \tau \rho c_p/C = 33 \text{ s m}^{-1}$$

Linacre also measured time constants for leaves sprayed with water. Analysis of this case can proceed by replacing T by the equivalent temperature $T + e/\gamma$, assuming that resistances for heat and vapor transfer are the same. For a transpiring leaf with $r_v > r_H$, it would be possible to determine by trial the value of γ^* needed to obtain the same time constant as for a non-transpiring leaf and the stomatal resistance could then be calculated as

$$r_v - r_H = r_H \left(\gamma^*/\gamma - 1\right) \tag{15.11}$$

Animals

Grigg et al [1979] exposed specimens of the Eastern Water Dragon to step changes of temperature and measured time constants (for heating) which increased from about 4 to 8 min as body mass increased from 140 to 590 g. In conflict with elementary theory, the time constant for cooling was longer by about 30%, a difference accounted for by a more realistic model in which an isothermal body core is surrounded by an insulating layer.

The theory that describes the response of an organism to step changes of temperature is particularly relevant to the thermal regime of an animal which "shuttles" between a cool and a hot microclimate in order to obtain food. Bakken and Gates [1975] and Porter et al [1973] have explored the thermal implications of shuttling by lizards in the desert. Willmer et al [2005] provided a comprehensive review of how animals adapt through behavior and physiology to extreme environments.

Streams

Increases in the water temperature of streams occur when removal of riparian forests increases radiation loads. As a stream enters a cleared area from a dense forest there is a step change in net radiation. Such change may influence fish habitat because fish survival, growth and reproduction are all influenced by water temperature [Fagerlund et al, 1995]. Sinokrot and Stefan [1993] estimated that the time constant of streams was about 40 h per meter of water depth, indicating that water temperatures do not adjust rapidly to a step change in energy input. Brown [1969] compared the heat balance components of small streams shaded by Douglas fir forest, and running through a clearcut. The maximum rate of temperature increase in water traveling through a clearcut area was about 9°C per hour, whereas stream temperature in a shaded reach increased by less than 1°C per hour. If a stream is treated for simplicity as a uniform, well-mixed flow of constant cross-section and velocity V, and turbulent dispersion in the downstream direction x is ignored, the heat transport equation is

$$\frac{\partial T}{\partial t} = -V\frac{\partial T}{\partial x} + \frac{\mathbf{S}_a + \mathbf{S}_b}{\rho_w c_p d} \tag{15.12}$$

where T is water temperature, d is stream depth, $\mathbf{S}_a$ and $\mathbf{S}_b$ are the rates of transfer into the stream water of heat exchanged across the upper and lower interfaces (with the atmosphere and stream bed respectively), and $\rho_w c_p$ is the volumetric heat capacity of water. The heat budget at the stream-atmosphere interface is

$$\mathbf{R}_{na} = \mathbf{C}_a + \lambda\mathbf{E} + \mathbf{S}_a \tag{15.13}$$

and at the water-stream bed interface is

$$\mathbf{R}_{nb} = \mathbf{S}_b + \mathbf{G} \tag{15.14}$$

where $\mathbf{R}_n$ and $\mathbf{C}$ are net radiation and sensible heat fluxes at the upper (a) and lower (b) interfaces, respectively, $\lambda\mathbf{E}$ is the latent heat flux to the atmosphere, and $\mathbf{G}$ is the heat flux conducted into the stream bed. In a shallow, clear stream, reflection at the surface and absorption of shortwave radiation by the water are small (absorption of total solar radiation in clear

water is about 4% per metre depth), so net radiative absorption directly
by the stream is unlikely to differ much from the net longwave radiation,
typically around -100 W m^{-2} under cloudless sky and about zero under
a tree canopy. Brown [1969] reported that C_a and λE were small terms
in a forest environment, but they may be large, and in opposing directions
when unshaded streams run through hot dry landscapes. The rate of solar radiation absorption at the stream bed may be large, depending on the
clarity of the stream and the shortwave reflection coefficient of the stream
bed material. Brown [1969] and Sinokrot and Stefan [1993] found that
the rate of conduction G of heat into the bedrock of shallow, clear streams
was an important term in the energy budget. A step change in radiation is
therefore the driving term influencing the temperature response of a stream
emerging from or entering shade, but the relative importance of the terms
G and S_b in Equation 15.14, that are likely to strongly influence the rate of
increase of stream temperature, must depend on the thermal conductivity
of the stream bed and the boundary layer resistance at the bedrock-water
interface. The heat budget of streams is a topic more environmental physicists should consider dipping in to.

RAMP CHANGE

When the rate of change of air temperature (or effective temperature) is α,
so that the temperature at time t is

$$T(t) = T(0) + \alpha t \qquad (15.15)$$

the solution of Equation 15.8 is

$$T_0(t) = T(0) + \alpha t - \alpha \tau \{1 - \exp(-t/\tau)\} \qquad (15.16)$$

where the heating rate is

$$\partial T_0/\partial t = \alpha \{1 - \exp(-t/\tau)\} \qquad (15.17)$$

This equation shows that the rate of heating of the system is zero initially,
increasing to α when t/τ is large. To be more specific, when t/τ exceeds
3, the exponential term is less than 0.05 and can therefore be neglected.
Equation 15.16 then becomes

$$T_0(t) = T(0) + \alpha(t - \tau) \qquad (15.18)$$

showing that the system heats at the same rate as its environment (Equation
15.15) but with a time-lag of τ and therefore with a temperature lag of αT.

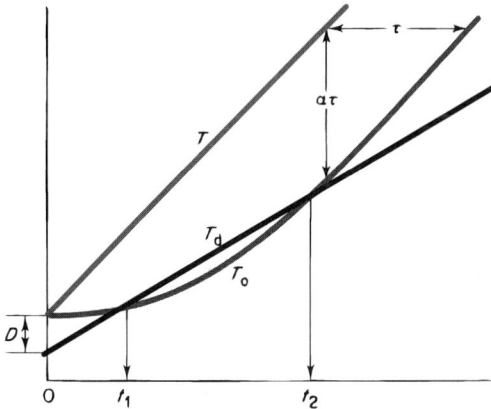

FIGURE 15.3 Idealized representation of the increase of air temperature (T), dew-point temperature (T_d) and pod temperature (T_0) when the dew-point depression is D at sunrise ($t = 0$). Condensation begins at t_1 and stops at t_2. For other symbols, see text (from Monteith and Butler, 1979).

Equation 15.16 has been used to investigate the phenomenon of condensation on cocoa pods in Bahia, Brazil, occurring after sunrise as a consequence of thermal inertia [Monteith and Butler, 1979]. In humid weather, the increase in pod temperature after dawn is initially much slower than the increase in air temperature (1.5 to 2.5 K h^{-1}), and pod surface temperature can therefore fall below the dew-point of the air, which increases at about 1 to 2 K h^{-1}. Condensation then begins and continues until the temperature of the pod surface eventually rises above the dew-point, after which the film of dew evaporates (Figure 15.3).Values of τ determined experimentally ranged from 0.5 to 1.5 h depending on pod size and windspeed.

By calculating the rate of condensation from the heat balance of a pod, it was possible to show that the mean thickness of the wet layer would often be between 10 and 20 μm, persisting for several hours. The existence of such a layer may play an important role in the germination of zoospores of the Black Pod fungus responsible for large losses of yield in some areas.

The same type of condensation must often occur on the trunks of trees but is rarely visible because of the rough nature of the surface. Unsworth et al [2004] observed that air in an old-growth Douglas fir forest canopy became saturated on summer mornings shortly after sunrise, and speculated that condensation on tree trunks could be an important factor for bark-dwelling organisms in the dry summers of the Pacific Northwest.

HARMONIC CHANGE

A harmonic change of air temperature has the form

$$T = \bar{T} + A \, \sin 2\pi t / P \qquad (15.19)$$

where $\bar{T}$ is a mean temperature and T oscillates between $\bar{T} + A$ and $\bar{T} - A$ (see Figure 15.1) with a period P, expressed in the same units as time. For an isothermal system whose surface temperature is governed by Equation 15.8, the solution of the equation is

$$T_0 = \bar{T} + A' \sin \{(2\pi t / P) - \phi\} \qquad (15.20)$$

where the amplitude of the surface temperature is

$$A' = A \cos \phi \qquad (15.21)$$

and ϕ, known as a "*phase lag*," has a value of

$$\phi = \tan^{-1} (2\pi \tau / P) \qquad (15.22)$$

In the natural environment, the diurnal change of temperature, though not a true sine wave, may be treated as one to demonstrate principles. For $P = 24$ h, ϕ is small and $A' \approx A$ unless τ is at least several hours, a figure appropriate for tree trunks or large fruit. In such cases, however, it is unrealistic to assume the system is isothermal, and although simple theory may give an approximate solution for conditions at the surface, it is not appropriate elsewhere. Herrington [1969] solved more complex equations for the radial flow of heat in a tree trunk and compared predictions with measurements in a pine tree. For the trunk surface temperature, A'/A was about 0.75 compared with 0.63 from Equation 15.21 but, at the center of the trunk, A'/A was much larger than the estimated value.

15.3   Heat Flow in Soil

The vertical flow of heat in soil provides a relevant and very important example of how a transient heat balance can be established in a system where temperature is a function of position and changing harmonically with time. As an introduction to the subject, the dependence of soil thermal properties on water content and mineral composition will be considered before deriving the appropriate differential equation for a soil in which thermal properties are assumed uniform with depth.

SOIL THERMAL PROPERTIES

By use of the symbol ρ for density and c for specific heat, the solid, liquid, and gaseous components of soil can be distinguished by subscripts s, l, and g. If the volume fraction x of each component is expressed per unit volume of bulk soil

$$x_s + x_l + x_g = 1 \qquad (15.23)$$

For a completely dry soil ($x_l = 0$), x_g is the space occupied by pores. In many sandy and clay soils x_g is between 0.3 and 0.4, and it increases with organic matter content, reaching 0.8 in peaty soils.

The *bulk density* of a soil ρ' is found by adding the mass of each component; i.e.,

$$\rho' = \rho_s x_s + \rho_l x_l + \rho_g x_g = \sum (\rho x) \qquad (15.24)$$

Because ρ_g for the soil atmosphere is much smaller than ρ_s or ρ_l, the term $\rho_g x_g$ can be neglected. When ρ_s and x_s are constant, soil bulk density increases linearly with the liquid fraction x_l, but, in a soil that swells when it is wetted, the relation is not strictly linear.

The *volumetric specific heat* (J m^{-3} K^{-1}) is the product of bulk density ρ' (kg m^{-3}) and *bulk specific heat* c' (J kg^{-1} K^{-1}). It can be found by adding the heat capacity of soil components to give

$$\rho' c' = \rho_s c_s x_s + \rho_l c_l x_l + \rho_g c_g x_g = \sum (\rho c x) \qquad (15.25)$$

and this quantity increases linearly with water content in a non-swelling soil. The bulk specific heat is therefore

$$c' = \sum \rho c x / \sum \rho x \qquad (15.26)$$

Thermal properties of soil constituents and of three representative soils are listed in Table 15.1. Quartz and clay minerals, which are the main solid components of sandy and clay soils, have similar densities and specific heats. Organic matter has about half the density of quartz but about twice the specific heat. As a result, most soils have volumetric specific heats between 2.0 and 2.5 MJ m^{-3} K^{-1}. As the specific heat of water is 4.18 MJ m^{-3} K^{-1}, the heat capacity of a dry soil increases substantially when it is saturated with water.

The dependence of *thermal conductivity* k' (W m^{-1} K^{-1}) on water content is more complex. The thermal conductivity of a very dry soil may increase by an order of magnitude when a small amount of water is added because relatively large amounts of heat can flow through the soil by the evaporation and condensation of water in the pores. For a sandy soil, for example, k' may increase from 0.3 to 1.8 W m^{-1} K^{-1}

Table 15.1 Thermal properties of soils and their componenets (after van Wijk and deVries, 1963).

	Water content x_l	Density ρ 10^6 g m^{-3}	Specific heat c J g^{-1}K^{-1}	Thermal conductivity k' W m^{-1}K^{-1}	Thermal diffusivity κ' 10^{-6} m^2 s^{-1}
(a) Soil components					
Quartz		2.66	0.80	8.80	4.18
Clay minerals		2.65	0.90	2.92	1.22
Organic matter		1.30	1.92	0.25	0.10
Water		1.00	4.18	0.57	0.14
Air (20°C)		1.20×10^{-3}	1.01	0.025	20.50
(b) Soils					
Sandy soil (40% pore space)	0.0	1.60	0.80	0.30	0.24
	0.2	1.80	1.18	1.80	0.85
	0.4	2.00	1.48	2.20	0.74
Clay soil (40% pore space)	0.0	1.60	0.89	0.25	0.18
	0.2	1.80	1.25	1.18	0.53
	0.4	2.00	1.55	1.58	0.51
Peat soil (80% pore space)	0.0	0.26	1.92	0.06	0.10
	0.4	0.66	3.30	0.29	0.13
	0.8	1.06	3.65	0.50	0.12

when x_1 increases from zero to 0.2. With a further inçrease of x_1, from 0.2 to 0.4, the corresponding increase of k' is much smaller because the diffusion of vapor becomes increasingly restricted as more and more pores are filled with water. The conductivity of very wet soils is therefore almost independent of water content.

When water is added to a very dry soil, k' increases more rapidly than $\rho'c'$ initially so that the *thermal diffusivity* $\kappa' = k'/\rho'c'$ also increases with water content. In a very wet soil, however, the increase of k' with water content is much less rapid than the increase of $\rho'c'$ so that κ' decreases with water content. Between these two regimes, κ' reaches its maximum at a point where an increase of water content is responsible for equal fractional increases of κ' and $\rho'c'$. Table 15.1 shows that sandy soils tend to have larger thermal diffusivities than other soil types because quartz has a much larger thermal conductivity than clay minerals. Peat soils have the smallest diffusivities because the conductivity of organic matter is relatively small.

FORMAL ANALYSIS OF HEAT FLOW

At a depth z below the soil surface, the downward flux of heat can be written

$$\mathbf{G}(z) = -k'(z)(\partial T/\partial z) \qquad (15.27)$$

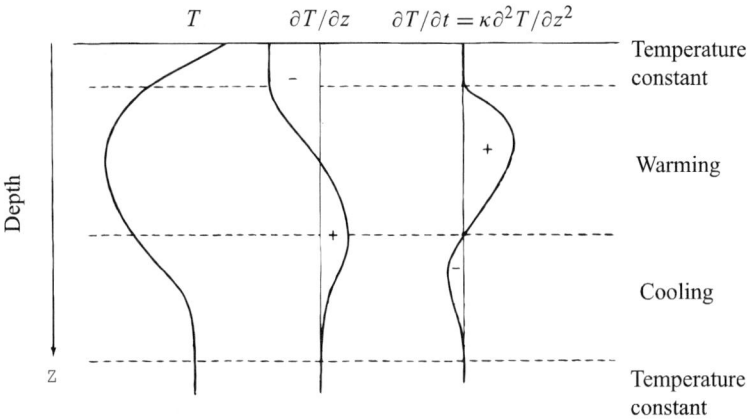

FIGURE 15.4 Imaginary temperature gradient in soil (left-hand curve), and the corresponding first and second differentials of temperature with respect to depth; i.e., $\partial^2 T/\partial z^2$. The second differential is proportional to the rate of temperature change $\partial T/\partial t$.

(The negative sign on the right-hand side of this equation is a convention that ensures that **G** is positive when temperature declines as depth increases.) In any thin layer of thickness Δz, say, the difference between the flux entering the layer at level z and leaving at $z + \Delta z$ is $\mathbf{G}(z) - \mathbf{G}(z + \Delta z)$ or, in the notation of calculus, $\Delta z(\partial \mathbf{G}(z)/\partial z)$. The sign of this quantity determines whether there is net gain of flux or "*convergence*" in the layer producing a local increase of soil temperature, or a net loss of flux or "*divergence*" producing a fall in temperature. In general, the rate at which the heat content of the layer changes can be written $\partial(\rho'c'T\,\Delta z)/\partial t$, and this quantity must be equal to the change of flux with depth; i.e.,

$$\frac{\partial \mathbf{G}(z)}{\partial z}\Delta z = \frac{\partial}{\partial z}\left(-k'\frac{\partial T}{\partial z}\right) = -\frac{\partial\left(\rho'c'T\right)}{\partial t}\Delta z \qquad (15.28)$$

For the special case in which the physical properties of the soil are constant with depth, the equation of heat conduction (15.28) reduces to

$$\frac{\partial T}{\partial t} = \kappa'\frac{\partial^2 T}{\partial z^2} \qquad (15.29)$$

Figure 15.4 is a graphical demonstration of this equation in terms of an imaginary temperature profile in a soil with constant diffusivity. Figure 15.5 shows the observed change of temperature beneath a bare soil surface

Local time

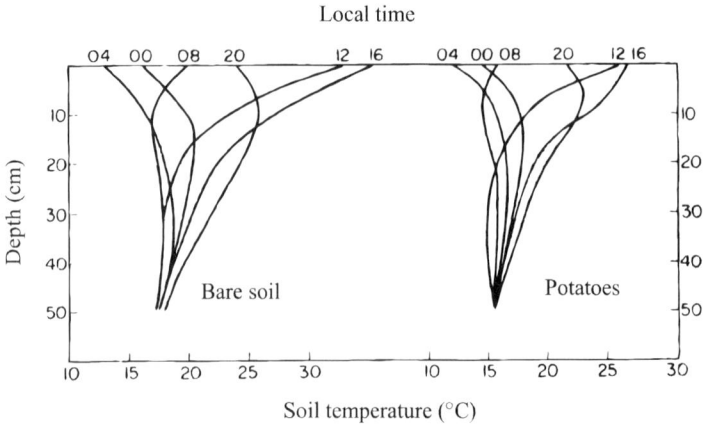

FIGURE 15.5   Diurnal change of soil temperature measured below a bare soil surface and below potatoes (from van Eimern, 1964).

and beneath a crop. Observed changes of temperature at different depths can be compared with the changes predicted from temperature gradients.

In most soils, composition, water content, and compaction (density) change with depth, and in cultivated soils, substantial changes often occur near the surface. Precise measurements of the bulk density and thermal conductivity of soils are therefore difficult to acquire *in situ*, although recent instrumentation simplifies conductivity measurement [Bristow et al, 1994]. The theory of heat transfer in soils has nevertheless been used to determine average thermal properties from an observed temperature regime and for predicting daily and seasonal changes of soil temperature. In most analyses, κ' is assumed independent of depth, but McCulloch and Penman [1956] derived a solution of Equation 15.29 when κ' was a linear function of depth.

If temperature at depth z and time t is $T(z, t)$, the boundary condition which describes an harmonic oscillation of temperature at the surface about a mean value $\bar{T}$ can be written as

$$T(0, t) = \bar{T} + A(0) \sin \omega t \qquad (15.30)$$

where $A(0)$ is the temperature amplitude at the surface and $\omega = 2\pi/P$ (P is the period) is the angular frequency of the oscillation; i.e., for daily cycles $\omega = (2\pi/24)$ h^{-1} and for annual cycles $\omega = (2\pi/365)$ d^{-1}.

The solution of Equation 15.29 satisfying this boundary condition is

$$T(z, t) = \bar{T} + A(z) \sin(\omega t - z/D) \qquad (15.31)$$

where the amplitude at depth z is

$$A(z) = A(0) \exp(-z/D) \tag{15.32}$$

and D is a depth defined by

$$D = (2\kappa'/\omega)^{0.5} \tag{15.33}$$

Several important features of conduction in soils can be related to the value of D as follows:

(i) At a depth $z = D$ (often called the "*damping depth*") the amplitude of the temperature wave is $\exp(-1)$ or 0.37 times the amplitude at the surface. Similar calculations can be made for 2 and 3 times the damping depth, providing a useful indication of the depth to which the daily or annual temperature wave can be detected in a soil.

(ii) The position of any fixed point on a temperature wave is specified by a fixed value of the phase angle ($\omega t - z/D$); e.g., the maximum temperature occurs when the phase angle is $\pi/2$, and the minimum is when the phase angle is $-\pi/2$. Differentiation of the simple equation $\omega t - z/D = \text{constant}$ gives $\partial z/\partial t = \omega D$, and this is the velocity with which temperature maxima and minima appear to move downward into the soil.

(iii) At a depth $z = \pi D$, the phase angle is π less than the angle at the surface; i.e., the temperature wave is exactly out of phase with the wave at the surface. When the surface temperature reaches a maximum, the temperature at depth πD reaches a minimum and vice versa.

(iv) By differentiating Equation 15.31 with respect to z and putting $z = 0$, it can be shown that the heat flux at the surface at time t is

$$\mathbf{G}(0, t) = \frac{\sqrt{2}A(0)k' \sin(\omega t + \pi/4)}{D} \tag{15.34}$$

The maximum heat flux is $\sqrt{2}A(0)k'/D$, which is the flow of heat that would be maintained through a slab of soil with thickness $\sqrt{2}D$ if one face were maintained at the maximum and the other at the minimum temperature of the surface (i.e., a temperature difference of $A(0)$). The quantity $\sqrt{2}D$ can therefore be regarded as an effective depth for heat flow. (Note that the flux reaches a maximum $\pi/4$ or one-eighth of a cycle before the temperature; i.e., 3 hours for the diurnal wave and 1.5 months for the annual wave.)

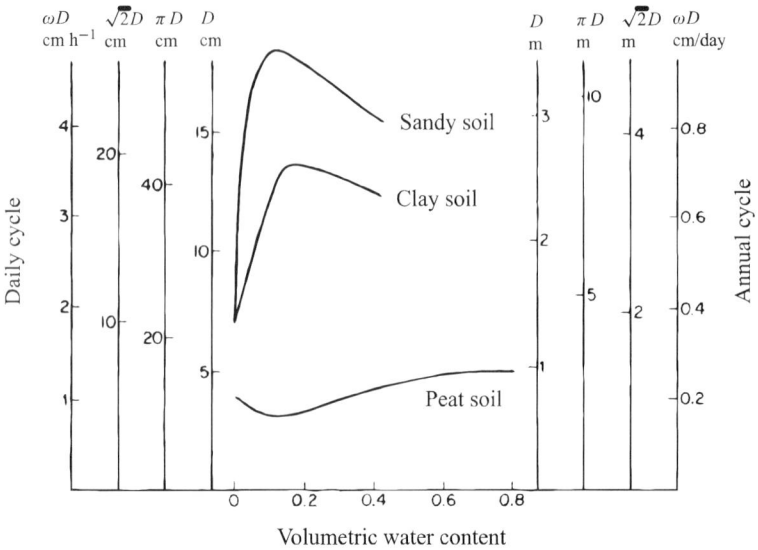

FIGURE 15.6   Change of damping depth and related quantities for three soils over a wide range of water contents. Left-hand axes refer to a daily and right-hand axes to an annual cycle (data from van Wijk and de Vries, 1963).

(v) The amount of heat flowing into the soil during one half cycle is found by integrating $\mathbf{G}(0, t)$ from $\omega t = -\pi/4$ to $+3\pi/4$ and is $\sqrt{2}D\rho'c'A(0)$. This is the amount of heat needed to raise through $A(0)$ degrees Kelvin a layer of soil equal to the effective depth $\sqrt{2}D$.

Values of D for three types of soil are plotted in Figure 15.6 for daily and annual cycles and as a function of volumetric water content. In sandy and clay soils, D increases rapidly when x_l increases from 0.0 to 0.1, reaching values between 12 and 18 cm, for the daily cycle. For the peat soil, D lies between 3 and 5 cm over the whole range of water content, consistent with the slow heating or cooling of organic soil in response to changes of radiation or of air temperature (see (v) above). Corresponding values for the out-of-phase depth πD, the effective depth, and the rate of penetration of the temperature wave can be read from the appropriate axes.

Figure 15.7 shows the type of record from which a damping depth and heat fluxes can be estimated. From the range of the temperature waves at 2.5 and 30 cm, it can be shown (e.g., using Equation 15.32) that the surface amplitude $A(0)$ was about 20 K. The value of D is 10 cm giving $\kappa' = 0.36 \times 10^{-6}$ m^2 s^{-1} from Equation 15.33 and, as the soil was a

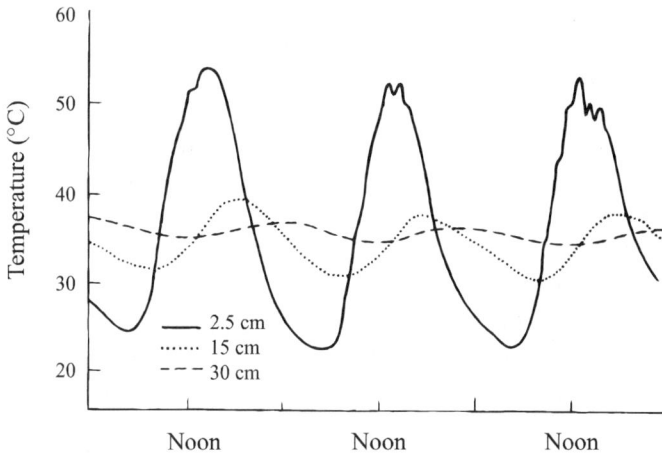

FIGURE 15.7   Diurnal course of temperature at three depths in a sandy loam beneath a bare uncultivated surface; Griffith, New South Wales, 17–19 January 1939 (from Deacon [1969] after West).

sandy loam, these values imply that it was very dry (see Figure 15.6 and Table 15.1). The volumetric heat capacity for a dry sandy soil is about 1.6 MJ m^{-3} K^{-1} and the thermal conductivity $k' = \kappa' \rho' c'$ would therefore be about 0.6 W m^{-1} K^{-1}. If the effective depth is taken as 14 cm ($\sqrt{2D}$), the maximum heat flux into the soil is $\sqrt{2} \times 20 \times 0.6/0.1 = 170$ W m^{-2} (see (iv) above), and the amount of heat stored in the soil during a half cycle is $\sqrt{2}D\rho' c' A(0) = 4.6$ MJ m^{-2}.

In this analysis, the surface amplitude $A(0)$ has been treated as the independent variable in the system. In practice, $A(0)$ depends on the heat balance of the soil surface and on the relative thermal properties of soil and atmosphere. To compare the behavior of different soils exposed to the same weather, it is necessary to begin by calculating the amplitude of the heat flux. The surface amplitude and the soil temperature distribution can then be derived as a function of D. Campbell [1985] published programs for simulating heat and mass transfer in soils. In practice, because density, water content and thermal properties vary with depth in soils, it is usually easier to measure temperature and soil heat flux directly than to rely on calculations based on an idealized model of a uniform soil.

MODIFICATION OF SOIL THERMAL REGIMES

Many biological processes depend on soil temperature: the metabolism and behavior of microorganisms and many invertebrates; the germination of seeds and extension of root systems; shoot extension of seedlings. Because it is difficult to observe the behavior of an undisturbed root system or of the fauna and flora in a natural soil, relatively little is known about physiological responses to changes of soil temperature or about the behavioral significance of soil temperature gradients. Ignorance of fundamental processes has not prevented agronomists, horticulturists, and foresters from developing empirical methods of modifying the thermal regime of soils to help the establishment and growth of crops and trees. Their methods include: mulching soils with layers of organic matter in the form of peat or straw to reduce heat losses in winter; covering peat soils with a layer of sand to inhibit evaporation and to reduce the risk of frost at the soil-air interface; irrigating dry soils in spring to increase conductivity and reduce frost risks; covering the soil with polythene sheets to increase soil surface temperature in spring; and the use of black or white powders to raise or lower the temperature of the surface by changing its reflectivity.

Apart from direct intervention by man, the temperature regime of any soil is profoundly modified by the growth of vegetation because the surface becomes increasingly shaded as the canopy develops. The presence of shade reduces soil heat fluxes, reducing the maximum and increases the minimum temperature at any depth, and there is usually a small decrease in average temperature. Although this effect is well documented, many implications for root and rhizosphere activity remain to be explored.

15.4 Problems

1. Calculate the time constant at $20°C$ for (i) a leaf with surface area 50 cm^2 and thickness 1 mm, and (ii) an apple, treated as a sphere with diameter 10 cm. Assume the heat capacity per unit volume of both organs is $3.0 \text{ MJ m}^{-3} \text{ K}^{-1}$.

2. A stream, 1 m deep, traveling at 0.3 m s^{-1}, emerges from a forest and flows for 1 km through a clear-cut area. On a day of bright sunshine, the stream temperature increased by $0.5°C$ as it traveled through the clear-cut. Estimate the rate $(W \text{ m}^{-2}$ per unit surface area of the stream) at which heat was being stored in the stream water. What are likely sources of this energy?

3. A soil consists of solid material (60% clay, 40% sand) and the pore space is 35%. Given that the thermal conductivity of the dry soil is 0.30 W m^{-1}K^{-1}, and of the wet soil is 1.60 W m^{-1}K^{-1}, find

   (i) The density and specific heat of (a) completely dry soil, and (b) saturated soil

   (ii) The thermal diffusivity and damping depth of soils in conditions (a) and (b)

4. A dry soil has density 1.60 Mg m^{-3}, specific heat 0.90 J g^{-1} K^{-1}, thermal conductivity 0.30 W m^{-1} K^{-1}, and thermal diffusivity 0.22 × 10^{-6} m^2 s^{-1}. Plot a graph of the diurnal temperature wave at the surface and at D assuming the surface amplitude of the wave was 10°C.

5. An oven-dry clay soil has a volumetric specific heat of 1.28 MJ m^{-3}K^{-1}. When the soil is saturated with water the volumetric specific heat is 2.96 MJ m^{-3}K^{-1}. Estimate (i) the fractional pore space in the dry soil, (ii) the volumetric specific heat of the solid material.

6. On a clear night when windspeed was very low, the following temperatures were measured in a soil with thermal conductivity 1.50 W m^{-1}K^{-1}:

Depth (mm)	10	20	50
Temperature (°C)	−2.2	−1.4	1.0

   Assuming that the net loss of radiant energy was equal to the soil heat flux, calculate the incoming long-wave irradiance, stating any assumptions you make.

Chapter 16

Micrometeorology (i) Turbulent Transfer, Profiles, and Fluxes

The physical principles governing the transfer of radiant energy, momentum, and mass in the atmosphere converge in the subject known as "*micrometeorology*," which may be defined as the study of weather on the scale of plants, including trees, and animals, including Man. This branch of environmental physics has progressed rapidly since the early 1960s with the advent of new instrumentation coupled to powerful systems for recording and processing field measurements.

Three major fields can be distinguished for the purposes of definition but they interact strongly.

1. *Hydrology.* In a given environment, how fast does vegetation lose water? How does the rate depend on air, soil, and plant factors and how can it be minimized?

2. *Physiology.* In a given environment, how fast does vegetation gain and lose carbon in the form of carbon dioxide; how do these rates depend on air, soil, and plant factors and how can they be maximized?

3. *Ecology.* What factors determine the regimes of temperature, humidity, wind, and carbon dioxide to which all organisms respond, and what part do they play in determining rates of growth and develop-

ment? How does the microclimate of leaves and soil determine rates of reproduction and activity for plants and animals covering a wide range of scales and diverse physiological responses?

Theme 1 has received much attention and has reached a stage where rates at which crops and forests transpire can be estimated from appropriate sets of environmental factors. Theme 2 was first tackled in the 1960s for agricultural crops. Short-term agricultural studies are numerous but only a few season-long studies of growth and carbon dioxide exchange have been published (e.g., Monteith, 1975 and Monteith et al, 1993). Since the mid 1990s, comparable measurements in forests have been stimulated by their potential role in absorbing a significant fraction of the carbon dioxide released by human activity [Baldocchi et al, 2002]. Studies of micrometeorological aspects of Theme 3 were retarded when it became clear that measurements of canopy transfer confined to vertical fluxes were inadequate. New methods are now available in principle (e.g., McNaughton and van den Hurk, 1995; Raupach, 1989; Katul and Albertson, 1999), but practical applications remain limited. Studies of the environmental physics of leaf-dwelling insects and pathogens have also received attention recently (e.g., Aylor, 1990, McCartney and Fitt, 1998), but there is scope for more work on this important topic.

This chapter deals primarily with the theory and methods needed to attack Themes 1 and 2; the interpretation of exchange measurements above and within canopies is considered in Chapter 17.

16.1 Turbulent Transfer

BOUNDARY LAYER DEVELOPMENT

In laminar flow, the movement of material is predictable, and results in the development of a boundary layer with well-defined profiles of velocity, concentration, and temperature (see p. 142). In contrast, turbulent flow is unpredictable both spatially and temporally. However, just as the depth of a *laminar* boundary layer above a flat plate increases with distance from the leading edge (p. 144), the depth of a *turbulent* boundary layer can be related to the "*fetch*" or distance of traverse (x) across a uniformly rough surface that generates turbulence by shear at the surface. Figure 16.1 illustrates the cross section of a level field of wheat adjacent to a level field of short grass. Because the rougher surface of the wheat exerts more drag than the grass, air is decelerated as it moves from the smoother to the rougher surface. The boundary layer that develops over the short grass

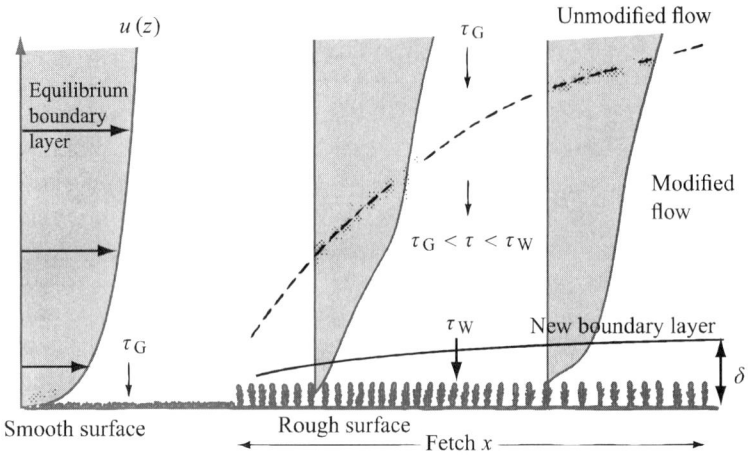

FIGURE 16.1   Development of a new equilibrium boundary layer when air moves from a relatively smooth to a rougher surface. The ratio of the vertical to the horizontal scale is about 20:1. The broken line is the boundary between unmodified flow in which the vertical momentum flux is τ_G and modified flow in which the flux is between τ_G and τ_W. The flux is τ_W below the height δ.

is disturbed by the change in roughness so that a layer of modified flow develops. The characteristics of the flow within it are intermediate be-tween those of the contrasting surfaces. Munro and Oke [1975] studied the development in depth δ of the new equilibrium wind profile following a change of roughness and they concluded that the commonly quoted fetch ratios (x/δ) in the range 100 to 200 were too large. Gash [1986] measured turbulent fluxes of momentum for an extreme transition from heathland (*ca.* 0.25 m high) to forest (*ca.* 10 m high). For the heath/forest transition, the flux measured at 3.5 m over the forest appeared to reach equilibrium about 120 m from the forest edge, and the fetch ratio x/δ was about 20 for this case; for the forest/heath transition, the value of x/δ was about 70 when equilibrium was reached, larger than for the reverse case because the smoother heath generated less turbulence.

Very close to the roughness elements of the crop the turbulent structure is influenced by wakes that they generate, thus establishing a *roughness sublayer* [Raupach and Legg, 1984]. Within this sublayer, the boundary layer structure depends on factors such as the distribution and structure of foliage elements and inter-plant spacing (Figure 16.2).Above the rough-ness sublayer is the *inertial sublayer*, within which fluxes are constant with

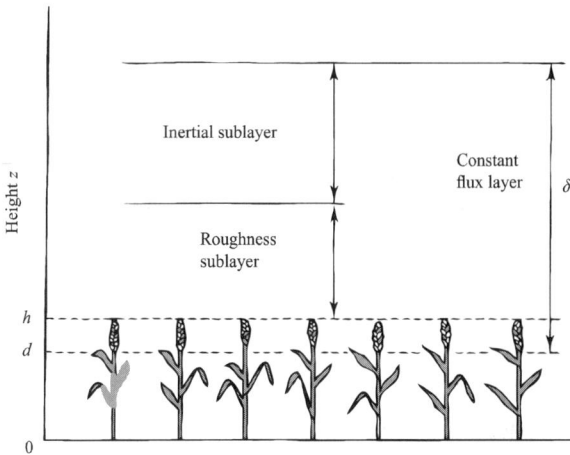

FIGURE 16.2 The constant flux layer and its sublayers. The depth δ is about 15% of the surface boundary layer.

height, and the structure of the boundary layer depends only on scales such as friction velocity (p. 311) and height. Usually, the constant flux layer constitutes only about the lowest 15% of the entire layer in which flow is influenced by the nature of the surface.

It is within the inertial sublayer that micrometeorological measurements can most readily be interpreted to deduce fluxes from turbulence measurements or profiles. The height z_l above the ground of the lower limit of the inertial sublayer is typically about 1.3–1.5 of the canopy height (but varies with aerodynamic roughness and canopy structure [Garratt, 1980]). For example, z_l would be typically 20–23 m for a forest plantation about 15 m high, and 3–3.5 m for a canopy of maize about 2.3 m high. Implications of taking measurements closer to the canopy; i.e., in the roughness sublayer, are discussed on p. 322.

PROPERTIES OF TURBULENCE

Turbulent three-dimensional motion transforms mechanical energy to internal energy (heat) through a cascade of rotating eddies of diminishing size. Maintaining this turbulence requires the supply of energy. Scales of time and length in turbulence extend over many orders of magnitude. In the inertial sublayer near the surface of bare soil or above vegetation, for example, eddies have time scales varying from about 10^{-3} to 10^4 s, and spatial scales from 10^{-3} to 10^4 m [Kaimal and Finnigan, 1994]. Within canopies, turbulence is enhanced by wakes of leaves and stems, as discussed later.

The unpredictable nature of turbulent motion can be analyzed in two ways. One employs statistical analysis. In the other, simplified descriptions of turbulent motion are used to deduce empirical relationships between fluxes and mean vertical gradients. The following treatment begins with the more fundamental statistical approach known as "*eddy covariance.*"

EDDY COVARIANCE
Reynolds Averaging
Reynolds introduced the idea of "decomposing" a time series of turbulent fluctuations into the sum of a mean parameter value (typically averaged over about 30 minutes) and random fluctuations. This type of "*Reynolds averaging,*" is useful only when the averaged quantities do not change significantly with time, so that the mean value of the fluctuating component over the averaging time is zero. In such cases, the atmospheric conditions are termed *stationary.* Thus, at any instant (t), for a wind velocity with components u, v, and w measured in right-handed Cartesian coordinates along mean wind direction (x), cross-wind (y), and vertically (z), respectively,

$$u(t) = \bar{u} + u'(t)$$
$$v(t) = \bar{v} + v'(t) \tag{16.1}$$
$$w(t) = \bar{w} + w'(t)$$

where $\bar{u}$, $\bar{v}$, and $\bar{w}$ are mean values of the velocity components over a long enough time to ensure that the mean values $\overline{u'} = \overline{v'} = \overline{w'} = 0$. Similarly, fluctuations of a scalar entity s (for example, temperature, gas concentration or humidity) may be expressed as the sum of a mean value and a turbulent fluctuation; i.e.,

$$s(t) = \bar{s} + s'(t) \tag{16.2}$$

Figure 16.3 shows typical fluctuations of wind components, temperature, specific humidity, and CO_2 concentration measured over a forest.

Eddy transfer
The principles of transfer by eddies acting as "carriers" were discussed in Chapter 3. This section focusses on the principles underlying the eddy

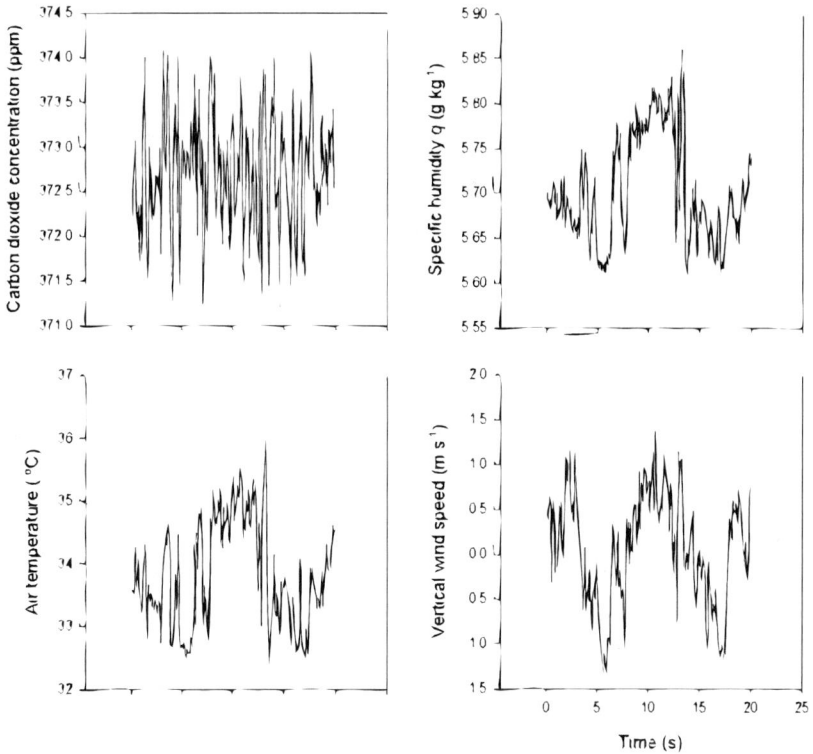

FIGURE 16.3 Fluctuations in wind components, temperature, humidity and carbon dioxide concentration measured over a young ponderosa pine forest in Oregon (Data courtesy of Larry Mahrt and Dean Vickers).

covariance (or eddy correlation) method for flux measurement applied to expanses of vegetation that are uniform, extensive and horizontal. Sakai et al [2001] and Finnigan et al [2003] extended the analysis to complex terrain and heterogeneous vegetation.

The eddy covariance method relies on measurements of the fluctuating components of wind within inertial sublayer (constant flux region) of the surface boundary layer, and of the associated fluctuations in temperature, humidity, or gas concentration. For an extensive horizontal surface, the mean wind at any height within the constant flux layer is horizontal, but, at an observation point, instantaneous values may be in any direction and generally have a vertical component that may be toward or away from the surface. The *mean* net vertical transport of dry air (density ρ_a) must be zero if the mean mass below the observation point is to be constant, so that

$$\overline{\rho_a w} = \overline{\rho_a}\,\overline{w} + \overline{\rho_a' w'} = 0 \qquad (16.3)$$

Rearranging Equation 16.3 shows that the mean vertical velocity is not zero, but is given by

$$\overline{w} = -\overline{\rho_a' w'}/\overline{\rho_a} \qquad (16.4)$$

Consider now the vertical transport of a substance S, assumed to be a minor constituent of the atmosphere so its presence does not significantly change the air density ρ, which can be written

$$\rho \simeq \rho_a + \rho_v \qquad (16.5)$$

where the subscripts denote dry air and water vapor, respectively.

For there to be *net transport* of S by vertical eddies, there must be fluctuations of S that are correlated to some extent with the fluctuations in vertical velocity w. A flux of S from the surface arises when, on average, eddies moving away from the surface contain air with a higher concentration of S than average, and those moving towards the surface contain lower than average concentrations.

At any instant, the flux $F(t)$ at a point of observation above the surface is

$$F(t) = \rho_s(t) w(t) \qquad (16.6)$$

where ρ_s is the density of S in air. The average vertical flux density is the time average of $F(t)$; i.e., $F = \overline{\rho_s w}$.Figure 16.4 illustrates a typical measurement arrangement over a forest. Assuming that the forest is uniform and horizontal, there are no net horizontal fluxes of S in the measuring volume. Provided that the measurements are made in the inertial sublayer, and if there is no storage of S in the volume, the vertical flux at the top of the volume is identical to the total flux of S arising from the forest and soil surface.

Using Reynolds averaging,

$$\begin{aligned} F &= \overline{(\overline{\rho_s} + \rho_s')(\overline{w} + w')} \\ &= \overline{\rho_s}\,\overline{w} + \overline{\rho_s' w'} + \overline{\rho_s'\overline{w}} + \overline{\overline{\rho_s} w'} \end{aligned} \qquad (16.7)$$

The last two terms in Equation 16.7 are zero because they include the time average of fluctuations. The first term is sometimes taken as zero, arguing that $\overline{w}$ is zero, in which case the equation reduces to $F = \overline{\rho_s' w'}$; i.e., the flux is given by the time average of the eddy covariance term. However, Equation 16.4 shows that it is generally incorrect to assume $\overline{w} = 0$. The following analysis explores some of the implications of this conclusion.

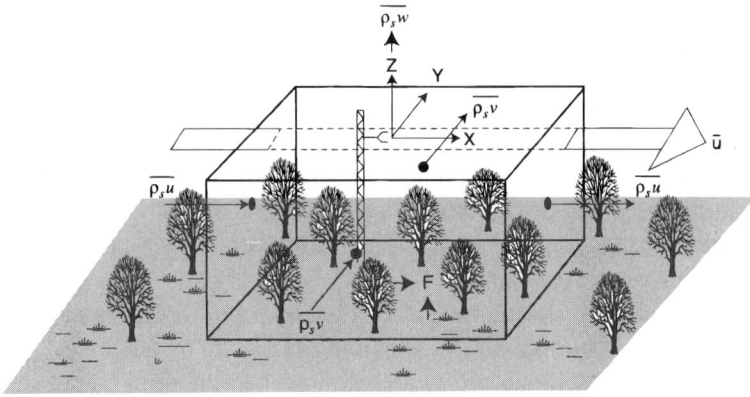

FIGURE 16.4 Flux relationships in and above a uniform, horizontal forest. The coordinate axes are aligned with the mean wind vector, which is parallel to the ground. In the absence of horizontal advection and storage in the sample volume, the vertical flux measured by instruments on the tower above the forest is identical to the flux from the soil and vegetation (after Finnigan et al, 2003).

By writing $F = \overline{\rho w q_s}$, where $q_s = \rho_s/\rho$ (the specific concentration), and expressing mean values of the quantities as $\overline{\rho w}$ and $\overline{q_s}$, and their fluctuations as $\rho w'$ and q_s', it follows that

$$F = \overline{(\overline{\rho w} + (\rho w)')(\overline{q_s} + q_s')} = \overline{\rho w}\,\overline{q_s} + \overline{\overline{\rho w}\, q_s'} + \overline{(\rho w)'\overline{q_s}} + \overline{(\rho w)'q_s'}$$
(16.8)

The final term on the right hand side of Equation 16.8 is the eddy covariance or eddy flux. The second and third terms are zero because fluctuations associated with mean values can make no net transport over the averaging interval. The first term is not zero, since it is the vertical flux of dry air $\overline{\rho_a w}$, not $\overline{\rho w}$ that is zero. However, using Equations 16.3 and 16.5 gives $\overline{\rho w} = \overline{\rho_v w} = E$, where E is the evaporation rate, demonstrating that F calculated from Equation 16.8 is determined not only by fluctuations of the property S, and by the water vapor flux.

Webb et al [1980] showed more generally that fluxes determined from measurements of density fluctuations require corrections arising from both heat and water vapor fluxes. Qualitatively this can be understood as follows. Parcels of air that are moister than average are also less dense than average (because the density of water vapor is less than the density of dry air, p. 16. If the water vapor flux is upwards, then rising parcels of air are on average moister and less dense than descending parcels. Since there must be zero mean vertical mass flow of air, it follows that there must exist a small mean upward velocity component (see Equation 16.4). Thus, when the flux of a substance S is measured using fluctuations of the density of S and of vertical wind speed w about its mean $\overline{w}$, the contribution to the flux of S associated with $\overline{w}$ is missed, and an additional term having the same sign as the water vapor flux must be added to the measured flux. A similar argument applies to heat fluxes. It will be shown later that the magnitude of the additional terms may be very important when measuring fluxes of trace gases such as carbon dioxide and atmospheric pollutants by the eddy covariance method.

As an alternative to Equation 16.8, the flux F can be expressed in pure eddy covariance form (with no remaining terms for mean flow) by the introduction of the mixing ratio $r_s = \rho_s / \rho_a$; i.e., the mass of substance per unit mass of dry air, using the constraint of Equation 16.3. Thus, $F = \overline{\rho_a w r_s} = \overline{\rho_a w}\ \overline{r_s} + \overline{(\rho_a w)' r_s'}$, and since $\overline{\rho_a w} = 0$,

$$F = \overline{(\rho_a w)' r_s'} \tag{16.9}$$

Equation 16.9 is not practical, since $(\rho_a w)'$ is not readily measurable, but it can be shown that the pure covariance of w' with r_s' gives a close approximation to F, as follows.

From the definition of r_s, we see that, for small fluctuations of r_s, $\delta r_s = (\partial r_s / \partial \rho_s)\delta \rho_s + (\partial r_s / \partial \rho_a)\delta \rho_a$. Hence, by differentiation of $r_s = \rho_s / \rho_a$, recognizing that small fluctuations may be regarded as identical to the δ terms,

$$r_s' = (1/\overline{\rho_a})\rho_s' - (\overline{\rho_s}/\overline{\rho_a}^2)\rho_a'$$

Hence,

$$\overline{w' r_s'} = (1/\overline{\rho_a})\overline{\rho_s' w'} - (\overline{\rho_s}/\overline{\rho_a}^2)\overline{\rho_a' w'} \tag{16.10}$$

Rearranging Equation 16.10, and substituting from Equation 16.4 yields

$$\overline{\rho_a}\ \overline{w' r_s'} = \overline{\rho_s' w'} + \overline{\rho_s}\ \overline{w} \tag{16.11}$$

Writing

$$F = \overline{\rho_s}\ \overline{w} + \overline{\rho_s' w'}, \tag{16.12}$$

Equation 16.11 shows that

$$F = \overline{\rho_a}\ \overline{w' r_s'} \tag{16.13}$$

Equation 16.13 shows that F could be measured directly by calculating the covariance of w' and r_s' if detectors for windspeed and mixing ratio with sufficiently fast responses were available, and if their signals could be sampled sufficiently rapidly and averaged over time to detect the full range of eddies in the turbulent flow. Sonic anemometers provide suitably fast responses for w' [Guyot, 1998], but most fast response gas analyzers (e.g., open-path infrared gas analyzers for CO_2 or water vapor) measure density rather than mixing ratio, so equations similar to Equation 16.8 must generally be used for eddy covariance calculations, and these include terms associated with density variations caused by heat and water vapor fluxes. Some researchers have avoided the need for density corrections by drawing air through tubes into closed path infrared gas analyzers so that the measurements of density are made at constant pressure and temperature (e.g., Goulden et al 1996), but in this case corrections are necessary for signal loss and distortion in the sampling tube. Leuning and Judd [1996] discussed the relative merits of open- versus closed-path gas analyzers for eddy covariance studies.

The necessary response time of the sensors used for eddy covariance measurements depends on the range of eddy sizes that carry the flux. Eddy sizes grow with height over the surface (see Figure 16.9) and increase with increasing surface roughness and windspeed. Consequently sensors capable of detecting fluctuations between 0.1 and 10 Hz would often be adequate for use several metres above a rough forest canopy, whereas a frequency response of 0.001 Hz might be needed for eddy flux measurements close to a smooth surface. To avoid artifacts in sampling, readings must be taken at a time interval that is not more than half the fastest sensor response time. Examples of the eddy covariance method are given in the next chapter. Leuning et al [1982], Goulden et al [1996], Finnigan et al [2003], Baldocchi [2003] and Dolman and Gash [2004] describe some of the practical and theoretical difficulties of the method.

The eddy covariance method can be applied to measure fluxes of any scalars for which fast sensors exist. When fast sensors are not available (e.g. for stable isotopes), a technique known as *Relaxed Eddy Accumulation* may be used [Guenther et al, 1996]. In this technique, air is sampled into separate reservoirs from upward- and downward-moving eddies at rates proportional to the vertical velocity. Analysis of the concentration difference between the reservoirs, using relatively slow response analyzers, allows the flux to be determined.

16.2 Flux-Gradient Methods

An alternative way of measuring fluxes by micrometeorological methods uses empirical relations between fluxes and mean gradients of quantities measured in the inertial sub-layer. The empirical relations are a simplified description of the complexities of turbulence. The gradients are averaged over periods long enough to include the time scales of eddies responsible for transporting the flux, typically about 30 minutes.

In Chapter 3, a general equation was derived for the vertical transfer of entities in a gas by "carriers" that could be molecules or eddies. For turbulent transport it was stated that the flux F of an entity S was given by

$$F = -\overline{\rho w l_s}\,(d\overline{r_s}/dz)$$

where ρ is air density, r_s is the mixing ratio of S in air, and l_s is called the "*mixing length*" for S. The quantity $\overline{w l_s}$ is called the "*turbulent transfer coefficient*" (or sometimes the *eddy diffusivity*) K_s, so that

$$F_{=} - \rho K_s \frac{d\overline{r_s}}{dz} \tag{16.14}$$

Equation 16.14 will be used to develop empirical equations relating fluxes of momentum, heat, water vapor, and mass to their respective gradients.

PROFILES

The change of potential of an entity with height above or within a crop canopy is called the *profile* of that entity. With sufficiently precise instrumentation, profiles of windspeed, temperature, and gas concentrations can be measured. Figure 16.5 shows some idealized profiles representative of a cereal crop that would typically grow to a height of $h = 1$ m with most of its green foliage between $h/2$ and h. The shape of the profiles above the canopy is determined partly by the turbulent eddies that are produced by the drag of the crop elements on the wind blowing over it and partly by the fluxes to the crop. In Chapter 3 it was shown that in *laminar* boundary layers the transfer of momentum, mass, and heat was determined by gradients of potential (profiles) and by diffusivities associated with molecular agitation. In the *turbulent* boundary layers above crops, the same principles apply, but the diffusivities are associated with turbulent eddies as defined previously. The size of turbulent eddies decreases as the surface is approached until finally they merge with molecular agitation. The change of eddy size with height, and the dependence of turbulent mixing on windspeed cause the shapes of profiles to be influenced both by windspeed and by the surface properties that generate turbulence.

MOMENTUM TRANSFER

Equation 16.8 demonstrated that the vertical flux of an entity S transported by turbulence can be represented by the mean value of the product $\rho \overline{w' q_s'}$.

When the entity is horizontal momentum, q'_s becomes the fluctuation of the horizontal velocity u', and the vertical flux of horizontal momentum is given by $\overline{\rho u' w'}$. Provided that the vertical flux is constant with height, this quantity can be identified as the force per unit ground area, known as the *shearing stress* (τ); i.e.,

$$\tau = \overline{\rho u' w'} \tag{16.15}$$

The ratio $\left(\dfrac{\tau}{\rho}\right)^{1/2}$ has the dimensions [velocity], and is called the *friction velocity* u_*. Consequently, it follows that $u_*^2 = \overline{u' w'}$, showing that the friction velocity is a measure of mean eddy velocities.

By comparison with Equation 16.14, the shearing stress may also be written as

$$\tau = \rho K_M \frac{\partial u}{\partial z} \tag{16.16}$$

where K_M is a turbulent transfer coefficient for momentum with dimensions $L^2 T^{-1}$. From experience, both windspeed and turbulent mixing increase with height, and a simple dimensional argument will now be used to obtain the functional relations.

The simplest assumption for the height dependence of K is

$$K_M = az \tag{16.17}$$

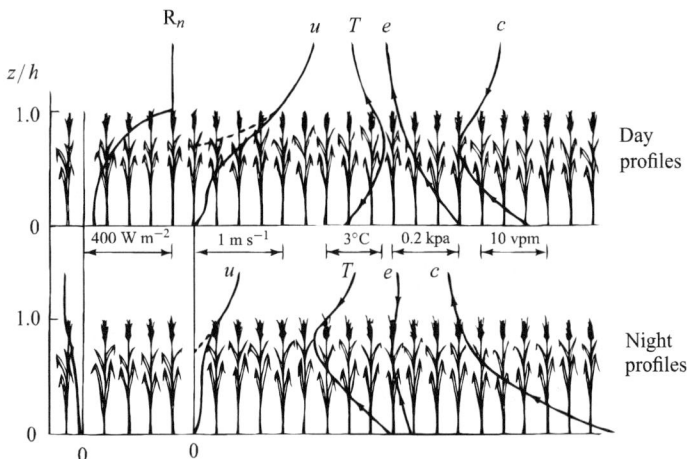

FIGURE 16.5 Idealized profiles of net radiation (R_n), windspeed (u), air temperature (T), vapor pressure (e), and CO_2 concentration (c) in a field crop growing to a height h plotted as a function of z/h. The wind profiles represent an extrapolation of the logarithmic relation between u and $(z - d)$ above the canopy (see p. 313).

where a is a quantity with the dimensions of velocity. Similarly, the simplest assumption for the wind gradient is

$$du/dz = b/z \qquad (16.18)$$

where b is a second constant with dimensions of velocity. Putting Equations 16.17 and 16.18 in Equation 16.16 gives

$$\tau = \rho a b \qquad (16.19)$$

implying that $ab = u_*^2$. Furthermore, because a and b are both velocities, Equation 16.19 implies that

$$a = k u_* \quad \text{and} \quad b = u_*/k$$

where k is a constant. Substitution in Equations 16.17 to 16.19 now gives

$$K_M = k u_* z \qquad (16.20)$$

and

$$du/dz = u_*/(kz) \qquad (16.21)$$

$$\tau = \rho u_*^2 \qquad (16.22)$$

Integration of Equation 16.21 between the limits z_0 and z yields

$$u = (u_*/k)\ln(z/z_0) \qquad (16.23)$$

where z_0 is a constant termed the *roughness length*, such that $u = 0$ when $z = z_0$. This does not imply that the real wind speed is zero at height z_0, because the assumptions made in deriving Equation 16.23 may fail near the boundary, so the limit $u = 0$ should be regarded as a mathematical convenience.

Equation 16.23 is the logarithmic wind-profile equation, which is found to be valid over many types of uniform surface provided that the following conditions are satisfied:

1. The surface is uniform, extensive, and horizontal.

2. Turbulence is generated only by shear stress at the surface (i.e., not by convection or upwind obstructions).

3. Measurements are made in the equilibrium boundary layer associated with the surface (i.e., the part of the inertial sublayer where the shearing stress is constant with height).

4. The windspeed is averaged over a sufficiently long time interval to include all scales of eddies contributing to the eddy flux.

Measurements confirm that k is a constant—the *von Karman constant,* named after a famous aerodynamicist—and it is usually assigned a value of 0.41, as determined by experiment.

A more general form of Equation 16.23 takes account of the height of the surface elements by assuming the vertical displacement of the zero plane to a height d known as the *zero plane displacement* such that the distribution of shearing stress over the elements is aerodynamically equivalent to the imposition of the entire stress at height d. The wind profile equation then becomes

$$\partial u/\partial z = u_*/[k(z - d)] \tag{16.24}$$

or

$$u = (u_*/k) \ln \{(z - d)/z_0\} \tag{16.25}$$

and Equation 16.20 becomes

$$K_M = ku_*(z - d) \tag{16.26}$$

These equations are valid in the air above the surface elements where the assumption that shearing stress is independent of height is satisfied. Because of the absorption of momentum below the tops of the elements, equation 16.23 is not valid within the canopy, nor is it valid in the roughness sublayer immediately above tall aerodynamically-rough vegetation (Figure 16.2) where the relation between flux and gradient defined by Equation 16.16 breaks down. Within a canopy, windspeed may be larger than predicted by extrapolating measurements above the canopy using the logarithmic equation. In stands that have less foliage near the ground than at higher levels—forests, for example—windspeed may increase towards the soil surface because the movement of air is impeded mainly by stems of trunks offering relatively little obstruction to airflow.

It is nevertheless possible to use Equation 16.25 to obtain an extrapolated theoretical value of u, which tends to zero at a height given by $z = z_0 + d$. To recap, d is an equivalent height for the absorption of momentum (a "center of pressure"), and $(d + z_0)$ is an equivalent height for zero windspeed. In general, the shape of the wind profile above tall crops and forests implies that the absorption of momentum by the surface of the ground is trivial compared with absorption by the vegetation.

Stanhill [1969] found that the average value of d for a range of primarily agricultural vegetation was close to 0.63 of the mean vegetation height

Table 16.1 Characteristic values
of roughness length, z_0.

Type of surface	z_0(m)
Ice	0.001
Open water	0.002–0.006
Bare soil (untilled)	0.005–0.020
(tilled)	0.002–0.006
Grass 0.01 m high	0.001
0.1 m high	0.023
0.5 m high	0.05–0.07
Wheat 1 m high	0.10–0.16
Coniferous forest	1.0

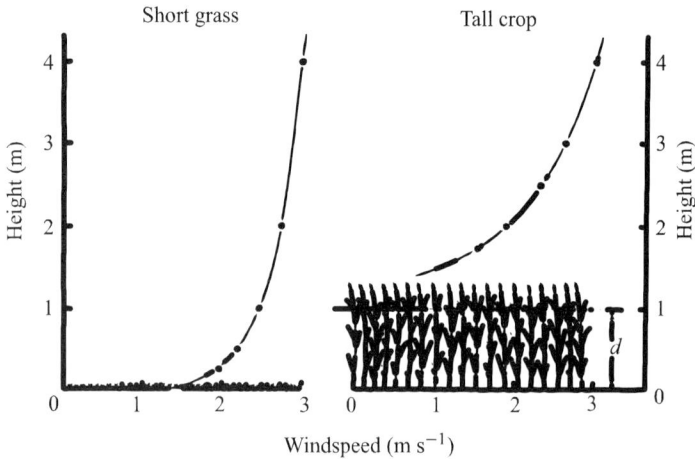

FIGURE 16.6 Wind profiles over short grass and a tall crop when windspeed at 4 m above the ground is 3 m s^{-1}. The filled circles represent hypothetical measurements from sets of anemometers.

h. Table 16.1, derived from a summary by Campbell and Norman [1998] of work by Hansen [1993], provides characteristic values of z_0 for a wide range of surfaces.

To demonstrate the relation between windspeed and height for contrasting surfaces, Figure 16.6 contains profiles over short grass ($d = 7.0$ mm, $z_0 = 1.0$ mm) and a tall crop ($d = 0.95$ m, $z_0 = 0.20$ m) when the windspeed at 4 m above the ground is 3 m s^{-1}. Figure 16.7 shows the equivalent logarithmic plots of u as a function of $\ln(z-d)$. Because height appears (by convention) on the vertical axis, the slope is k/u_*, giving values of $u_* = 0.15$ and 0.46 m s^{-1} for the grass and tall crop respectively: $\tau = 0.027$ and 0.25 N m^{-2}: and 0.25 and 0.58 m^2 s^{-1} for the transfer coefficient K_M at a height of 4 m.

When windspeed is measured over heights much larger than d (or in other situations where d is negligible), this type of analysis requires values u_1 and u_2 at a minimum of two heights z_1 and z_2 so u_* can be eliminated initially to give

$$\ln z_0 = (u_2 \ln z_1 - u_1 \ln z_2)/(u_2 - u_1) \qquad (16.27)$$

When the value of d is significant, at least three heights are needed so that both u_* and z_0 can be eliminated initially to give

$$\frac{u_1 - u_2}{u_1 - u_3} = \frac{\ln(z_1 - d) - \ln(z_2 - d)}{\ln(z_1 - d) - \ln(z_3 - d)} \qquad (16.28)$$

This equation allows d to be found either by iteration in a computer program or graphically by plotting the right hand side as a function of d.

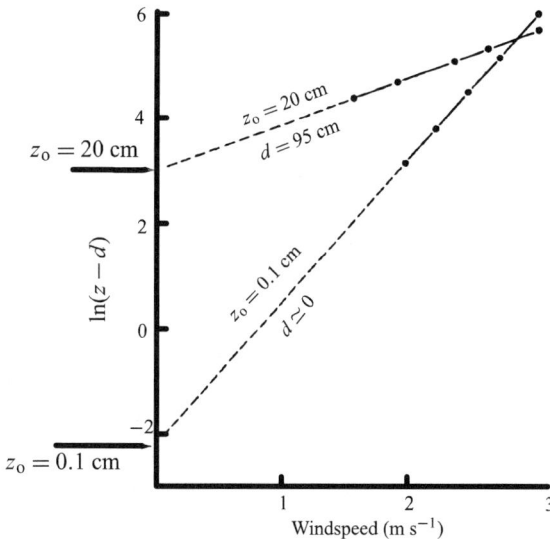

FIGURE 16.7 Relationships between windspeed and $\ln(z - d)$ for the wind profiles in Figure 16.6.

Behavior of z_0 and d with Vegetation Height and Structure

The dependence of z_0 and d on the height and structure of roughness elements has been examined both experimentally and theoretically. Lettau [1969] analyzed the classic experimental studies of Kutzbach, where the roughness over a frozen lake was modified with large arrays of "bushel baskets," concluding that z_0 increased proportionally to obstacle height and to the ratio of silhouette area to plot area. Shaw and Pereira [1982] found that attempts to explain the exchange of momentum within canopies in terms of traditional mixing-length models were unsatisfactory, and they developed second-order equations for turbulent mixing. In their numerical model, leaf area index was replaced by a plant area index (P) and the area per unit height was assumed to increase linearly with height from zero at the top of the canopy to a maximum at a height z_m, below which it decreased linearly to the surface. By assuming a uniform drag coefficient c_d within the canopy, and by computing wind profiles, they were able to predict how z_0/h and d/h should depend on the parameter $c_d P$.

As $c_d P$ increased, the ratio d/h also increased, and for representative values of $c_d = 0.5$ and $z_m/h = 0.5$, d/h was between 0.5 and 0.7, consistent with field experience. The dependence of z_0/h on $c_d P$ was more complex. When there were relatively few roughness elements per unit ground area, any increase in number increased the drag but had relatively little effect on d, so z_0/h increased (Figure 16.8). With a further increase of P, however, a point was reached where an increase in the area of roughness elements tending to increase drag was offset by an increase in the height of the zero plane which reduced the effective depth of the canopy for momentum exchange. Beyond this point, z_0/h decreased as $c_d P$ increased (Figure 16.8).

As a corollary, when vegetation is *sparse*, drag is greatest when most plant material is near the top of the canopy (z_0/h large) but *dense* vegetation is least rough when z_0/h is large because the canopy then presents a relatively smooth surface to the air passing over it. Below the point of maximum roughness, the value of z_0/h predicted by the model is approximately $0.29(1 - d/h)$, consistent with measurements, but above the maximum, z_0/h depends on $c_d P$, z_m/h, and on d/h.

Combining field evidence with predictions from the model, when the maximum density of foliage is approximately at half the height of the canopy, z_0/h is expected to be between 0.08 and 0.12 and d/h between 0.6 and 0.7. However, both ratios depend to some extent on windspeed, and for flexible stands of cereals there are many reports that the ratios

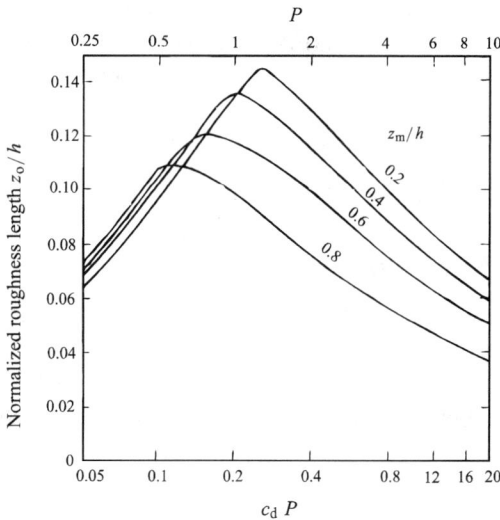

FIGURE 16.8 Normalized roughness length as a function of $c_d P$ (and against P assuming that $c_d = 0.2$). Curves are labeled according to the height at which foliage density reaches a maximum (from Shaw and Pereira, 1982).

decrease with windspeed as a consequence of three factors working in the same direction:

(i) decrease in the drag coefficient of individual leaves at a fixed angle to the wind (Figure 9.5)

(ii) decrease in the drag coefficient of leaves moving into a more streamlined position

(iii) decrease in the drag coefficient of the whole canopy as stems bend

Legg et al [1981] found that small changes of d/h and z/h with windspeed were not statistically significant for field beans, but for potatoes at the beginning of the season, d/h decreased with windspeed.

Over surfaces such as sand and water, the force exerted by wind can detach elements which carry momentum upward into a thin layer of air immediately above the surface before falling back again. In a classic study of the movement of sand in the desert (*saltation*), Bagnold [1941] suggested that the initial vertical velocity of a detached sand grain should be proportional to the friction velocity u_*, a measure of the mean velocity of eddies (p. 311). A grain moving upward with an initial velocity u_* will come to rest at a height u_*^2/g where g is the gravitational acceleration. It can therefore be argued on dimensional grounds that the depth z_0 of the roughness layer within which horizontal momentum is absorbed should be proportional to u_*^2/g. Chamberlain [1983] pointed out that the relation

$$z_0 = 0.016u_*^2/g \qquad (16.29)$$

appears to be valid for sand, snow, and sea.

AERODYNAMIC RESISTANCE

Equation 16.16, which expresses momentum flux in terms of the gradient of horizontal momentum per unit volume, can be rewritten in the general form of Ohm's Law by introducing an aerodynamic resistance to momentum transfer between heights z_1 and z_2 where windspeeds are u_1 and u_2. Then if

$$\tau = \rho(u_2 - u_1)/r_{aM} \qquad (16.30)$$

the resistance can be evaluated as

$$r_{aM} = (u_2 - u_1)/u_*^2 = \ln\{(z_2 - d)/(z_1 - d)\}/ku_* \qquad (16.31)$$

The resistance for momentum transfer between a single height where the windspeed is $u(z)$ and the level $d + z_0$ where the extrapolated value of u is zero can be written in several equivalent forms; e.g.,

$$r_{aM} = \frac{u(z)}{u_*^2} = \frac{\ln\{(z-d)/z_0\}}{ku_*} = \frac{\ln\{(z-d)/z_0\}^2}{k^2 u(z)} \qquad (16.32)$$

These resistance equations, like all others in the chapter, apply to momentum transfer in neutral stability; i.e., when the change in temperature with height is equal to the adiabatic lapse rate (p. 11). A discussion of the more general non-adiabatic cases follows discussion of fluxes of other entities in neutral conditions.

FLUXES OF HEAT, WATER VAPOR AND MASS

By analogy with Equation 16.16, flux-gradient equations in neutral stability can be written for heat, water vapor, and mass concentrations, to give

$$\begin{aligned}
\mathbf{C} &= -K_H/[\partial(\rho c_p T)/\partial z] \\
\mathbf{E} &= -K_V/(\partial\chi/\partial z) \qquad (16.33) \\
\mathbf{F} &= -K_S/(\partial s/\partial z)
\end{aligned}$$

In neutral conditions, eddies transport all entities equally effectively, so $K_H = K_V = K_S = K_M$. This similarity in transfer coefficients is the

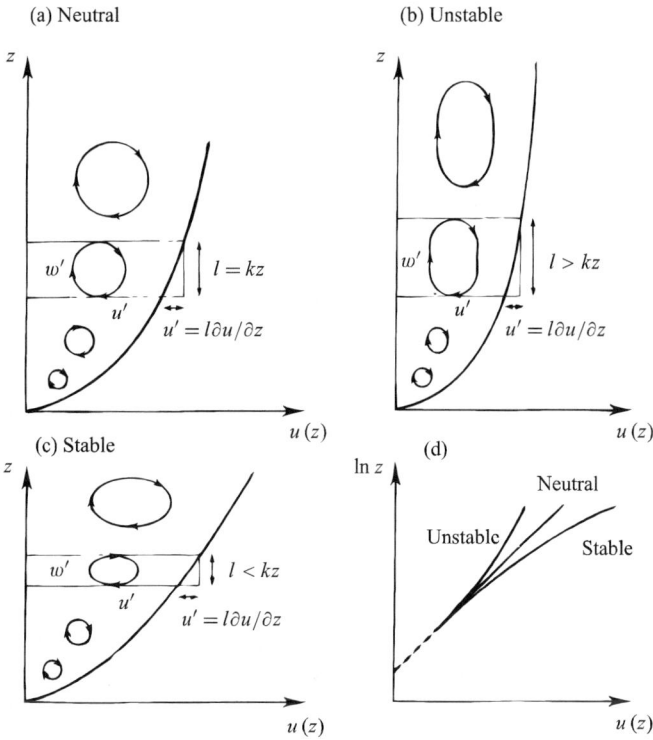

FIGURE 16.9 Windspeed profiles and simplified eddy structures characteristic of the three basic stability states in air flow near the ground (from Thom, 1975).

basis of the *Similarity Theory* developed by Monin–Obhukov. Equations 16.16 and 16.33 provide the basis of the aerodynamic method for measuring fluxes (p. 324).

Profiles and Stability

A simple interpretation of the relation between mean wind profile and turbulence is illustrated in Figure 16.9. In neutral stability (Figure 16.9a) the eddy structure can be envisaged as a set of circular eddies with diameters increasing with height and given by the mixing length $l = kz$, rotating with tangential speed equal to the friction velocity u_*, i.e.,

$$u' = w' = u_* = l\partial u/\partial z$$

where w' and u' represent the vertical and horizontal velocity fluctuations respectively [Thom, 1975].

In unstable (lapse) conditions, which occur when the surface is strongly heated, vertical motion is enhanced by buoyancy. The amount of enhancement increases as the wind shear (depending on viscosity) decreases, and this is illustrated in Figure 16.9b, where the eddies are progressively stretched vertically. Thus w' exceeds u', where u' is still given by $l\partial u/\partial z$ but l is greater than kz.

Conversely, in stable conditions (inversion), for example on a clear night with light winds, vertical eddy velocities are damped, and so (Figure 16.9c) $w' < u'$, with $u' = l\partial u/\partial z$ but $l < kz$.

The qualitative effect of stability on the shape of the wind profiles is apparent in Figure 16.9a–c, and is summarized in semi logarithmic form in Figure 16.9d. In each of the examples in Figure 16.9 the momentum flux transmitted to the surface is assumed the same, and so u_* is constant. Since, by differentiating Equation 16.25,

$$u_* = k\partial u/\partial[\ln(z - d)],$$

this requires the gradient of each profile at the lowest levels to be the same. As height increases, velocity gradients become smaller in unstable conditions and larger in stable conditions than those for the neutral case. The differential wind profile (Equation 16.27) can therefore be written in generalized form as

$$\frac{\partial u}{\partial z} = \frac{u_*}{k(z - d)}\Phi_M \tag{16.34}$$

where Φ_M is a dimensionless *stability function* with a value of unity in neutral stability and larger or smaller than unity in stable or unstable conditions, respectively.

Using the relation between momentum flux and gradient

$$\tau = \rho u_*^2 = K_M \partial(\rho u)/\partial z$$

it can be readily shown that

$$K_M = ku_*(z - d)\Phi_M^{-1} \tag{16.35}$$

Stability functions can be defined for other entities by

$$K_H = ku_*(z - d)\Phi_H^{-1} \tag{16.36}$$

and similarly for Φ_V and Φ_S.

The relations between K_M, K_H, K_V, and K_S (or equivalent functions Φ_M, Φ_H etc.) have been a source of considerable argument in micrometeorology and a number of empirical relationships have been proposed. In neutral stability, all entities are transported equally effectively, all profiles are logarithmic in the constant flux layer, $\Phi_M = \Phi_H = \Phi_V = \Phi_S$, and $K_M = K_H = K_V = K_S$. In unstable conditions, K_H exceeds K_M because there is preferential upward transport of heat. Measurements (reviewed by Dyer, 1974) support the view that $K_H = K_V = K_S$ in unstable conditions. In slightly to moderately stable conditions, Dyer inferred that $K_M = K_H = K_V$, but as stability increases turbulence is increasingly damped, and the concepts underlying similarity theory become invalid.

The dependence of the functions Φ on stability is generally expressed as a function of parameters that depend on the ratio of the production of energy by buoyancy forces to the dissipation of energy by mechanical turbulence. The two best established parameters are the gradient Richardson number Ri, calculated from gradients of temperature and windspeed, and the Monin–Obukhov length L, which is a function of fluxes of heat and momentum. In symbols

$$Ri = (gT^{-1}\partial T/\partial z)/(\partial u/\partial z)^2 \qquad (16.37)$$

and

$$L = \frac{-\rho c_p T u_*^3}{kg\mathbf{C}} \qquad (16.38)$$

where T = absolute temperature (K), g = gravitational acceleration, and $\mathbf{C}$ is sensible heat flux.

When measurements are made over rough surfaces where gradients are small, or at heights of more than a few meters over any vegetated surface, it is important to allow for the decrease in temperature with height that arises from adiabatic expansion when a parcel of air rises; i.e., the dry adiabatic lapse rate Γ which is approximately -0.01 K m^{-1} (see Chapter 2). Temperature gradients $\partial T/\partial z$ in Equation 16.37 (and in other heat flux equations elsewhere in this chapter) should be replaced by potential temperature gradient $\partial\theta/\partial z = (\partial T/\partial z) - \Gamma$. In a neutral atmosphere θ remains constant with height.

From the definitions of Ri and L it can be shown that

$$(z - d)/L = (\Phi_M^2/\Phi_H)Ri \qquad (16.39)$$

(i) Unstable Conditions
In unstable conditions, Dyer and Hicks [1970] concluded that

$$\Phi_M^2 = \Phi_H = \Phi_V = [1 - 16(z - d)/L]^{-0.5} \qquad (16.40)$$

i.e.,

$$\Phi_M^2 = \Phi_H = \Phi_V = (1 - 16Ri)^{-0.5} \text{ for } Ri < -0.1 \qquad (16.41)$$

For slightly unstable conditions when $16(z - d)/L$ is only slightly less than zero, Equation 16.40 can be written as

$$\Phi_M \simeq [1 + 4(z - d)/L] \qquad (16.42)$$

(ii) Stable Conditions

From measurements in stable and slightly unstable conditions, Webb (1970) deduced the empirical relation

$$\Phi_M = \Phi_H = \Phi_V = [1 + 5(z - d)/L] \qquad (16.43)$$

i.e.,

$$\Phi_M = \Phi_H = \Phi_V = (1 - 5Ri)^{-1} \qquad (16.44)$$

for $-0.1 \leqslant Ri \leqslant 1$.

As $\Phi_V/\Phi_H = K_H/K_V$ (from Equations 16.35 and 16.36, equality of Φ_V and Φ_H in both stability states implies that turbulent exchanges of water and heat are always similar to each other, and presumably also to any other entity entrained in the atmosphere.

For correcting flux measurements it will be seen later (p. 326) that it is useful to define the product $(\Phi_V\Phi_M)^{-1} \equiv (\Phi_H\Phi_M)^{-1} = F$ where F is called a *generalized stability factor* [Thom, 1975]. From Equations 16.41 and 16.44

$$F - (1 - 5Ri)^2 \qquad -0.1 \leqslant Ri \leqslant 1 \qquad (16.45)$$

and

$$F = (1 - 16Ri)^{0.75} \qquad Ri < -0.1 \qquad (16.46)$$

Figure 16.10 shows these relationships plotted against Ri on a logarithmic scale.

When $-0.01 < Ri < +0.01$, F is within 10% of unity, and this range is often taken to define "fully forced" convection. As Ri approaches $+0.2$, F tends to zero and turbulent exchange is completely inhibited. When Ri is less than -1, it is generally assumed that "free" convection dominates and the value of F then exceeds 8.

The relations shown in Figure 16.10 were all derived from measurements over extensive, and relatively smooth, flat surfaces such as short grass, and at heights where $(z - d)/z_0$ was generally 10^2–10^3, i.e., the measurements were well above the division between the roughness sublayer and the inertial sublayer. Several studies over aerodynamically rough

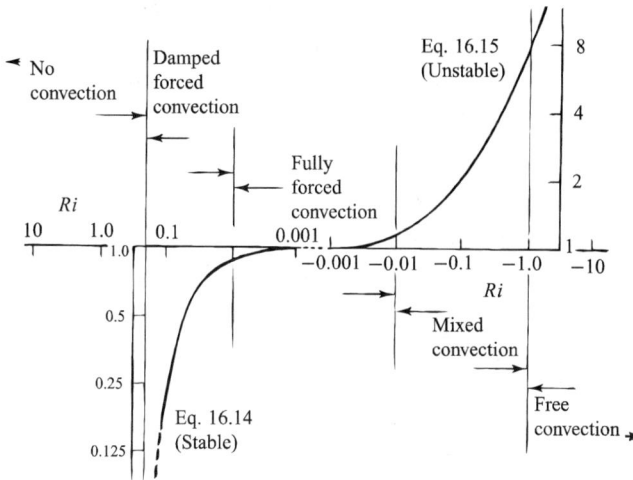

FIGURE 16.10   The "stability factor" F plotted logarithmically against the Richardson number Ri. Fluxes calculated in non-neutral conditions, using profile-gradient equations valid for neutral conditions must be multiplied by F (from Thom, 1975).

surfaces (forests and scrub) have suggested that, in unstable and neutral conditions, values of F measured with $(z - d)/z_0$ in the range 10–50 were up to twice the values in Figure 16.10 [Garratt, 1978; Thom et al, 1975]. The causes of the discrepancy seem to be the wakes generated by the roughness elements, and thermals rising between the elements; e.g., between the trees in a forest [Raupach 1995]. In stable conditions the discrepancy appears much less. Such problems arise when measurements are made in the roughness sublayer rather than the inertial sublayer, but it is often difficult to satisfy the strict fetch requirements over tall, rough crops.

16.3 Methods for Indirect Measurements of Flux above Canopies

It was shown earlier that eddy covariance provides a method for *direct* measurements of turbulent fluxes of entities between vegetation canopies and the atmosphere. There are two *indirect* methods for deducing fluxes from measurements taken in the constant flux layer above extensive canopies. The indirect methods rely on the measurement of mean potentials and their gradients in the atmosphere, and are usually referred to as the "aerodynamic" and "Bowen ratio" methods. Although, with the advent of commercially available eddy covariance sensors, the indirect "gradient" methods are less commonly applied than previously, they can be valuable because they require less sophisticated instrumentation and can be used for entities for which fast detectors are not available.

AERODYNAMIC METHOD

The method relies on the existence of relations between fluxes and gradients of the form described earlier (Equations 16.16 and 16.33)

$$\tau = K_M \rho \partial u / \partial z$$
$$\mathbf{C} = -K_H / [\partial(\rho c_p T)/\partial z]$$
$$\mathbf{E} = -K_V / (\partial \chi / \partial z)$$
$$\mathbf{F} = -K_S / (\partial s / \partial z)$$

The similarity hypothesis states that, in neutral stability,

$$K_M = K_H = K_V = K_S$$

Consequently,

$$\frac{-\rho c_p \partial T / \partial z}{\mathbf{C}} = \frac{\rho \partial u / \partial z}{\tau} \tag{16.47}$$

Similar equalities may be written between $\mathbf{E}$ and $\mathbf{C}$ or $\mathbf{F}$ and τ. By rearranging Equation 16.47 and setting $\tau = \rho u_*^2$ it can be shown that

$$\mathbf{C} = -c_p(\partial T / \partial u)\tau = -\rho c_p(\partial T / \partial u)u_*^2 \tag{16.48}$$

and it follows that
$$\mathbf{E} = -(\partial \chi / \partial u)u_*^2 \tag{16.49}$$

and
$$\mathbf{F} = -(\partial S / \partial u)u_*^2 \tag{16.50}$$

In neutral stability, u_* can be estimated from the wind profile alone, and so the aerodynamic method requires only two sets of profiles: temperatures or concentrations of water vapor or gas are measured at a series of heights above the crop, and windspeed is measured at identical heights. The friction velocity is found from the wind profile, and the gradient $\partial T / \partial u$ is found by plotting values of T against u (similarly for χ or S). Flux is then calculated from Equation 16.48 (or Equations 16.49–16.50) If a fast response anemometer is available, a hybrid eddy covariance/aerodynamic method may be used, replacing u_*^2 in Equations 16.48 to 16.50 with the measured value $\overline{u'w'}$, thus avoiding the empiricism of wind profile analysis.

An alternative way of applying the aerodynamic method eliminates u_* by differentiating the wind profile equation so that

$$\frac{\partial u}{\partial[\ln(z - d)]} = \frac{u_*}{k}$$

Substituting in Equation 16.48 for u_* gives

$$C = -\rho c_p k^2 \frac{\partial u}{\partial[\ln(z - d)]} \frac{\partial T}{\partial[\ln(z - d)]} \qquad (16.51)$$

The minimum number of heights over which the gradients may be determined is two. If the heights are distinguished by subscripts 1 and 2, Equation 16.51 becomes

$$C = \rho c_p k^2 \frac{(u_1 - u_2)(T_2 - T_1)}{\{\ln[(z_2 - d)/(z_1 - d)]\}^2} \qquad (16.52)$$

Similar equations can be written for **E** and **S**.

An equation of this form was first derived to calculate water vapor transfer by Thornthwaite and Holzman [1942] and has been used in many subsequent studies of transfer in the turbulent boundary layer. Its main defect is the dependence on wind and temperature (or humidity or mass) at two heights only, so that the estimate of flux is sensitive to the error in a single instrument or to local irregularities of the site. More accurate estimates of flux can therefore be obtained by using Equations 16.48 to 16.50 with temperature and windspeed measured at four or more heights than by using Equation 16.52 with only two heights.

In non-neutral conditions it is necessary to know profiles of u and T to estimate u_* from wind profile analysis (though it could still be measured directly by eddy covariance), and equality of K_M, K_H, K_V and K_S cannot be assumed. It can be shown that Equation 16.52 takes the generalized form

$$C = \rho c_p k^2 \frac{(u_1 - u_2)(T_2 - T_1)}{\{\ln[(z_2 - d)/(z_1 - d)]\}^2} (\Phi_H \Phi_M)^{-1} \qquad (16.53)$$

where $(\Phi_H \Phi_M)^{-1}$ is the stability factor F defined earlier. Similar equations may be written for **E** and **S**. Equation 16.53 may then be evaluated using Equations 16.45 or 16.46.

An alternative approach [Biscoe et al, 1975], avoiding the errors inherent in "two point" profiles, makes use of Webb's [1970] relation for near neutral and stable conditions (i.e., $-0.03 < (z - d)/L < +1$) as given in Equation 16.43.

This equation allows three steps to be taken. First, equality of Φ_M and Φ_H means that the profiles of windspeed and temperature have the same shape, so a plot of temperature versus windspeed will give a straight line with slope $\partial T/\partial u$. Second, Φ_M is a linear function of $(z - d)/L$, and so Equation 16.34) can be integrated to give the wind profile equation

$$u = (u_*/k)\{\ln[(z - d)/z_0] + 5(z - d - z_0)/L\} \qquad (16.54)$$

Third, because $\mathbf{C} = -\rho c_p (\partial T/\partial u)u_*^2$, the Monin–Obukhov length (Equation 16.38) may be written $L = u_* T/[kg(\partial T/\partial u)]$, and so Equation 16.54 may be written

$$u_c = (u_*/k) \ln[(z - d)/z_0] \qquad (16.55)$$

where

$$u_c = u - [5(z - d - z_0)(\partial T/\partial u)gT^{-1}]$$

Equation 16.55 can be used to find u_*/k as the slope of the straight line defined by plotting u_c against $\ln(z - d)$. When u_* is known,

$$\mathbf{C} = -\rho c_p (\partial T/\partial u)u_*^2$$

Similar expressions can be used to find fluxes of water vapor and other entities in stable and slightly unstable conditions. Paulson [1970] derived profile equations for the more complicated case when the atmosphere is more unstable.

Note: It was stressed on p. 321 that temperature gradients in flux equations should strictly be replaced by potential temperature gradients. Furthermore, humidity should be expressed as mixing ratio r (mass per unit mass of dry air) because E is strictly proportional to $\partial r/\partial z$, and r is conserved when pressure changes with height. The specific humidity q (mass per unit mass of moist air) is also conserved, and is often used in meteorological literature, but, as discussed on p. 307, Webb et al [1980] showed that r is strictly the correct ratio for flux calculations because it defines fluxes in a coordinate system fixed relative to the surface (i.e., there is no net vertical flux of dry air). However, for consistency with earlier chapters, and because the errors involved are usually small, temperature and vapor pressure are retained in the analysis that follows.

BOWEN RATIO METHOD

The Bowen ratio formula for flux measurement is derived from the energy balance of the underlying surface, which can be rewritten in the form

$$\lambda \mathbf{E} = \frac{\mathbf{R}_n - \mathbf{G}}{1 + \beta} \tag{16.56}$$

where β is the Bowen ratio $\mathbf{C}/\lambda \mathbf{E}$. Measurements of the net radiation ($\mathbf{R}_n$) and soil heat flux ($\mathbf{G}$) are needed to establish ($\mathbf{R}_n - \mathbf{G}$), and β is found from measurements of temperature and vapor pressure at a series of heights within the constant flux layer. Assuming the transfer coefficients of heat and water vapor are equal, it can be shown that

$$\beta = \mathbf{C}/\lambda \mathbf{E} = \gamma \, \partial T / \partial e \tag{16.57}$$

and $\partial T / \partial e$ is found by plotting the temperature at each height against vapor pressure at the same height.

The Bowen ratio method can be generalized by writing the heat balance equation as

$$\mathbf{R}_n - \mathbf{G} = -K \rho c_p (\partial T / \partial z) - K \rho c_p \gamma^{-1} \partial e / \partial z = K \rho c_p (\partial T_e / \partial z) \tag{16.58}$$

where K is a turbulent transfer coefficient assumed identical for heat and water vapor and T_e is the equivalent temperature $T = T + (e/\gamma)$. The Bowen ratio Equation 16.57 is derived from Equation 16.58 by writing $T_e = T + (e/\gamma)$.

By writing the sensible heat flux as $\mathbf{C} = -K \rho c_p (\partial T / \partial z)$ and forming similar expressions relating the latent heat flux $\lambda \mathbf{E}$ to $\gamma^{-1} (\partial e / \partial z)$ and the flux $\mathbf{F}$ of any other gas to $\partial S / \partial z$, it can be shown that, combining each expression in turn with Equation 16.58

$$\mathbf{C} = (\mathbf{R}_n - \mathbf{G})(\partial T / \partial T_e)$$
$$\lambda \mathbf{E} = (\mathbf{R}_n - \mathbf{G})(\partial e / \partial T_e) \gamma^{-1}$$
$$\mathbf{F} = (\mathbf{R}_n - \mathbf{G})(\partial S / \partial T_e)(\rho c_p)^{-1}$$

The aerodynamic and Bowen ratio methods of flux determinations are usually applied to potentials that have been averaged for periods of a half to one hour. Fluctuations in the potentials, especially on a day of intermittent cloud, often preclude the estimation of mean fluxes for shorter periods. On the other hand, diurnal changes make time-averaging undesirable for periods of more than two hours, particularly near sunrise and sunset when conditions could no longer be treated as stationary.

16.4 Relative Merits of Methods of Flux Measurement

The eddy covariance technique has the advantages of elegance and a sound theoretical basis, but it requires fast-response sensors and rapid data acquisition. It is widely used over crops, forests, and natural vegetation in studies to understand relationships between vegetation and the atmosphere. Over the last decade the technology has become sufficiently reliable to allow continuous measurements for periods of months to years [Goulden et al, 1996], opening up prospects for studying seasonal and annual variability in plant-atmosphere exchange. There remain difficulties in applying the method over tall, heterogeneous canopies, sloping surfaces, and when atmospheric conditions change rapidly. The method also can be subject to error if the size of the instrument sensing path is larger than the dominant eddy size; this can occur close to aerodynamically smooth surfaces. Lee et al [2005] reviewed the theoretical principles and practical difficulties of the eddy covariance method

The aerodynamic method is straightforward in neutral stability, when profiles of only windspeed and the entity in question are required. It is unreliable at low windspeeds when mechanical anemometers may stall for part of the measuring period, but one- or two-dimensional sonic anemometers can avoid this limitation. Empirical corrections for stability necessitate measurements of temperature profiles, and are not well defined in strongly stable conditions (e.g., clear nights with low windspeeds): the accepted corrections appear valid over short vegetation, but the aerodynamic method appears to underestimate fluxes seriously over tall rough vegetation unless it is possible to make measurements well above the roughness sublayer but still within the inertial sublayer.

The Bowen ratio method, which assumes that $K_H = K_V$, does not require stability corrections and so is often the preferred of the two gradient techniques, but it becomes indeterminate when $\mathbf{R}_n - \mathbf{G}$ tends to zero, and is generally difficult to apply at night or in other conditions when net radiation is small.

Stannard [1997] analyzed theoretically the fetch requirements for eddy covariance (EC) and Bowen ratio (BR) measurements for the minimal case where only two heights are used to determine Bowen ratio. Assuming that the EC instrumentation was placed at a height corresponding to the top level for the BR measurements, the BR measurement required less fetch than EC. Differences in fetch requirements were largest over smooth surfaces. Fetch for BR measurements can be significantly reduced by reduc-

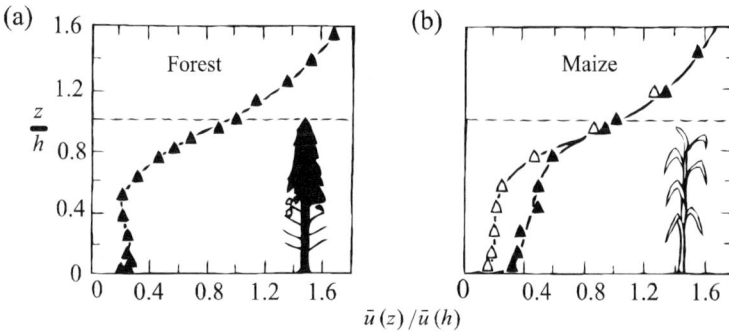

FIGURE 16.11 (a) Profile of mean windspeed in a pine forest canopy ($h = 16$ m); data averaged from 18 very-near- neutral one-hour runs. (b) Profiles of mean windspeed in a maize canopy ($h = 2.1$ m) during periods of light wind [$\bar{u}(h) = 0.88$ m s^{-1}, ▲] and strong wind [$\bar{u}(h) = 2.66$ m s^{-1}, Δ] (from Raupach and Thom, 1981).

ing the height of the lower measurement level and lowering the upper sensors, whereas issues of the scale of turbulence often limit the lower level for EC measurement. A particular *caveat* is the potential source of error when measurements are taken in the roughness sublayer [Stewart and Thom, 1973].

16.5 Turbulent Transfer in Canopies

Turbulence in plant canopies differs in several important ways from turbulence in boundary layers above canopies. Before the mid-1970s it was generally thought that the observed high levels of turbulence in canopies were caused by eddies generated by flow over leaves, branches and other structures. Since such turbulence would be fine scale in relation to canopy height, it was believed that turbulent transport in canopies could be described by flux gradient equations similar to Equation 16.33, and analysis by these methods was described in the first two editions of this book. But through the 1970s and 1980s experimental evidence emerged to demonstrate that mass and momentum were often transported in canopies in opposite directions to the mean gradients of the entities. Figure 16.11 illustrates one of these situations for a forest canopy.

The vertical flux of momentum must be downward throughout a canopy because there are no sources of mean momentum in the canopy. Consequently $\frac{\partial u}{\partial z}$ should decrease monotonically throughout the canopy, but in

Figure 16.11 the wind profile has a secondary maximum in the trunk space. If flux was proportional to mean gradient, there would need to be a source of momentum associated with this secondary maximum.

Finnigan and Brunet [1995] reviewed the emerging understanding of canopy flow from wind tunnel and field experiments. In the inertial sub-layer above rough surfaces, turbulent kinetic energy (TKE), produced by shear at the surface, "cascades" in a continuous spectrum from large- to small-scale eddies, and eventually is converted to heat. Within canopies, TKE may also be produced in the wakes of plants or by waving and flutter-ing leaves, contributing to the spectrum of energy. Additionally, Baldocchi and Meyers [1988] identified a "spectral shortcut" process whereby eddies generated from shear production are "chopped up" into finer scale eddies when they are intercepted by foliage, and are rapidly converted to heat. It appears however that eddies directly generated in the wakes of foliage are much less important than large-scale, intermittent turbulent eddies in transporting momentum and scalars in canopies. Full details of how these eddies develop and are related to canopy structure are not yet clear, but Finnigan and Brunet [1995] speculated that the mechanism was similar to that generated in the turbulent layer formed when two streams with differ-ent velocities mix in a wind tunnel. The resulting mean windspeed profile is strikingly similar to the shape of the wind profile in and above a canopy (e.g., Figure 16.11). They concluded, by analogy with the turbulent struc-ture observed in wind tunnel mixed layer experiments, that the point of inflection in the wind profile near the top of a vegetation canopy might trigger instability that generates large scale turbulent structures, with the eddy size related to $(h - d)$, where h is canopy height and d is zero plane displacement, and with the rate of eddy production being proportional to the shear magnitude at the inflection point.

Consistent with this theory, Baldocchi and Meyers [1991] found that the dominant time scale for turbulent fluxes in a deciduous forest canopy was 200–300 s. At about this frequency, large scale eddies swept down into the trunk space, resulting in ejection of air from the canopy. These brief events were followed by relatively quiescent periods during which humidity and respired carbon dioxide built up again. As a result of this quasi-periodic process, evaporation from the forest floor and understory was more closely coupled to the above-canopy saturation deficit than to the in-canopy environment (see also p. 348).

Although it is seldom appropriate to use flux-gradient methods to mea-sure fluxes within plant canopies, eddy covariance methods are successful provided that sampling durations are over long enough periods to measure

most of the turbulent events that contribute to the fluxes [Baldocchi and
Meyers, 1999].

16.6   Density Corrections to Flux Measurements

Earlier (p. 308), attention was drawn to problems that arise when eddy
covariance fluxes of an atmospheric constituent are measured with sensors
that detect density rather than mixing ratio. In such cases it was shown
that it is usually necessary to take into account the simultaneous fluxes of
other entities, in particular heat and water vapor. These issues also apply
to fluxes measured by gradient methods. Sources of heat or water vapor
result in expansion of the air and so affect the density of a constituent
(but not its mixing ratio r). For example, if gradients of CO_2 above a
crop were measured by drawing air through a gas analyzer from various
heights sequentially without altering the temperature or humidity of the air
sample, the apparent CO_2 gradient would be in error, because the density
of CO_2 in the air at each height would be influenced by the content of
water vapor and by the temperature at that height. Correction would not be
necessary if the flux was evaluated from measurements of the mixing ratio
of the constituent (mass of constituent per unit mass of dry air), or if the
samples were allowed to reach a constant temperature and pressure before
analysis. Webb, Pearman, and Leuning [1980] derived expressions relating
the correction δF required to allow for fluxes of sensible and latent heat
when a mass flux is deduced from density measurements. Their analysis
shows that, for a typical situation with $T_a = 20°C$ and $e = 1.0$ kPa, when
C and λE are in W m^{-2}, and δF is in kg m^{-2}s^{-1}

$$\delta F = (\overline{\rho_c/\rho_a})(0.65 \times 10^{-6}\lambda E + 3.36 \times 10^{-6}C). \tag{16.59}$$

The mean density of the constituent is ρ_c, and ρ_a is the density of dry air.
Equation 16.59 shows that the correction for sensible heat flux is typically
about five times larger than that for an equivalent flux of latent heat. For
CO_2, with a mean content of 330 vpm in dry air, $\rho_c/\rho_a = 0.502 \times 10^{-3}$
so that, when $\lambda E = C = 250$ W m^{-2}, $\delta F = 0.5$ mg m^{-2} s^{-1}, a value
that is comparable with typical CO_2 fluxes over crops of 1–2 mg m^{-2} s^{-1}.
The correction is therefore important if appropriate methods of avoiding
density gradients of CO_2 (e.g., allowing sampled air to reach a standard
temperature) have not been adopted. The situation is exacerbated in semi-
arid regions where sensible heat fluxes may be much larger than latent
heat fluxes. Equation 16.59 must also be applied to analysis of other trace
gases. For example, for SO_2 at 0.010 vpm, $\overline{\rho_c/\rho_a} = 0.22 \times 10^{-6}$ and

so $\delta \mathbf{F} = 0.2 \ \mu\mathrm{g \ m}^{-2} \ \mathrm{s}^{-1}$, comparable with reported fluxes [Fowler and Unsworth, 1979].

<h1>16.7   Problems</h1>

1. Mean windspeeds $u(z)$, averaged over 30 minutes, were measured simultaneously at several heights z above the canopy of an extensive poplar plantation, height $h = 5.0$ m, in conditions when a logarithmic wind profile was expected. Results were:

Height above ground z (m)	6.0	7.0	9.0	12.0	18.0	30.0
Windspeed u (m s^{-1})	0.832	0.994	1.198	1.386	1.610	1.858

   (i) Estimate the roughness length of the canopy z_0, and the friction velocity u_*. (Assume the zero plane displacement d was 4.0 m, and von Karman's constant $k = 0.41$.)

  (ii) Give some possible reasons why the values for d and z_0 are different for this canopy than values you would estimate from equations in this chapter.

 (iii) Estimate the aerodynamic resistance for momentum transfer r_{aM}, relative to a reference height 12.0 m above the ground.

2. At 40 m above the ground, over a coniferous forest with a canopy height (h) of 25 m, mean windspeed u was 5.0 m s^{-1} in conditions when a logarithmic profile was expected.

   (i) Assume the value of the zero plane displacement d was 0.60 h and the roughness length z_0 was 0.10 h. Hence, calculate the friction velocity u_* (von Karman's constant k is 0.41).

  (ii) Calculate the wind speed at 30 m above the ground and estimate the aerodynamic boundary layer resistance for momentum (r_{aM}) between the height 30 m and the height $(d + z_0)$.

3. The following measurements of temperature and vapor pressure were made simultaneously above a crop of barley 0.8 m high when net radiation above the canopy was 450 W m^{-2} and downward soil heat flux was 20 W m^{-2}.

Height z (m)	2.00	0.90
Temperature ($^\circ$C)	20.00	21.00
Vapor pressure (kPa)	1.500	1.633

(i) Calculate the Bowen ratio and the sensible and latent heat fluxes from the crop.

(ii) Assuming temperature and vapor pressure varied logarithmically with height, the zero plane displacement was 0.60 m and the roughness length was 0.10 m, estimate graphically or otherwise the values of the temperature T_0 and vapor pressure e_0 at the height of the apparent sink for momentum.

(iii) If heat fluxes from the crop can be treated as analogous to those from a "big leaf" situated at the height of the apparent sink for momentum, calculate the vapor pressure within this "leaf" and hence calculate the resistances for sensible and latent heat transfer between the "big leaf" and the height $z = 2.00$ m.

4. Profiles of windspeed u and carbon dioxide concentration c were measured in neutral stability above a wheat crop, height 0.80 m. Results were:

z (m)	1.10	1.30	1.60	2.10	2.46
u (m s^{-1})	1.68	1.93	2.19	2.49	2.65
c (vpm)	324.5	326.2	327.9	330.0	331.1

(i) Find graphically or otherwise, values of the zero plane displacement, roughness length, and friction velocity.

(ii) Calculate the momentum flux density to the crop.

(iii) Calculate the aerodynamic resistance to momentum transfer between the height 2.10m and the crop.

(iv) Calculate the flux density of CO_2 to the crop in g CO_2 m^{-2} h^{-1} (assume 330 vpm CO_2 is 605 mg CO_2 m^{-3}).

5. Profiles of windspeed u and ozone concentration S were measured at several heights z in neutral stability over an extensive field of beans 0.30 m high. Results were:

z(m)	0.35	0.50	0.90	1.75	3.20
u(m s^{-1})	0.95	1.23	1.61	1.99	2.31
$S(\mu g\,m^{-3})$	83.0	87.0	90.6	96.0	99.5

By plotting appropriate graphs, or otherwise, determine (i) the zero plane displacement, (ii) the roughness length, (iii) the friction velocity, (iv) the momentum flux density, (v) the ozone flux density, and (vi) the deposition velocity for ozone, referred to height $z = 1.75$ m.

6. For an atmosphere with temperature T and vapor pressure e at a reference height over an open water surface with temperature T_s, explain with the aid of a diagram how Penman derived the relationship $e'_{(T_s)} = e'_{(T)} + \Delta(T_s - T)$, showing clearly the meaning of the term Δ. ($e'_{(T)}$ is the saturation vapor pressure at T). Hence, show that the Bowen ratio β is given by:

$$\beta = \gamma \frac{(T_s - T)}{\left\{ (e'_{(T)} - e) + \Delta(T_s - T) \right\}}$$

where γ is the psychrometer constant. On a day with air temperature $T = 25°C$ and vapor pressure $e = 2.00$ kPa, the Bowen ratio over a flooded field of rice was 0.10. By considering the relationship between the Bowen ratio and T_s, find graphically or otherwise the temperature of the water surface.

Chapter 17

Micrometeorology (ii) Interpretation of Measurements

17.1 Resistance Analogues

Measurements of fluxes by micrometeorological methods are of relatively little value to the ecologist, agricultural or forest scientist unless they can be associated with some factor or group of factors that describes how the canopy or landscape controlled or responded to the flux. A useful way of extending the study of transfer from single leaves to complex canopies, and of pointing out the shortcomings of certain approaches, is to consider the canopy as an electrical analogue as in Figure 17.1. The rate of exchange (flux density) of an entity between a *single leaf* and its environment can be estimated when (a) the potential of the entity (e.g., the vapor pressure or CO_2 concentration) is known at the leaf and in the surrounding air and (b) the relevant resistances (e.g., stomatal and leaf boundary layer) can be measured or estimated. In the same way, the bulk exchange of any entity between a canopy and the air above it can be estimated by measuring the potentials at two or more heights above the canopy (z_1, z_2, etc.) if the resistances across these potentials are also known. Within the canopy, resistances corresponding to the stomata and boundary layers of individual leaves have a clear physical significance, but the validity of describing transfer in the air within the canopy by Ohm's Law analogues (as shown in Figure 17.1) is discussed later in this chapter.

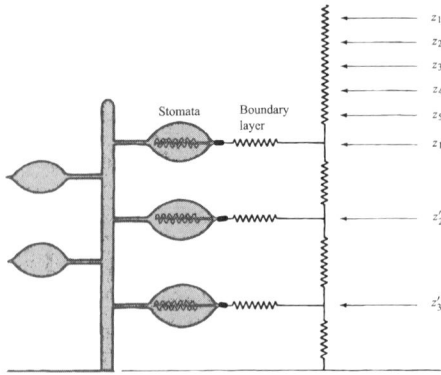

FIGURE 17.1　Resistance model for a plant in a stand of vegetation.

CANOPY RESISTANCE

It is possible to derive a parameter playing the same part in equations for the water vapor exchange of a canopy as the stomatal resistance plays in similar equations for a single leaf, thus describing the canopy by a "big leaf model" (p. 250). The parameter will be given the symbol r_c, where the subscript denotes canopy, crop, or cover.

It was shown in Chapter 16 that the rate of sensible heat loss from a surface can be written in the form

$$\mathbf{C} = -\rho c_p u_*^2 \, (\partial T / \partial u)$$

where T is a linear function of u. As a special case, the gradient $\partial T / \partial u$ can be written $\{T(z) - T(0)\}) / \{u(z) - 0\}$ where $T(0)$, obtained by the extrapolation shown in Figure 17.2, is the air temperature at the height where $u = 0$, i.e., $z = d + z_0$ (see p. 313). The above equation may therefore be written as

$$\mathbf{C} = -\rho c_p u_*^2 \, \{T(z) - T(0)\} / u(z)$$
$$= -\rho c_p \, \{T(z) - T(0)\} / r_{aH} \tag{17.1}$$

where $r_{aH} = u(z)/u_*^2 = [\ln(z - d) - \ln(z_0)]/u_* k$ can be regarded as an aerodynamic resistance to sensible heat transfer between a fictitious surface at the height $d + z_0$, and the height z. Similarly, it can be shown that

$$\lambda \mathbf{E} = \frac{-\rho c_p}{\gamma} \frac{\{e(z) - e(0)\}}{r_{aV}} \tag{17.2}$$

where $e(0)$ is the value of the vapor pressure extrapolated to $u = 0$ and the aerodynamic resistance r_{aV} to water vapor transfer is assumed identical to that for sensible heat transfer, i.e., $r_{aV} = r_{aH} = u(z)/u_*^2$. The diffusion of water vapor between the intercellular spaces of leaves in a canopy and the atmosphere at height z can now be described formally by the equation

$$\lambda E = \frac{-\rho c_p}{\gamma} \frac{\{e(z) - e(T(0))\}}{r_{aV} + r_c} \tag{17.3}$$

This relation defines the *canopy resistance* r_c and is identical to the corresponding equation for an amphistomatous leaf, with r_c replacing the stomatal resistance r_s (p. 244) and r_{aV} replacing leaf boundary layer resistance r_H. Introducing the energy balance equation, writing $r_{aH} = r_{aV} = r_a$ and eliminating $T(0)$ (see p. 238) gives

$$\lambda E = \frac{\Delta (R_n - G) + \rho c_p \{e_s(T(z)) - e(z)\}/r_a}{\Delta + \gamma^*} \tag{17.4}$$

where $\gamma^* = \gamma (r_a + r_c)/r_a$ Values of r_c for a given stand can therefore be derived either directly from profiles of temperature, humidity and windspeed using Equations 17.1, 17.2, and 17.3, or indirectly from the

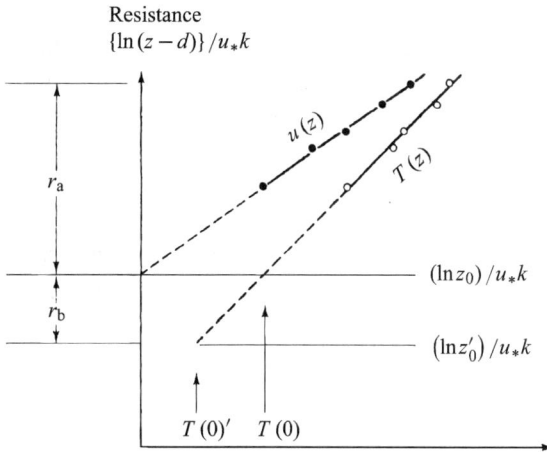

FIGURE 17.2 Diagram showing how $T(0)$ and $T(0')$ are determined by plotting temperature against $\ln(z - d)/u_*k$ (Equation 17.6). $T(0)$ is the value of temperature extrapolated to $z - d = z_0$, and $T(0')$ is the value extrapolated to $z - d = z_0'$. The significance of r_a and r_b is shown on the left-hand axis and is discussed on p. 339.

Penman–Monteith Equation 17.4 when the relevant climatological parameters are known and λE is measured or estimated independently.

Two objections can be raised to this apparently straightforward method of separating the aerodynamic and physiological resistances of a crop canopy. In the first place, the values of r_c derived from measurements are not unique unless the sources (or sinks) of sensible and latent heat have the same spatial distribution. In a closed canopy, fluxes of both heat and water vapor are dictated mainly by the absorption of radiation by the foliage. Provided that the stomatal resistance of leaves does not change sharply with depth in the part of the canopy where most of the radiation is absorbed, distributions of heat and vapor sources will usually be *similar* but will seldom be *identical.* Conversely, anomalous values of r_c are likely to be obtained in a crop with little foliage if evaporation from bare soil beneath the leaves makes a substantial contribution to the total flux of water vapor. In the second place, the analysis cannot yield values of r_c, which are strictly independent of r_a unless the apparent sources of heat and water vapor, as determined from the relevant profiles, are at the same level $(d + z_0)$ as the apparent sink for momentum. This is a more serious restriction. Form drag, rather than skin (molecular) friction (p. 145), is often the dominant mechanism for the absorption of momentum by vegetation, so that the resistance r_{aM} to the exchange of momentum between a leaf and the surrounding air is smaller than the corresponding resistances to the exchange of heat and vapor, which depend on molecular diffusion alone. It follows that the *apparent* sources of heat and water vapor will, in general, be found at a lower level in the canopy than the *apparent* sink of momentum, say at $z = d + z_0'$ rather than at $z = d + z_0$ where z_0' is smaller than z_0. Atmospheric resistances to heat and mass transfer may therefore be described in terms of r_{aM}, the aerodynamic resistance to momentum transfer, and r_b, an additional resistance, assumed to be the same for heat and water vapor (see Figure 17.2). In Chapter 16 it was shown, from the definition of r_{aM} and the wind profile equation in neutral stability, that

$$
\begin{aligned}
r_{aM} &= \rho u\left(z\right)/\tau = u\left(z\right)/u_*^2 \\
&= \left\{\ln\left[\left(z - d\right)/z_0\right]\right\}/ku_* \\
&= \left\{\ln\left[\left(z - d\right)/z_0\right]\right\}^2/k^2 u\left(z\right)
\end{aligned}
\qquad (17.5)
$$

In the derivation of this expression the effective height for the sink of momentum is $z = d + z_0$. If z_0 were independent of windspeed, Equation 17.5 shows that $1/r_{aM}$ would be proportional to $u(z)$. Figure 17.3 illustrates this for values of roughness lengths z_0 appropriate for short grass, cereal crops, and forests. In practice the roughness length of many crops decreases as windspeed increases (pp. 316–317) and $1/r_{aM}$ is approximately constant over a range of low windspeeds.

ADDITIONAL AERODYNAMIC RESISTANCE FOR HEAT AND MASS TRANSFER

In a similar manner to Equation 17.5, the aerodynamic resistance between a height z above the ground and the apparent source (or sink) of heat and vapor at a height $d + z_0'$ can be written as

$$r_a = \frac{\ln\left[(z - d)/z_0'\right]}{ku_*} = \frac{\ln\left[(z - d)/z_0\right]}{ku_*} + \frac{\ln\left[z_0/z_0'\right]}{ku_*} \qquad (17.6)$$

$$= r_{aM} + r_b$$

where r_b is the additional aerodynamic resistance. The implications of Equation 17.6 are shown in Figure 17.2.

Extending this analysis to non-neutral stabilitiy, in near-neutral and stable conditions the resistance r_{aM} derived from Equation 16.54 is

$$[ku_*]^{-1}\left[\ln\left(\frac{z-d}{z_0}\right) + \frac{5(z-d)}{L}\right] \qquad (17.7)$$

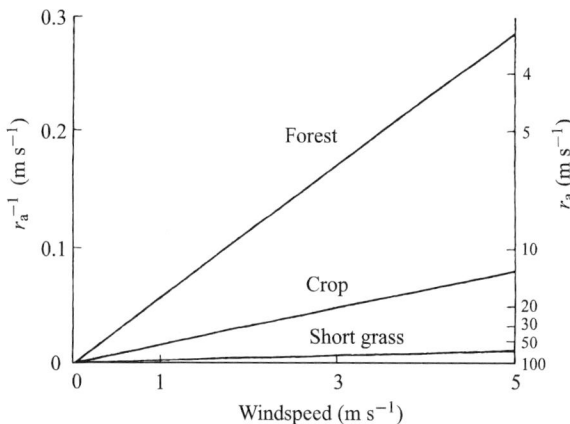

FIGURE 17.3 Calculated values (from Equation 17.5) of resistance r_a in relation to windspeed over surfaces with roughnesses characteristic of: short grass ($z_0 = 1$ mm, $d = 7$ mm); cereal crop ($z_0 = 0.2$ m, $d = 0.95$ m); forest ($z_0 = 0.9$ m, $d = 11.8$ m). Windspeeds are referred to a standard height $z - d = 5$ m for each surface.

provided that $z - d$ is at least an order of magnitude greater than z_0. When $[\ln(z - d)/z_0]/ku_*$ is identified as the aerodynamic resistance to momentum transfer in *neutral* stability, the additional component $+[5(z - d)/L]/ku_*$ can be regarded as a *stability resistance*. Similarly, Equation 16.42 can be used to show that the stability resistance in more unstable conditions (L negative) is approximately $[4(z - d)/L]/ku_*$. The tendency for $1/r_{aM}$ to become independent of windspeed as the speed decreases is more pronounced in unstable than in neutral or stable conditions because the decrease in turbulent energy associated with decreased friction is compensated by an increase in the supply of energy from buoyancy.

The resistance r_b is $[\ln(z_0/z_0')]/ku_*$ (Equation 17.6), and $ku_* r_b$ is identical to the parameter kB^{-1} ($= \ln(z_0/z_0')$), which a number of workers have used to analyze processes of exchange at rough surfaces. Values of kB^{-1} summarized by Massman [1999] and Su et al [2001] for natural surfaces cover a wide range. A dense forest had kB^{-1} close to zero, implying that the roughness lengths for heat and momentum were very similar and r_b was negligibly small. Over bare soil, kB^{-1} ranged up to 7. Su et al [2001] used models based on canopy structure and multiple heat sources to estimate kB^{-1} for canopies of cotton ($kB^{-1} = 4$), savanna (≈ 5), and grass (≈ 5). Campbell and Norman [1998] suggested that $z_0/z_0' = 5$ ($kB^{-1} = 1.6$) was a reasonable working assumption, but strictly the ratio should depend on windspeed.

The dependence of kB^{-1}, and therefore of r_b, on roughness and windspeed for real and for model vegetation was studied by Chamberlain [1966] and Thom [1972], and the results of their work can be summarized as follows:

(i) For a given value of u_*, kB^{-1} and r_b were almost constant over a wide range of surface roughnesses. For example, a set of measurements of evaporation from an artificial grass surface gave $z_0 = 1$ cm, $u_* = 0.25$ m s^{-1}, $kB^{-1} = 1.8$. At the same value of u_* achieved at a higher windspeed over towelling with $z_0 = 0.045$ cm, kB^{-1} was 1.9. The corresponding resistances r_b are 18 and 19 s m^{-1}

(ii) For a given value of z_0, kB^{-1} increased with windspeed and therefore with u_*. For a fourfold increase of u_* from 0.25 to 1.00 m s^{-1}, kB^{-1} increased by a factor of 1.3 for the grass and 1.7 for the towelling. For evaporation from a bean crop, Thom found that $kB^{-1}(= ku_* r_b) = Au_*^{0.33}$ where the constant A had the value 2.54 when u_* was in m s^{-1}. This implies that $r_b \propto u_*^{-0.67}$.

(iii) The value of z_0' and hence of r_b is expected to depend on the molecular diffusivity of the property being transferred. On the assumption that $r_b \propto (\text{diffusivity})^n$, values of n determined experimentally range from about -0.8 to about -0.3. For a stand of beans, n appeared to be about -0.66, implying that r_b for heat may be 10% greater than r_b for water vapor. This difference is often ignored, given the uncertainty with which r_b is known.

Values of r_b are seldom determined directly in crop micrometeorology, but are usually estimated from the results such as those above. Thom's empirical equation

$$r_b = 6.2 u_*^{-0.67} \tag{17.8}$$

is an adequate approximation for estimating r_b (in s m^{-1} when u_* is in m s^{-1}) with reference to heat and water vapor transfer to crops, at least over the typical range of u_*, 0.1–0.5 m s^{-1}. Values of r_b for a gas with diffusivity much different from the value for water vapor may be estimated from Equation 17.8 using the approximations $r_b \propto D^{-0.67}$ and $D_g = D_v (M_v / M_g)^{0.5}$ where M is molecular weight, D diffusivity and the subscripts g and v refer to the gas and water vapor respectively.

When vegetated surfaces are unusually rough or fibrous (e.g., pine needles), Equation 17.8 may not be a good approximation to estimating r_b. Thom [1972] discussed more detailed treatments, Wesely and Hicks [1977] suggested the expression

$$r_b = 2(ku_*)^{-1}(\kappa/D)^{0.67}$$

or the models of Massman [1999] or Su et al [2001] may be applied.

For vapor transfer to *rigid* rough surfaces, Figure 17.4 shows that the expression

$$B^{-1} = 7.3 \, \text{Re}_*^{0.25} \, \text{Sc}^{0.5} - 5.0$$

proposed by Chamberlain et al [1984] on the basis of analysis by Brutsaert [1982] fits observations well, and may be useful for estimating r_b for ploughed fields or urban areas; Re_* is the roughness Reynolds number $u_* z_0 / v$ and Sc is the Schmidt number v/D. The expression is applicable also to the diffusive transfer of particles in the size range where impaction and sedimentation are unimportant, i.e., radius $r \lesssim 0.5$ μm (see Chapter 12).

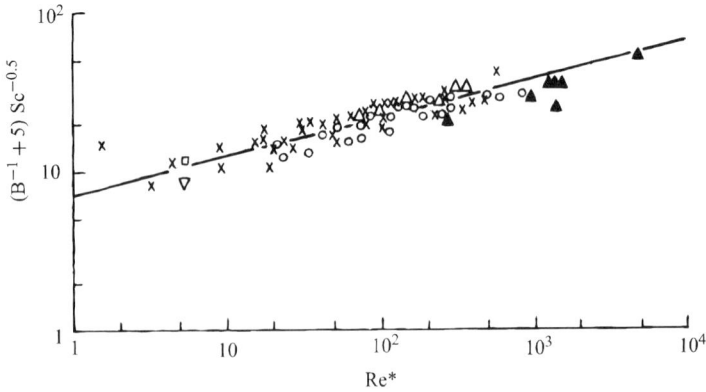

FIGURE 17.4 Diffusive transport to rough surfaces (from Chamberlain et al, 1984). Symbols refer to different surface structures, vapors and particles: ^{212}Pb vapor $\triangledown$, $\triangle$, $\times$; ^{123}I vapor $\blacktriangle$; water vapor $\bigcirc$; Aitken nuclei ($r = 0.08$ μm) $\square$. The straight line has slope 0.25 and intercept 7.3.

"APPARENT" AND "TRUE" CANOPY RESISTANCES

The canopy resistance of vegetation has sometimes been determined from relations such as

$$\lambda E = \rho c_p \frac{[e_s\,(T(0)) - e]}{\gamma\,(r_{aM} + r_c)} \tag{17.9}$$

where $r_{aM} = u/u_*^2$, r_c is the "apparent" canopy resistance, and $T(0)$ the "apparent" surface temperature at height $z_0 + d$. More rigorous analysis (cf. Figure 17.2) gives

$$\lambda E = \rho c_p \frac{[e_s\,(T(0)') - e]}{\gamma\,(r_{aM} + r_c')} \tag{17.10}$$

where r_c' is the "true" resistance allowing for the existence of the additional boundary layer resistance r_b, and $T(0)'$ is the "true" surface temperature. By manipulating these equations and using the relation $C = \rho c_p(T(0)' - T(0))/r_b$, it can be shown that the error in calculating r_c without allowing for r_b is

$$\delta r_c = r_c' - r_c = r_b \left(\frac{\Delta}{\gamma} \frac{C}{\lambda E} - 1 \right) \tag{17.11}$$

This error is zero when the Bowen ratio β ($= C/\lambda E$) is equal to γ/Δ. For a well-watered crop growing in a temperate climate, the average value of β is usually about 0.1 and when $(\Delta/\gamma) = 2.0$, $\delta r_c = -0.8r_b$. The absolute magnitude of this error is less important than the fact that it may change in size and sign during the day as the Bowen ratio changes. For example, if β decreases from $+0.3$ in the early morning to -0.3 in the late afternoon and r_b is 20 s m^{-1}, the value of r_c will change during the day from $(r'_c - 8)$ to $(r'_c + 32)$ s m^{-1}.

Because there is a good correlation between the "true" canopy resistance r'_c and the Bowen ratio β, owing to stomatal control of the transpiration flux, the *relative* importance of the error in r_c (Equation 17.11) depends principally on the magnitude of r_b, which is inversely proportional to u_*. For a given windspeed, u_* is larger over tall vegetation than short, so that r_c is generally a better estimate of the true canopy resistance for tall, rough vegetation than it is for short.

CANOPY RESISTANCES FOR TRANSFER OF POLLUTANT GASES

Resistances of canopies and other surfaces to the uptake of pollutant gases have also been determined from equations analogous to Equation 17.9; e.g.,

$$F = \frac{S - 0}{r_{aM} + r_b + r_c} \qquad (17.12)$$

assuming that there are sinks within the crop canopy where the gas is absorbed and hence where the gas concentration S may be assumed zero. Examples of this analysis are given later.

17.2 Case Studies

WATER VAPOR AND TRANSPIRATION

One of the earliest practical applications of micrometeorology was for measuring the water use of agricultural crops. The relatively simple instrumentation required for the Bowen ratio or aerodynamic methods, and the importance of efficient planning of irrigation in many parts of the world ensure that this remains an active area of research. Water use of forests has also been studied, often to assess the consequences for water resources of changing land use. Gradients are much smaller over tall, rough forest canopies than over agricultural crops, and the small aerodynamic resistance between forest canopies and the atmosphere has consequences that are discussed later for the evaporation rates of intercepted rainfall (i.e., water remaining on the canopy during and after rain, p. 246). Over forests, there are advantages in using the eddy covariance technique rather than a gradient method because the large aerodynamic roughness that creates small, hard-to-measure, gradients generates large scale turbulence during the day that can readily be measured by eddy covariance instruments.

In a relatively early example of forest micrometeorology, Stewart and his colleagues [Stewart and Thom, 1973; Thom et al, 1975] used the aerodynamic and Bowen ratio techniques to study evaporation from a forest of Scots and Corsican pine at Thetford in southeast England. In determining the available energy for the Bowen ratio method it was necessary to allow for the storage of heat in the trunks and branches and in the air within the canopy; this term $\mathbf{J}$ (W m^{-2}) was about $18\delta T$ where δT is the rate of temperature change of air in the canopy (K h^{-1}). The maximum value of $\mathbf{J}$ was ± 55 W m^{-2}.

Comparison of flux measurements at Thetford by the aerodynamic and Bowen ratio methods identified the discrepancy discussed on p. 322, requiring a large empirical correction to aerodynamic estimates. Bowen ratios on fine days ranged from near 1 to 4 or more.

Stewart and Thom [1973] analysed the flux measurements at Thetford using resistance analogues. The aerodynamic resistance r_{aM}, derived from wind profiles, was about 5–10 s m^{-1}, and the additional resistance r_b (Equation 17.6) was estimated as 3–4 s m^{-1}. They then used the canopy form of the Penman–Monteith equation (Equation 17.4) to deduce the canopy resistance, making the small correction (Equation 17.11) for additional boundary layer resistance.Figure 17.5 shows the variation of canopy resistance on a fine day, and compares it with values found at other forest sites. Near dawn, the foliage was wet with dew and so r_c was small. Once foliage had dried, r_c was about 100–150 s m^{-1} in the middle part of the day, implying that the average stomatal resistance of needles in the canopy (which had leaf area index about 10) was about 1000–1500 s m^{-1}. Later in the day r_c increased, probably as a result of stomatal closure in response to water stress.

Figure 17.5 indicates that minimum canopy resistances of other mature forests are also typically about 100 s m^{-1}. The ratio r_c/r_a for forests is often about 10–50 when the foliage is dry. When forest canopies are wet with rain, so that r_c is effectively zero, Equation 17.4 can be used to show that evaporation of intercepted rainfall on foliage proceeds much faster than transpiration from dry forest canopies exposed to the same weather. This contrasts with the situation for many agricultural crops for which minimum values of r_c are also typically 100 s m^{-1} but r_c/r_a is often close to unity. Consequently, forests in regions where rain is frequent tend to use more water by evaporation from foliage and transpiration than shorter crops growing nearby [Calder 1977, Shuttleworth 1989].

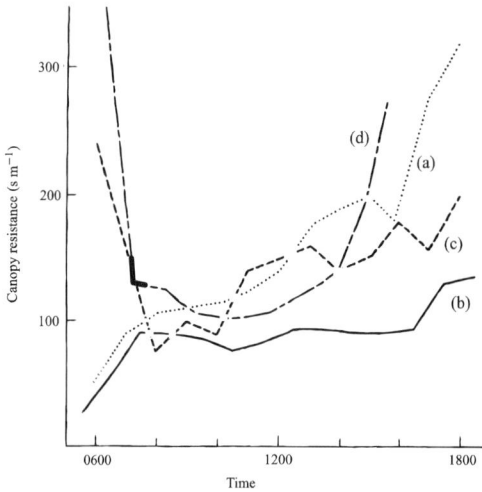

FIGURE 17.5 Diurnal variation of the canopy resistance of forests (after Jarvis, James, and Landsberg, 1976): (a) Scots pine (*Pinus sylvestris*), Thetford, UK [Stewart and Thom, 1973], (b) Sitka spruce (*Picea sitchensis*), Fetteresso, UK [Jarvis, James, and Landsberg, 1976], (c) Douglas fir (*Pseudotsuga menziesii*), British Columbia [McNaughton and Black, 1973], and (d) Amazonian forest, Brazil [Shuttleworth et al, 1984].

Kelliher et al [1995] summarized the minimum canopy resistances (maximum conductances) reported in the literature for major vegetation types when canopies were dry, and some values are given in Table 17.1.

Table 17.1 Minimum canopy resistances r_c for evaporation from various vegetation types (derived from Kelliher et al 1995). The only statistically significant difference was between natural vegetation and agricultural crops, so a simple approximate generalization is r_c (minimum) $\simeq 60$ s m^{-1} (natural vegetation) and 30 s m^{-1} (crops).

Vegetation Type	r_c (minimum) (s m^{-1})
Temperate grassland	59
Coniferous forest	47
Eucalypt forest	59
Temperate deciduous forest	48
Tropical rain forest	77
Cereal crops	31
Broadleaved herbaceous crops	32

Black and his colleagues [Black et al, 1996] used eddy covariance instrumentation at 39.5 m and 4 m to measure water vapor and carbon dioxide fluxes above and below the overstorey of an aspen stand, 21 m high, in the boreal forest region of Saskatchewan, Canada. They also used mini-lysimeters (trays of soil weighed at regular intervals) to estimate soil evaporation. Leaves of the aspen overstorey emerged about a month earlier than those of the hazlenut understorey, with the canopies reaching leaf area indices of 1.8 and 3.3, respectively, by mid-July. Figure 17.6a shows the daily evaporation and latent heat fluxes for an almost complete growing season, and Figure 17.6b illustrates estimates of the cumulative evaporation, including the soil evaporation and measured precipitation. When measurements began in early April (Day of year DOY≈90), snow was on the ground and no foliage had emerged; the 24-hour ratio r of below- to above-overstorey evaporation was 0.43. Just after snow melt in the second half of April, when leaves were beginning to emerge, r increased to 0.84. The ratio was much higher after snow melt because the vapor pressure at the surface of the rapidly warming forest floor was much larger than that over the snow surface, the temperature of which was thermodynamically limited to 0°C or less. For the period from full leaf (15 June, DOY 166) to senescence in early September (DOY ≈ 250), r averaged 0.22. Over

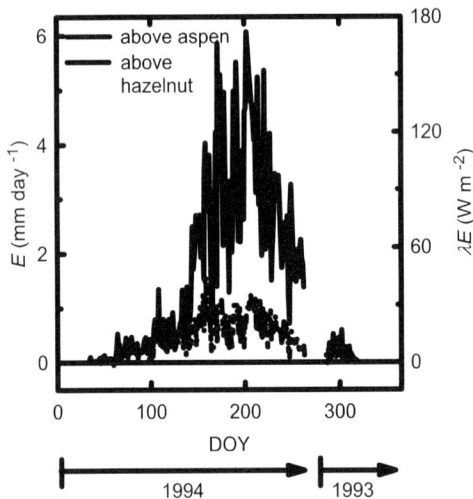

FIGURE 17.6   (a) Comparison of daily average evaporation rates and latent heat fluxes measured using eddy covariance techniques above an aspen overstorey and above a hazelnut understorey.

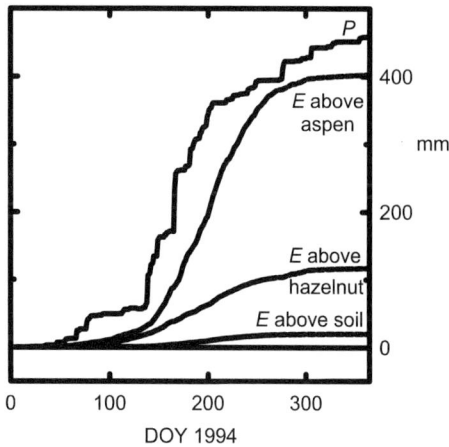

FIGURE 17.6 (b) Cumulative evaporation (E) over a year above an aspen overstorey, above a hazelnut understorey, and above the soil surface. The dashed sections are estimated. Also shown is the cumulative precipitation (from Black et al. 1996).

a two-week period in midsummer when lysimeters were operational the proportions of evaporation from the various sources were: aspen 78%; hazelnut 17%; soil 5%. At a nearby site with a more open canopy of Jack pine, Baldocchi, Vogel and Hall [1997] found that evaporation from the ground cover of shrubs and lichens accounted for 42% of forest evaporation when the soil was moist after several days of rain. This large proportion was largely due to the effectiveness of turbulent gusts penetrating the relatively permeable canopy airspace and transporting water vapor through the overstorey to the free atmosphere.

Baldocchi and Meyers [1991] treated understorey evaporation in a deciduous forest as a non-steady state phenomenon, and demonstrated that the frequency of canopy "flushing" by gusts penetrating the canopy from above was a critical feature in determining understorey evaporation (see also p. 330). Large scale eddies, with sufficient energy to penetrate deep into the canopy, typically occurred every 50–100 s, when air swept rapidly into the trunk space, ejecting the residual air. Then followed a quiescent period during which water evaporated from the soil/litter surface into the trunk space. To demonstrate how soil moisture influenced evaporation, they developed a simple box model of the atmosphere above the soil surface in which the saturation deficit D decreased with time during quiescent periods as a consequence of surface evaporation. The time rate of change of D depends on energy partitioning at the soil surface, and is

$$-\frac{dD(t)}{dt} = \frac{\Delta(\mathbf{R_n} - \mathbf{G}) - (\Delta + \gamma)\lambda\mathbf{E}(t)}{\rho c_p h} \quad (17.13)$$

where h is the height of the box and the other symbols have their usual meanings. Solving the differential equation for $\lambda\mathbf{E}$ yields

$$\lambda\mathbf{E}(t) = \lambda\mathbf{E}(0)\exp\left(\frac{-t}{\tau}\right) + \frac{\Delta}{\Delta + \gamma}(\mathbf{R_n} - \mathbf{G})\left[1 - \exp\left(\frac{-t}{\tau}\right)\right] \quad (17.14)$$

where $\lambda\mathbf{E}(0)$ is the latent heat flux at the beginning of the quiescent period (when the saturation deficit is equal to that of air above the canopy). The term $\Delta(\mathbf{R_n} - \mathbf{G})/(\Delta + \gamma)$ is the equilibrium latent heat flux $\lambda\mathbf{E_q}$ (Equation 13.37), so Equation 17.14 may be written

$$\mathbf{E}(t) = \mathbf{E}(0)\exp\left(\frac{-t}{\tau}\right) + \mathbf{E_q}\left[1 - \exp\left(\frac{-t}{\tau}\right)\right] \quad (17.15)$$

The time constant τ is

$$\tau = r_{aV}h[\Delta + \gamma(1 + \frac{r_g}{r_{aV}})]/(\Delta + \gamma) \quad (17.16)$$

where r_g is the soil surface resistance and r_{aV} is the boundary layer resistance for water transfer from the soil surface to the box. Equation 17.15 shows that the evaporation rate from the soil is initially related to the saturation deficit of air from above the canopy, but tends to $\mathbf{E_q}$ when $t \gg \tau$. Baldocchi and Meyers [1991] reported that values of r_{aV} ranged from 50 to 100 s m^{-1}. For dry soils, r_g was between 500 and 3000 s m^{-1}; for wet soils r_g approached zero. Substituting these values into Equation 17.16 for a box 2 m deep demonstrates that when the soil surface was dry, the time constant was typically between about 1500 and 5000 s, so that air would be renewed by intermittent turbulence before equilibrium evaporation rates were achieved. In contrast, evaporation from wet soils would be sufficiently rapid (τ typically 100 to 200 s) to allow an approach to equilibrium between the arrival of large eddies. Thus, when soils are dry, the forest floor is often closely coupled to the saturation deficit of the above-canopy environment.

The extent to which different layers in a forest contribute to its total water use therefore depends on several physical and biological factors: the relative leaf areas in the layers and their stomatal resistances; properties of the soil surface such as albedo, water phase and water content; and canopy structure through its influence on the frequency of large eddies that penetrate to the forest floor (see p. 360).

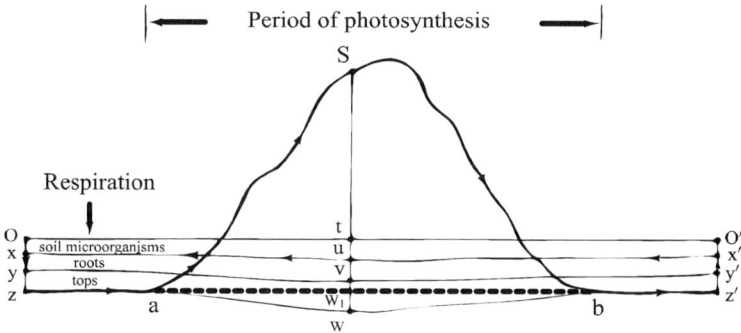

FIGURE 17.8 The diurnal change of CO_2 flux above a stand of vegetation, shown by the bold line zasbz'. For the significance of other components, see text.

CARBON DIOXIDE AND GROWTH

The measured flux of CO_2 from the air to a soil-plant ecosystem describes the net exchange of CO_2 between the ecosystem and the atmosphere. Figure 17.7 (see color plates) illustrates the various components that make up the carbon cycle of an ecosystem, depicted here as a forest.

Gross photosynthesis (uptake of CO_2) by trees and understorey species is responsible for the *gross primary productivity* (GPP) of the ecosystem. Some of the carbon gained is respired both night and day from foliage, tree boles, and living roots, forming the *autotrophic respiration* (R_a). The *net primary productivity* (NPP) is

$$NPP = GPP - R_a$$

An additional flux of respired CO_2, *heterotrophic respiration* (R_h) arises from the decay and consumption of dead carbon-containing material (e.g., dead roots, soil carbon) by fungi, micro-organisms, and other living components of the ecosystem, so that the *net ecosystem productivity* (NEP) is

$$NEP = NPP - R_h = GPP - R_a - R_h$$

Conventionally, biologists treat NEP as a positive term, since it is the net *gain* of carbon by the ecosystem from the atmosphere, whereas atmospheric scientists often plot the flux as negative, calling it *net ecosystem exchange* (NEE), i.e., NEE $= -$NEP, because it is the net *loss* of carbon dioxide from the atmosphere.

Figure 17.8 shows how the components of the CO_2 exchange of a crop are likely to alter over a period of 24 hours [Monteith, 1962]. The axis $00'$ represents zero flux. The line zasbz' represents the net ecosystem exchange (NEP or $-$NEE) between the atmosphere and a crop. The net CO_2 flux is directed upwards during the night when there is a respiratory loss of CO_2 from the system, and downwards during the day when the rate of photosynthetic uptake of CO_2 exceeds the daytime rate of respiration so there is a net carbon gain by the crop. Respiration from both heterotrophic and autotrophic sources contributes to the respired flux of CO_2. Total respiration between midnight and sunrise is represented by za. At sunrise (a), the photosynthetic system begins to assimilate some of the respired CO_2 and the net upward flux decreases to zero when solar irradiance reaches the light compensation point for the stand, usually about 1 to 2 hours after sunrise over actively growing vegetation. After the irradiance exceeds the light compensation point, there is a net downward flux of CO_2 representing the atmospheric contribution to photosynthesis. Shortly before sunset (b), the compensation point is reached again, and after sunset the rate of respiration is shown by bz'.

Components of the CO_2 balance at an instant during the day are given by segments of the line sw:

st $=$ net uptake of CO_2 from the atmosphere

tw $=$ respiration of CO_2 from plant (autotrophic) and soil

(heterotrophic) sources

sw $=$ gross assimilation of CO_2

uw $=$ plant (autotrophic) respiration

su $=$ net photosynthesis

Only one of these quantities, st, can readily be measured by micrometeorology. The respiration during the day is not known but can be estimated from the average flux at night adjusted for higher temperatures in the day [Goulden et al, 1996], or more simply, as Monteith [1962] did, by drawing a straight line through za and bz' intersecting sw at w_1. The segment ww_1 represents the increase in total respiration as a result of the higher soil and air temperature during the day. The proportion of total respiration attributable to soil organisms (heterotrophic respiration) is very difficult to establish experimentally because the presence of plant roots stimulates microbial activity in the rhizosphere. If β is the ratio of autotrophic respiration to the total respiration of the system, the instantaneous rate of net photosynthesis is $(sw_1) - \beta(tw_1)$. During the life of an agricultural crop, the value of β will increase from zero at germination to a maximum, usually between 0.5 and 0.9, when the crop is mature. The integrated rates of photosynthesis for the 24-hour period are

> gross photosynthesis: area asbwa $\approx$ asbw$_1$a
> plant respiration: area xux$'$z$'$wzx
> net photosynthesis area zsz$'$x$'$uxz

In practice, the net photosynthesis for a 24-hour period is found from the area zsz$'$O$'$Oz plus the nocturnal respiration from soil organisms which is $(1 - \beta)$ times the total nocturnal respiration.

Measurement of the net flux $\mathbf{P}_a$ of CO_2 from the atmosphere to crops has been possible since about 1960 using gradient methods of micrometeorology and sensitive infrared gas analyzers. During the 1980s, reliable fast-response CO_2 analyzers were developed to allow eddy covariance measurements over crops [Ohtaki, 1984; Anderson and Verma, 1986], and this more direct technique is now comonly used in short- and long-term micrometeorological studies (see the review by Baldocchi 2002).

Agricultural Crops

Biscoe, Scott and Monteith [1975] used aerodynamic and Bowen ratio methods to measure the net CO_2 flux $\mathbf{P}_a$ to barley throughout a complete growing season, and supplemented their study with measurements of soil and root respiration to enable net photosynthesis to be calculated on an hourly basis. Figure 17.9 shows the relationship between net photosynthesis rate and irradiance for five consecutive weeks from the stage of maximum green leaf area up to a time approaching harvest. At the beginning of the period, photosynthesis rate increased with irradiance, even in strong sunlight. Later, as the foliage senesced, maximum rates of photosynthesis declined and were achieved at steadily lower irradiances. The decrease was partly a consequence of decreasing green leaf area, but changes in the photosynthetic activity of individual organs also affected the crop photosynthesis.

Figure 17.10 shows the result of summing the hourly fluxes of carbon dioxide over a period of eight days when the canopy was fully established but before sensescence. The accumulation of carbon progresses in cycles corresponding to the succession of light and dark periods when the crop gains and loses carbon. In this example, where light was the principal factor limiting productivity, large differences in photosynthesis from day to day were well correlated with the daily insolation shown in the histogram.

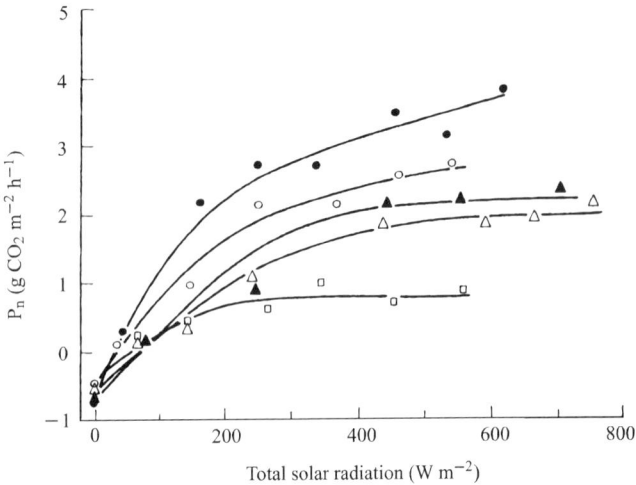

FIGURE 17.9   The relation between net CO_2 fixation of a barley crop and irradiance for the five weeks after anthesis in 1972. Dates, total green leaf area indices, and symbols are as follows: 28 June, 5.95 (●), 5 July, 5.69 (○), 12 July, 5.59 (▲), 19 July, 4.02 (△), 26 July, 2.68 (□) (from Biscoe, Scott and Monteith, 1975).

There have been relatively few studies of the CO_2 exchange of crops over periods long enough to demonstrate, as in Figure 17.10, the importance of short-term changes in photosynthetic activity in determining crop growth over periods of a week, the shortest time for which growth can be determined by conventional destructive methods.

Forests

As illustrated previously, micrometeorological methods allow studies on a time scale that enables physiological responses to the weather to be studied in the field. Anthoni et al [1999] used the eddy covariance method to study the CO_2 and energy exchange of an open-canopied ponderosa pine ecosystem in central Oregon where summers are very dry. Figure 17.11 shows the variations in net carbon uptake (NEP), latent heat flux (LE) and ecosystem respiration (R_e), with several weather factors over about 40 days in July–August 1997. The total solar radiation record (S_r) shows that the sky was generally cloudless apart from three periods when cooler, moister air masses from the Pacific penetrated the region. As high pressure re-established after these incursions, air temperature (T_a) and va-

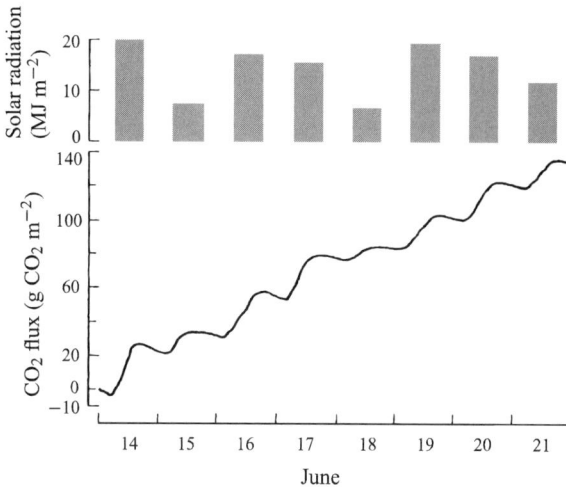

FIGURE 17.10 The hourly rates of net CO_2 fixation by a barley crop summed over the period 14–21 June 1972. The histogram shows the total solar radiation for each day (from Biscoe, Scott and Monteith, 1975).

por pressure deficit (VPD) steadily increased as a consequence of the regional energy balance. Respiration, which is strongly influenced by air and soil temperature, varied in synchrony with air temperature variations. However, net carbon dioxide uptake (NEP) was generally close to zero, increasing substantially only when VPD was below about 2 kPa. More detailed analysis of the carbon dioxide fluxes in response to radiation and temperature revealed that the decline in NEP on hot days with large VPD occurred primarily because gross photosynthesis was only about half the value it reached on humid days, rather than because respiration increased significantly. The decline in photosynthesis was probably because stomata closed to avoid excessive water loss.

The relative constancy of LE in Figure 17.11 while temperature and VPD vary substantially is striking, and demonstrates how stomata adjust when soil is dry and evaporative demand is large. The adjustment matches the evaporation rate from the foliage to the rate at which water can be transported from the soil through roots and stems. If this homeostasis did not occur, water potential gradients between roots and foliage would become so large that water columns in the plant xylem would break (*cavitate*), with potentially damaging consequences for the water balance [Sperry 1995].

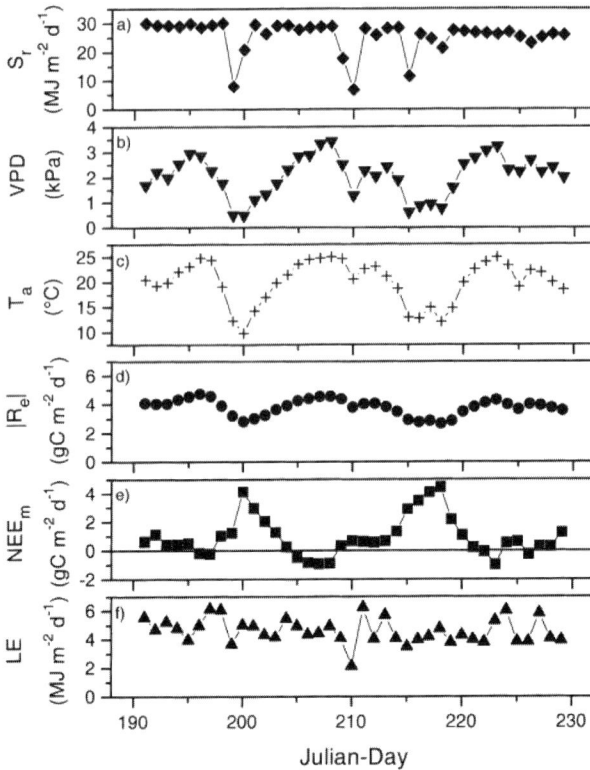

FIGURE 17.11   Day-to-day variation over an open-canopied ponderosa pine ecosystem in central Oregon of (a) daily total solar insolation S_r, (b) mean daylight vapor pressure deficit VPD, (c) mean daily air temperature T_a, (d) daily total ecosystem respiration R_e (absolute value), (e) daily net ecosystem exchange of CO_2 NEE_m (measured using eddy covariance methods and plotted as positive when CO_2 is being taken up by the ecosystem), and (f) daily latent heat flux LE (measured using eddy covariance) (from Anthoni et al, 1999).

In a groundbreaking study beginning in 1992, Wofsy and his colleagues established long-term micrometeorological measurements of the CO_2 exchange of Harvard Forest (42.5°N, 72.2°W). By careful attention to calibrations and data analysis techniques, they generated a consistent and continuous record of the forest's net carbon accumulation over several years, and supported the data with biometric measurements [Barford et al 2001]. Figure 17.12 illustrates the annual variation of the components of carbon exchange of the ecosystem. NEE was observed directly using eddy

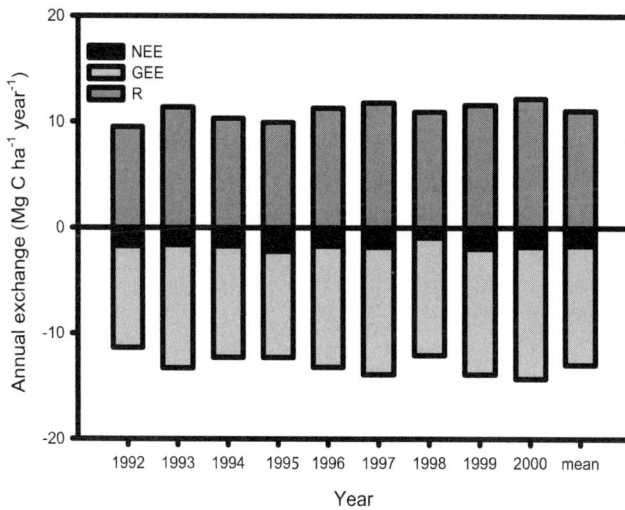

FIGURE 17.12 Components of the annual exchange of CO_2 between the atmosphere and a northern temperate hardwood forest (Harvard Forest, USA) over the period 1992–2000. Net ecosystem exchange NEE (negative sign indicates loss from the atmosphere) was measured using eddy covariance. Respiration R was inferred from night-time NEE measurements used to estimate daytime respiration. Gross ecosystem exchange is calculated as NEE-R. Data are summed from 28 October of the previous year to 27 October of the year plotted. Data bars are not cumulative; i.e., in 1992 NEE was −2.0 and GEE, was −11.4 Mg C ha^{-1} year^{-1} (replotted from data in Barford et al, 2001).

covariance instrumentation; respiration was interpreted as the eddy flux observed at night when friction velocity was above a minimum threshhold of 0.2 m s^{-1}, and was extrapolated for daytime on the basis of soil-air temperature variation to get daily values for R. Gross ecosystem exchange was calculated as NEE − R. The detailed information from the micrometeorological sytem and accompanying physical and biological measurements enable causes of annual variability to be determined. For example, low values of gross carbon uptake (GEE) in 1998 were caused by low temperatures and excessive cloudiness in early summer, which depressed photosynthesis; net uptake (NEE) was large in 1995 because dry soil in summer suppressed respiration. Measurements such as this in ecosystems around the world provide information on net carbon accumulation (*sequestration* of CO_2) that can be used to test global and regional models of the carbon cycle [Baldocchi et al, 2001].

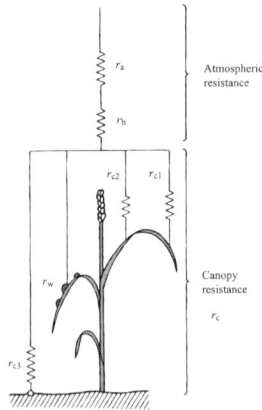

FIGURE 17.13 Resistance analogue of dry deposition of SO_2 to a wheat canopy, showing aerodynamic resistance, r_a, additional boundary layer resistance, r_b, and resistances in canopy, r_{c1}, to stomatal uptake, r_{c2}, to surface deposition, r_{c3}, to uptake by soil, and, r_w, to uptake by surface moisture (from Fowler and Unsworth, 1979).

SULPHUR DIOXIDE AND POLLUTANT FLUXES TO CROPS

In the same way that carbon dioxide is transported from the atmosphere to crops, to be absorbed at sites of photosynthesis, pollutant gases are absorbed within a crop canopy. This process is sometimes referred to as *dry deposition* to distinguish it from the *wet deposition* of pollutants in rain and snow.

In general, pollutant gases can be absorbed (or adsorbed) at various sites in the canopy, depending on their solubility and affinity for materials on the surface of, and within, leaves, and in soils. Resistance analogues can be used to establish the importance of these several pathways, applying the analysis to micrometeorological measurements of fluxes by gradient or eddy covariance techniques.

Figure 17.13 shows a resistance analogue of dry deposition of sulphur dioxide to a wheat canopy [Fowler and Unsworth, 1979]. Resistances to turbulent transfer within the canopy were much smaller than resistances associated with the sinks and were ignored. There are four possible sinks in the canopy: (i) SO_2 may diffuse through stomata, dissolve in the substomatal cavity and ultimately be used as sulphate in plant metabolism. The canopy resistance component for the stomatal pathway r_{c1} is therefore similar to the canopy resistance for water vapor loss, but a correction is required for the smaller diffusivity of SO_2; (ii) SO_2 may be absorbed or adsorbed onto the surface of leaves; the controlling resistance r_{c2} probably depends on the surface structure, and on any deposited particles, dust, other gaseous pollutants, etc.; (iii) drops of water on leaf surfaces absorb SO_2; the resistance r_w is influenced by other soluble substances, and increases as the liquid increases in acidity, eventually halting any further uptake of SO_2; (iv) SO_2 transported through the canopy can be absorbed by the soil; the resistance r_{c3} is smaller for chalky soils than for clays.

The flux F_s of SO_2 to a crop can be described by Equation 17.12,

$$F_s = \frac{S}{r_{aM} + r_b + r_c}$$

where S is the SO_2 concentration at a reference height above the crop, and r_c, the canopy resistance, is the resultant of the resistances in Figure 17.13 acting in parallel.

Fowler and Unsworth [1979] made measurements of F_s to a wheat crop in central England throughout a growing season, using the aerodynamic method with wet chemical techniques for measuring SO_2 concentration at five heights. Values of r_a ranged from 10–200 s m^{-1} and r_b was 20–100 s m^{-1} (the larger values in both cases being for light winds at night). Using Equation 17.12, values of r_c were derived, and the component resistances were estimated by interpreting diurnal changes. During the day, provided that the crop was dry and not senescent, r_c was dominated by the stomatal pathway; consequently, the minimum values of r_c during the day, 50–100 s m^{-1}, give an approximate value to r_{c1}. At night, when stomata closed, r_c increased to about 250–300 s m^{-1} and this is an estimate of r_{c2}. This value is about an order of magnitude lower than cuticular resistances for water vapor loss (allowing for a leaf area index of 4.5), indicating that there was an effective sink for SO_2 on leaf surfaces. Figure 17.14 shows an occasion when dew formed from about 0300 to 0600, and r_c decreased rapidly to about 100 s m^{-1}, indicating that r_w was the controlling resistance in this case. Analysis of flux measurements when the crop was senescent suggested that absorption of SO_2 by soil below the canopy was not significant. From a knowledge of the components of r_c and of the seasonal mean SO_2 concentration (50 μg m^{-3}), it was estimated that the wheat crop absorbed about 11 kg sulphur (ha)$^{-1}$ (1.1 g sulphur m^{-2}) in the period May–July; 5 kg ha^{-1} entering through stomata and 6 kg ha^{-1} deposited on leaf surfaces.

These methods for measuring and analysing SO_2 fluxes are applicable to other gaseous pollutants. For example, Wesely et al [1978] used an eddy covariance technique to study ozone deposition on maize, demonstrating that stomatal control formed the main component of r_c by day, but that absorption of ozone by soil below the canopy was also an important pathway. Fowler et al [1989] used resistance models to compare annual quantities of gaseous pollutants reaching grassland and forests with those deposited by wet deposition.

Just as net carbon dioxide fluxes may be bidirectional between leaves and the atmosphere, so may fluxes of other trace gases for which there is a source associated with plant metabolism. An example is ammonia (NH_3). Noting that ammonium arises in metabolic processes, and that in plant tissue it reaches equilibrium with gaseous NH_3, Farquhar et al [1980] postulated that there was a "compensation concentration" (χ_s) of NH_3 in stomatal cavities such that fluxes of NH_3 would be *into* leaves when atmospheric concentrations exceeded χ_s, but would be *from* leaves at atmospheric concentrations below χ_s. The magnitude of χ_s would be

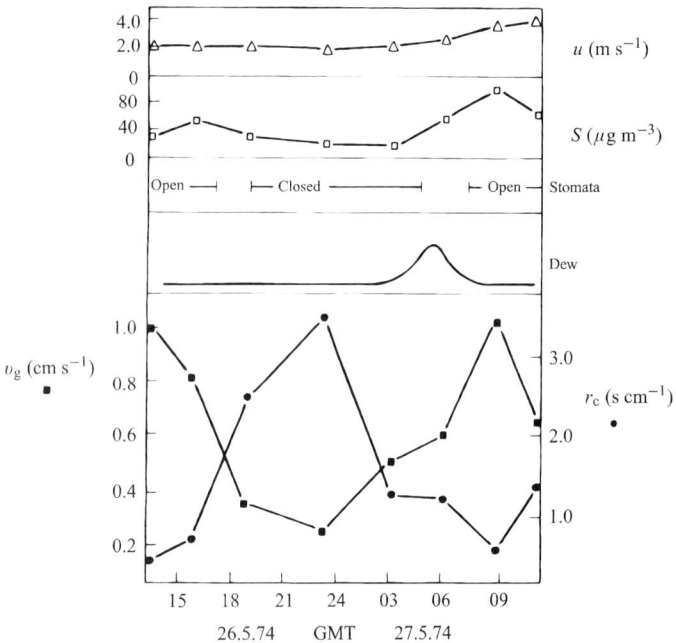

FIGURE 17.14   Diurnal variations of SO_2 deposition velocity v_g, canopy resistance, r_c, windspeed, u, and SO_2 concentration, S, over a wheat crop. Durations of dew deposition and of estimated stomatal opening are also shown. All height-dependent parameters are referred to 1 m above the zero plane (from Fowler and Unsworth, 1979).

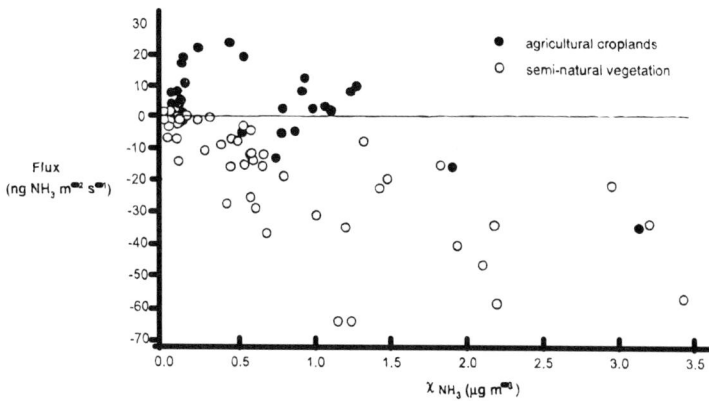

FIGURE 17.15 Variation of fluxes of ammonia (NH_3) between the atmosphere and two types of vegetation canopies with ammonia concentration. (Fluxes are plotted as positive when the flux of ammonia was from the vegetation to the atmosphere, from Sutton et al, 1995).

expected to depend on the nitrogen (N) status of the plant tissue. Sutton et al [1993 a,b] used the aerodynamic gradient method in conjunction with filter packs to measure ammonia exchange over unfertilized (low N status) semi-natural vegetation and fertilized (high N status) crops. As predicted by Farquhar, he found that the direction of the flux depended on atmospheric NH_3 concentration, and that the apparent canopy compensation point differed with the nitrogen status of the vegetation. Figure 17.15 (from Sutton et al, 1995) illustrates these findings. By modifying the resistance model of Figure 17.13 to include a compensation concentration of NH_3 within foliage, he developed a model that explained his observations and could be applied to estimated annual dry deposition of NH_3 to various ecosystems [Sutton et al, 1995].

17.3 Transport within Canopies

Much of the previous part of this chapter was concerned with the measurement and interpretation of vertical profiles or the eddy covariance of heat, mass, and momentum *above* plant canopies to deduce appropriate fluxes and to identify controlling resistances. A logical extension is to deduce fluxes within canopies by analysing profiles, and hence to investigate how different layers in the canopy contribute to the total flux. This problem has challenged environmental physicists for more than 30 years.

Initial research tried to relate observed local mean concentration gradients in the canopy to local turbulent fluxes using turbulent diffusivity theory (*K-theory*, see page 309). However numerous theoretical analyses and experimental measurements have demonstrated that scalar and momentum fluxes in canopies do not obey K-theory [Denmead and Bradley 1985, Finnigan 1985, Raupach 1989, Kaimal and Finnigan 1994]. Finnigan [1985] and others pointed out that it is possible to define a diffusivity as a simple function of the transporting mechanism (i.e., a flux-gradient relationship) *only* if the scale of the mechanism is much smaller than the scale of the gradients. Within canopies, it was believed for some time that turbulence (the transporting mechanism) was generated mainly in the wakes behind leaves and other within-canopy structures, which are much smaller than mean gradients of windspeed and other entities in canopies, so the concepts of a local eddy diffusivity and a flux-gradient relationship seemed appropriate. However, measurements reveal that, close to rough surfaces, large coherent eddies are the primary mechanisms of momentum and heat transfer, and their generation is more related to the gross roughness of the surface than to the fine structure of the elements. These large eddies, which penetrate into vegetation canopies, are responsible for much of the exchange between the canopy and the atmosphere, and violate the assumptions of K-theory.

Experimental evidence illustrating the failure of flux-gradient relationships in canopies and the role of large gusts in penetrating the canopy is shown in Figures 17.16 and 17.17. In Figure 17.16, the vertical temperature profile, averaged over one hour, has a maximum in mid-canopy, consistent with the likely zone of maximum absorption of solar radiation. Flux-gradient theory would therefore predict sensible heat fluxes upward and downward from this level, but the observed flux (H_6) below the canopy is "counter-gradient." Figure 17.17 demonstrates that transfer in this canopy was effected by large gusts, persisting for a minute or longer, that replaced canopy air with cooler air from above.

If K-theory cannot be used to derive fluxes and source strengths from observed mean concentrations in canopies, what other methods are available? Two approaches have been used with some success: Lagrangian and Eulerian advection-diffusion models. (In Lagrangian analysis, diffusion is computed following a "parcel" of air as it is advected by the mean wind; in Eulerian models, the solution is obtained relative to a fixed grid. Data that need to be specified *a priori* differ somewhat between the methods, but for both approaches it is necessary to specify or calculate the flow field

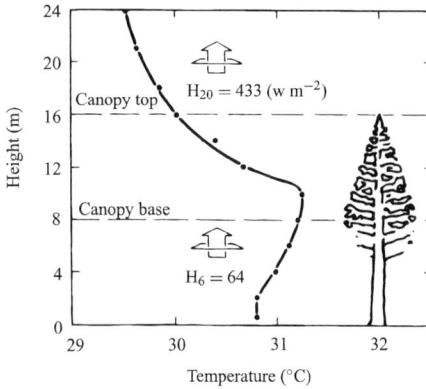

FIGURE 17.16 The vertical profile of air temperature (averaged over one hour) and simultaneous measurements of sensible heat flux (**H**) by eddy covariance above and below the canopy of an extensive Ponderosa pine forest (from Finnigan, 1985).

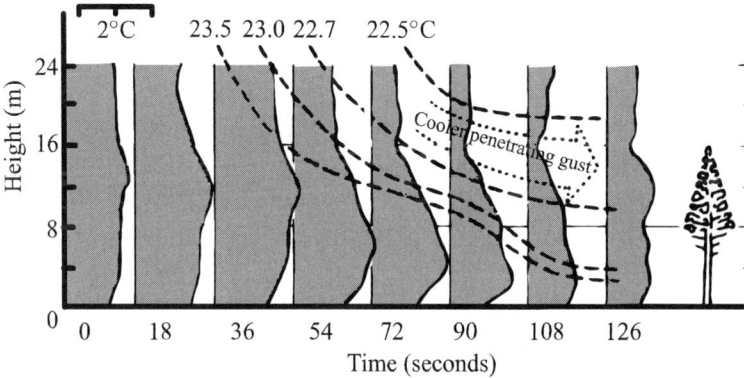

FIGURE 17.17 Profiles of temperature averaged over 10 seconds and taken during a short fraction of the hour depicted in Figure 17.16. The start time of each period was advanced by 18 s for each profile. Dashed lines are contours of constant temperature, and the base-line for the temperature profiles was $18.5°C$ (from Finnigan, 1985).

in the canopy.) Details of both methods are beyond the scope of this book but consulting the following examples would provide a good introduction.

Raupach [1989] developed a Lagrangian "Localized Near-field" (LNF) theory of dispersion in plant canopies, which has been applied with reasonable success by several groups. For example, Leuning et al [2000] used LNF theory to estimate source/sink distributions of heat, carbon dioxide, water vapor and methane in a rice canopy from observations of concentration fields and turbulence statistics (the so-called *inverse method*). In another application of Lagrangian models, Hsieh et al [2003] compared a two-dimensional Lagrangian stochastic dispersion model with a one-dimensional analytical Lagrangian dispersion model using data from a forest canopy, and found that both performed equally well for short ($\approx$ 100 m) fetches.

Katul et al [2001] tested an Eulerian model developed by Katul and Albertson [1999] with data from the rice canopy study of Leuning et al [2000] mentioned previously. The model coupled scalar and momentum transport together in the canopy to infer sources and sinks from mean concentration profiles. The model calculations agreed well with measured fluxes at the top of the canopy. However the model indicated a very different distribution of CO_2 source/sink distributions in the canopy than was inferred by Leuning et al [2000] using the LNF model. Katul et al [2001] concluded that, in general, estimating the source distribution from mean concentrations of scalars within a canopy is complicated by many processes that are not well resolved by both Eulerian and Lagrangian methods.

The development of inverse methods, using numerical models to estimate sources and sinks from observed concentrations and flow fields, is likely to continue to be an active area of research, but there are major challenges: atmospheric inverse modeling problems are mathematically ill-posed and inherently unstable [Styles et al, 2002], and small errors in observations can be transformed into large uncertainties in source distributions.

THE SYNTHESIS OF PROFILES IN CANOPIES

Although the nature of transport in crop canopies is not yet fully understood, it is still possible to develop mathematical models that generate realistic in-canopy microclimates from a known distribution of sources and sinks. This apparent inconsistency arises for two reasons: (i) as mentioned earlier, the source and sink strengths deduced from observed profiles are extremely sensitive to the shape of the measured profiles. Although this limits the scope for deducing the distribution of flux from measurements of profiles, it means that plausible profiles can be generated by integration from an approximate distribution of sources; (ii) the main restrictions on the transfer of water vapor, CO_2, and some pollutant gases between the atmosphere and leaves are associated with stomata, so that a realistic distribution of stomatal resistance in various layers of a canopy can be combined with a relatively uncertain model of atmospheric transfer between layers to produce canopy profiles. Several models of sensible and latent heat transfer in the literature follow the same procedure:

(i) The distribution of radiant energy in the canopy is expressed as a function of cumulative leaf area index.

(ii) The net radiation absorbed by each leaf is partitioned between sensible and latent heat, making assumptions about the distribution of stomatal and leaf boundary layer resistances.

(iii) Transfer within the canopy is estimated using diffusion models (K theory approaches in early work; Lagrangian or Eulerian models more recently).

Models of this type were developed by Waggoner and Reifsnyder [1968], Goudriaan [1977], Jarvis, Miranda and Muetzelfeldt [1985], McNaughton and van den Hurk [1995], and and Baldocchi and Meyers [1998]. As a generalization, the models demonstrate that the exchange of heat and water vapor between a canopy and the air above it depends much more critically on the behavior of stomata than on the structure of the microclimate within the canopy.

While models provide useful insight into the distribution of sources and sinks in canopies, they are often weak in simulating exchange between the atmosphere and the soil below the canopy, because the resistances associated with the soil boundary layer are difficult to define in a convenient form. This is seldom of major importance below dense canopies, but merits further work for sparse canopies (e.g., Baldocchi et al [2000]) and for early stages of crop development. An attempt to solve the heat balance equations for a canopy and soil divided into four discrete layers was described by Choudhury and Monteith [1988], and similar approaches have been used to estimate evaporation from sparse canopies [Shuttleworth and Wallace, 1985].

One application of models particularly useful for interpreting fluxes above and within canopies has been "footprint" modelling, to investigate the upwind area that contributes to fluxes sensed by instrumentation on a tower, and, more specifically, the probability that a flux contribution originating at a particular point upwind will be detected at the tower. Schmid [2002] reviewed this literature comprehensively, and Finnigan [2004] explored its mathematical basis. Flux footprints for measurements *above* canopies can be estimated reliably using Eulerian or Lagrangian diffusion

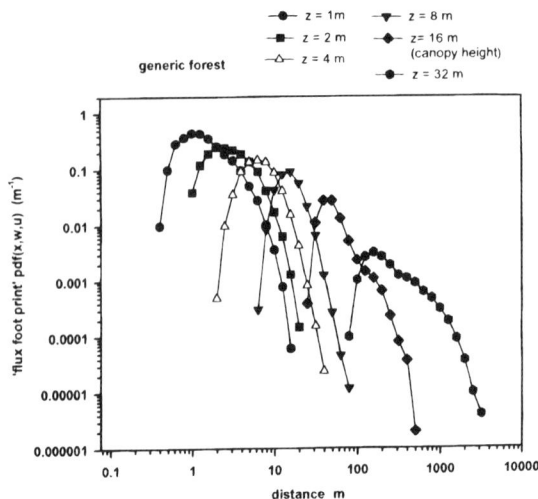

FIGURE 17.18   The source probability density function (pdf) of flux footprints for observations at several heights within and above a "generic" forest assuming near-neutral stability. The pdf is the probability that fluid elements released at numerous positions upwind will be observed at the defined height. The height (h) of the forest was assumed to be 16 m, zero-plane displacement d was $0.60\ h$, and roughness length z_0 was $0.10\ h$. An exponential wind profile within the canopy was assumed (from Baldocchi, 1997).

models, and provide greater insight into the relations between source area and measurement height than older rule-of-thumb "fetch" requirements.

Within canopies, the calculations are more problematic because the velocity, time and length scales of turbulence are heterogeneous and very different from above canopies. Figure 17.18 illustrates flux footprints calculated by Baldocchi [1997] for a typical forest, using a Lagrangian random walk model that could be applied above and below canopies. The figure demonstrates that measurements made at 32 m, twice the canopy height, would include significant flux contributions to about 3 km in neutral stability. In contrast, the probability functions for fluxes observed at 2 m height inside the canopy would peak at about 3 m upwind of the receptor, and very little flux information would be received from beyond 20 m. This indicates the difficulties of measuring representative below-canopy fluxes by micrometeorology when understories are heterogeneous.

17.4 Problems

1. A field of short grass has roughness length $z_0 = 1$ mm, and zero plane displacement $d = 7$ mm. Determine the variation of aerodynamic resistances to momentum transfer (r_{aM}) and the additional aerodynamic resistance (r_b) with windspeed in the range 0.5 to 3.0 m s^{-1}, assuming a measurement level of 5.0 m.

2. The friction velocity u_* over a crop canopy is 0.25 m s^{-1}, and the ratio Δ/γ is 2.0. Assume the parameter B^{-1} has a value of 4.0, and that the relationship between the canopy resistance r_c (in s m^{-1}) and the Bowen ratio β is given by $r_c = 100 \exp[3.22 - 4.6/(1 + \beta)]$ (as reported by Monteith 1965). Over the range $0 < \beta < 10$, determine the variation of: (i) r_c; (ii) the excess resistance $r_c' - r_c$; (iii) the "true" canopy resistance r_c'; (iv) the relative error made in using r_c rather than r_c' (i.e., $(r_c' - r_c)/r_c'$).

3. On a fine summer day, the measured latent heat flux λE at noon above a dense Douglas fir forest was 200 W m^{-2}. Net radiation $\mathbf{R_n}$ was 600 W m^{-2}, soil heat flux was 50 W m^{-2}, and air temperature and vapor pressure deficit at a reference level above the canopy were 28.0°C and 2.0 kPa, respectively. If the aerodynamic resistance r_{aM} between the canopy and the reference height was 10.0 s m^{-1}, estimate the canopy resistance r_c, stating any assumptions you need to make. If this resistance remained constant, how would λE vary with saturation deficit? In practice, would you expect to see this variation?

4. For the data in Problem 3, assume the observed concentration of sulphur dioxide (SO$_2$) at the reference height over the forest was 100 μg m^{-3}. If the canopy resistance for stomatal uptake of SO$_2$ was the same as for water vapor loss calculated in Problem 3, with the addition of a parallel resistance of 300 s m^{-1} for deposition of the gas on leaf surfaces, estimate the rate of SO$_2$ deposition from the atmosphere and the fraction of SO$_2$ entering the plant. Are the assumptions made likely to be valid?

References

Achenbach, E. (1977), The effect of surface roughness on the heat transfer from a circular cylinder, *International Journal of Heat and Mass Transfer, 20,* 359–369.

Anderson, D. E., and S. B. Verma (1986), Carbon dioxide, water vapor and sensible heat exchanges of a grain sorghum canopy, *Boundary-Layer Meteorology, 34,* 317–331.

Anderson, M. C. (1966), Stand structure and light penetration. II A theoretical analysis, *Journal of Applied Ecology, 3.*

Angstrom, A. (1929), On the atmospheric transmission of sun radiation and on dust in the air, *Geografiska Annaler, 11,* 156–166.

Anthoni, P. M., et al (1999), Carbon and water vapor exchange of an open-canopied ponderosa pine ecosystem, *Agricultural and Forest Meteorology, 95,* 151–168.

Ar, A., et al (1974), The avian egg; water vapour conductance, shell thickness and functional pore area, *Condor, 76,* 153–158.

Asrar, G., et al (1984), Estimating absorbed photosynthetic radiation and leaf area index from spectral reflectance in wheat, *Agronomy Journal, 76,* 300–306.

Aylor, D. E. (1975), Force required to detach conidia of Helminthosporium maydis, *Plant Physiology, 55,* 99–101.

Aylor, D. E. (1990), The role of intermittent wind in the dispersal of fungal pathogens, *Annual review of phytopathology, 28,* 73–92.

Aylor, D. E, and F. J. Ferrandino (1985), Rebound of pollen and spores during deposition on cylinders by inertial impact, *Atmospheric Environment, 19,* 803–806.

Bagnold, R. A. (1941), The Physics of Blown Sand and Desert Dunes, *Chapman and Hall.*

Bailey, B. J., and J. F. Meneses (1995), Modelling leaf convective heat transfer, *Acta horticulturae, 399,* 191–198.

Baker, C. J. (1995), The development of a theoretical model for the windthrow of plants, *Journal of Theoretical Biology*, *175*, 355–372.

Bakken, G. S, and D. M. Gates (1975), Heat transfer analysis of animals, in *Perspectives of Biophysical Ecology*, edited by D. M. Gates and R. G. Schmerl, Springer-Verlag, New York.

Bakwin, P. S., et al (1998), Measurements of carbon dioxide on very tall towers: results of the NOAA/CMDL program, *Tellus*, *50B*, 401–415.

Baldocchi, D. (1997), Flux footprints within and over forest canopies, *Boundary Layer Meteorology*, *85*, 273–292.

Baldocchi, D., et al (2001), FLUXNET: A New Tool to Study the Temporal and Spatial Variability of Ecosystem-Scale Carbon Dioxide, Water Vapor and Energy Flux Densities, *Bulletin of the American Meteorological Society*, *82*, 2415–2434.

Baldocchi, D., et al (1997), Seasonal variation of energy and water vapor exchange rates above and below a boreal forest jack pine canopy, *Journal of Geophysical Research, Atmosphere, In press*.

Baldocchi, D. D. (2002), Assessing ecosystem carbon balance: problems and prospects of the eddy covariance technique, *Global Change Biology*, *9*, 479–492.

Baldocchi, D. D. (2003), Assessing the eddy covariance technique for evaluating carbon dioxide exchange rates of ecosystems: past, present and future, *Global change biology*, *9*, 479–492.

Baldocchi, D. D., et al (1988), Measuring biosphere-atmosphere exchanges of biologically related gases with micrometeorological methods, *Ecology*, *69*, 1331–1340.

Baldocchi, D. D., and T. P. Meyers (1991), Trace Gas Exchange Above the Floor of a Deciduous Forest 1. Evaporation and CO_2 Efflux, *Journal of Geophysical Research*, *96*, 7271–7285.

Baldocchi, D. D., and T. P. Meyers (1998), On using eco-physiological, micrometeorological and biogeochemical theory to evaluate carbon dioxide and water vapor and gaseous deposition fluxes over vegetation: a perspective, *Agricultural and Forest Meteorology*, *90*, 1–25.

Baldocchi, D. D., and C. A. Vogel (1997), Seasonal variation of energy and water vapor exchange rates above and below a boreal jack pine forest canopy, *Journal of Geophysical Research*, *102*, 28, 939–928,951.

Barford, C. C., et al (2001), Factors controlling long- and short-term sequestration of atmospheric CO2 in a mid-latitude forest, *Science*, *294*, 1688–1691.

Baumgartner, A. (1953), Das Eindringen des Lichtes in den Boden, *Forstwissenschaftliches Zentralblatt*, *72*, 172–184.

Beament, J. W. L. (1958), The effect of temperature on the water-proofing mechanism of an insect, *Journal of Experimental Biology*, *35*, 494–519.

Becker, F. (1981), Angular reflectivity and emissivity of natural media in the thermal infrared bands, paper presented at Proceedings of Conference on Signatures Spectrales D'objects en Teledetection, Avignon, 8–11 September 1981.

Becker, F., et al (1981), An active method for measuring thermal infrared effective emissities: implications and perspectives for remote sensing, *Advanced Space Research*, *1*, 193–210.

Berman, A. (2004), Tissue and external insulation estimates and their effects on prediction of energy requirements and of heat stress, *Journal of Dairy Science*, *87*, 1400–1412.

Betts, A. K., et al (1998), Controls on evaporation in a boreal spruce forest.

Betts, R. A., et al (1997), Contrasting physiological and structural vegetation feedbacks in climate change simulations, *Nature*, *387*, 796–799.

Bird, R. B, and C. Riordan (1986), Simple solar spectral model for direct and diffuse irradiance on horizontal and tilted planes at the Earth's surface for cloudless atmospheres, *Journal of Climate and Applied Meteorology*, *25*, 87–97.

Bird, R. B., et al (1960), *Transport Phenomena*, John Wiley, New York.

Bird, R. E, and R. L. Hulstrom (1981), Simplified clear sky model for direct and diffuse insolation on horizontal surfaces, Solar Energy Research Institute, Golden, CO.

Biscoe, P. V. (1969), Stomata and the Plant Environment, Ph.D. thesis, University of Nottingham.

Biscoe, P. V., et al (1975a), Barley and its environment. I. Theory and practice, *Journal of Applied Ecology*, *12*, 227–257.

Biscoe, P. V., et al (1975b), Barley and its environment. III. Carbon budget of the stand, *Journal of Applied Ecology*, *12*, 269–293.

Black, T. A., et al (1996), Annual cycles of water vapor and carbon dioxide fluxes in and above a boreal aspen forest, *Global Change Biology*, *2*, 219–229.

Blanken, P. D., et al (1997), Energy balance and canopy conductance of a boreal aspen forest: Partitioning overstory and understory components, *Journal of Geophysical Research*, *102*, 28, 915–928, 927.

Blaxter, K. L. (1967), *The Energy Metabolism of Ruminants*, Second ed, Hutchinson, London.

Bolin, B. e. (1981), *Carbon Cycle Modelling*, New York.

Bonan, G. B. (1997), Effects of Land Use on the Climate of the United States, *Climatic Change*, *37*, 449–486.

Bonhomme, R, and P. Chartier (1972), The interpretation and automatic measurement of hemispherical photographs to obtain sunlit foliage area and gap frequency, *Israel Journal of Agricultural Research*, *22*, 53–61.

Bowers, S. A, and R. D. Hanks (1965), Reflection of radiant energy from soils, *Soil Scientist*, *100*, 130–138.

Bowling, D. R., et al (1999), Modification of the relaxed eddy accumulation technique to maximize measured scalar mixing ratio differences in updrafts and downdrafts, *Journal of Geophysical Research, 104*, 9121–9133.

Bristow, K. L, and G. S. Campbell (1986), Simulation of heat and moisture transfer through a surface residue-soil system, *Agricultural and Forest Meteorology, 36*, 193–214.

Bristow, K. L., et al (1994), Comparison of single and dual probes for measuring soil thermal properties with transient heating, *Australian Journal of Soil Research, 32*, 447–464.

Brown, G. W. (1969), Predicting temperatures of small streams, *Water Resources Research, 5*, 69–75.

Bruce, J. M, and J. J. Clark (1979), Models of heat production and critical temperature for growing pigs, *Animal Production, 28*, 353–369.

Brutsaert, W. H. (1982), *Evaporation into the Atmosphere*, D. Reidel Publishing Company, Dordrecht, Holland.

Buatois, A, and J. P. Crose (1978), Thermal responses of an insect subjected to temperature variations, *Journal of Thermal Biology, 3*, 51–56.

Burton, A. C, and D. G. Edholm (1955), *Man in a Cold Environment*, Edward Arnold, London.

Buss, I. O, and J. A. Estes (1971), The functional significance of movements and positions of the pinnae of the African elephant, *Loxodonta africana, Journal of Mammology, 52*, 21–27.

Calder, I. R. (1977), A model of transpiration and interception loss from a spruce forest in Plynlimon, cental Wales, *Journal of Hydrology, 33*, 247–265.

Calder, I. R, and C. Neal (1984), Evaporation from saline lakes: a combination equation approach, *Journal of Hydrological Sciences, 29*, 89–97.

Campbell, G. S. (1985), *Soil physics with BASIC: Transport models for soil-plant systems*, 150 pp, Elsevier, New York.

Campbell, G. S. (1986), Extinction coefficients for radiation in plant canopies calculated using an ellipsoidal inclination angle distribution, *Agricultural and Forest Meteorology, 36*, 317–321.

Campbell, G. S., et al (1980), Windspeed dependence of heat and mass transfer thorugh coats and clothing, *Boundary-Layer Meteorology, 18*, 485–493.

Campbell, G. S, and J. M. Norman (1989), The description and measurement of plant canopy structure, in *Plant canopies: their growth, form and function*, edited by G. Russell et al, pp. 1–19, Cambridge University Press, Cambridge.

Campbell, G. S., and J. M. Norman (1998), *Environmental Biophysics*, 2nd ed, 286 pp, Springer-Verlag, New York.

Campbell, G. S., and F. K. van Evert (1994), Light interception by plant canopies: efficiency and architecture, in *Resource Capture by Crops*, edited by J. L. Monteith et al, pp. 35–52, Nottingham University Press, Nottingham.

Carolus, R. L. (1971), Evaporative cooling techniques for regulating plant water stress, *HortScience, 6.*

Cena, K, and J. L. Monteith (1975a), Transfer processes in animal coats. I. Radiative transfer, *Proceedings of the Royal Society of London, B, 188,* 377–393.

Cena, K, and J. L. Monteith (1975b), Transfer processes in animal coats. II. Conduction and convection, *Proceedings of the Royal Society of London, B, 188,* 395–411.

Cena, K, and J. L. Monteith (1975c), Transfer processes in animal coats. III. Water vapour diffusion, *Proceedings of the Royal Society of London, B, 188,* 413–423.

Chamberlain, A. C. (1966), Transport of gases to and from grass and grass-like surfaces, *Proceedings of the Royal Society of London, A, 290,* 236–265.

Chamberlain, A. C. (1974), Mass transfer to bean leaves, *Boundary-Layer Meteorology, 6,* 477–486.

Chamberlain, A. C. (1975), The movement of particles in plant communtities, in *Vegetation and the Atmosphere,* edited by J. L. Monteith, pp. 155–203, Academic Press, London.

Chamberlain, A. C. (1983), Roughness length of sea, sand and snow, *Boundary-Layer Meteorology, 25,* 405–409.

Chamberlain, A. C., et al (1984), Transport of gases and particles to surfaces with widely spaced roughness elements, *Boundary-Layer Meteorology, 29,* 343–360.

Chamberlain, A. C, and P. Little (1981), Transport and capture of particles by vegetation, in *Plants and their Atmospheric Environments,* edited by J. Grace et al, pp. 147–173, Blackwell Scientific, Oxford.

Chandrasekhar, S. (1960), *Radiative Transfer,* Dover, New York.

Chauliaguet, C., et al (1979), *Solar Energy in Buildings,* 174 pp, John Wiley & Sons, Chichester.

Chen, J. (1984), Mathematical Analysis and Simulation of Crop Micrometeorology, Ph. D. thesis, Agricultural University, Wageningen, Netherlands.

Chen, J. M, and T. A. Black (1992), Foliage area and architecture of plant canopies from sunfleck size distributions, *Agricultural and Forest Meteorology, 60,* 249–266.

Chen, J. M., et al (1997), Radiation regime and canopy architecture in a boreal aspen forest, *Agricultural and Forest Meteorology, 86,* 107–125.

Choudhury, B. J, and J. L. Monteith (1988), A four-layer model for the heat budget of homogeneous land surfaces, *Quarterly Journal of the Royal Meteorological Society.*

Chrenko, F. A, and L. G. C. E. Pugh (1961), The contribution of solar radiation to the thermal environment of man in Antarctica, *Proceedings of the Royal Society of London, B, 155,* 243–265.

Church, N. S. (1960), Heat loss and the body temperature of flying insects, *Journal of Experimental Biology, 37,* 171–185.

Clapperton, J. L., et al (1965), Estimates of the contribution of solar radiation to the thermal exchanges of sheep, *Journal of Agricultural Science, 64,* 37–49.

Clark, J. A. (1976), Energy transfer and surface temperature over plants and animals, in *Light as an Ecological Factor*, edited by G. C. Evans., et al, pp. 451–463, Blackwell Scientific Publications, Oxford.

Clark, R. P, and N. Toy (1975), Natural convection around the human head, *Journal of Physiology, 244*, 283–293.

Coakley, J. A., et al (1987), Effect of Ship–Stack Effluents on Cloud Reflectivity, *Science, 237*, 1020–1022.

Colls, J. (1997), *Air Pollution: an introduction*, 1st ed, 341 pp, Chapman and Hall, London.

Coutts, M. P. (1986), Components of tree stability in Sitka spruce on peaty gley soil, *Forestry, 59*, 173–197.

Cowan, I. R. (1977), Stomatal behaviour and environment, *Advances in Botanical Research, 4*, 117–228.

Dawson, T. E. (1993), Hydraulic lift and plant water use: implications for water balance, performance and plant-plant interactions, *Oecologia, 95*, 565–574.

Deacon, E. L. (1969), Physical processes near the surface of the earth, in *World Survey of Climatology*, edited, Landsberg, H. E, Amsterdam.

DeBruin, H. A. R. (1983), A model for the Priestley–Taylor parameter a, *Journal of Climate and Applied Meteorology, 22*, 572–578.

Denmead, O. T, and E. F. Bradley (1985), Flux-gradient relationships in a forest canopy, in *The forest-atmosphere Interaction*, edited by B. A. Hutchinson and B. B. Hicks, pp. 421–442, D. Reidel, New York.

Di-Giovanni, F, and P. M. Beckett (1990), On the mathematical modelling of pollen dispersal and deposition, *Journal of Applied Meteorology, 29*, 1352–1357.

Digby, P. S. B. (1955), Factors affecting the temperature excess of insects in sunshine, *Journal of Experimental Biology, 32*, 279–298.

Dixon, M, and J. Grace (1983), Natural convection from leaves at realistic Grashof numbers, *Plant, Cell and Environment, 6*, 665–670.

Dollard, G. J, and M. H. Unsworth (1983), Field measurements of turbulent fluxes of wind-driven fog drops to a grass surface, *Atmospheric Environment, 17*, 775–780.

Dolman, A. J, and J. H. C. Gash (2003), Sonic anemometer (co)sine response and flux measurement: 1. The potential for(co)sine error to affect sonic anemometer-based flux measurements, *Agricultural and Forest Meteorology, 119*, 195–208.

Dyer, A. J. (1974), A review of flux-profile relationships, *Boundary-Layer Meteorology, 7*, 363–372.

Dyer, A. J, and B. B. Hicks (1970), Flux-gradient relationships in the constant flux layer, *Quarterly Journal of the Royal Meteorological Society, 96*, 715–721.

Ede, A. J. (1967), *An Introduction to Heat Transfer Principles and Calculations*, Pergamon Press, Oxford.

Ehleringer, J. R, and O. Bjorkman (1978), Pubescence and leaf spectral characteristics in a desert shrub, *Encelia farinosa, Oecologia, 36,* 151–162.

Ellington, C. P, and T. J. Pedley (Eds.) (1995), *Biological fluid dynamics,* 363 pp, Cambridge University Press, Cambridge.

Ennos, A. R. (1991), The mechanics of anchorage in Wheat (*Triticum aestivum* L.), *Journal of Experimental Botany, 42,* 1607–1613.

Fagerlund, U. H. M., et al (1995), Stress and tolerance, in *Physiological ecology of Pacific salmon,* edited by L. M. C. Groot, and W.C. Clarke, pp. 459–504, University of British Columbia Press, Vancouver, BC.

Farman, J. C., et al (1985), Large losses of total ozone in Antractica reveal seasonal ClOx/NOx interaction, *Nature, 315,* 207–210.

Farquhar, G. D., et al (1980), On the gaseous exchange of ammonia between leaves and the environment: determination of the ammonia compensation point, *Plant Physiology, 66,* 710–714.

Farquhar, G. D, and M. L. Roderick (2003), Pinatubo, Diffuse Light, and the Carbon Cycle, *Science, 299,* 1997–1998.

Finch, V. A., et al (1984), Coat colour in cattle, *Journal of Agricultural Science, 102,* 141–147.

Finnigan, J. J. (1985), Turbulent transport in flexible plant canopies, in *The Forest-Atmosphere Interaction,* edited by B. A. Hutchison and B. B. Hicks, pp. 443–480, D. Reidel, Dordrecht, Holland.

Finnigan, J. J. (2004a), The footprint concept in complex terrain, *Agricultual and Forest Meteorology, 127,* 117–129.

Finnigan, J. J. (2004b), A re-evaluation of long-term flux measurement techniques. Part II: Coordinate systems, *Boundary-Layer Meteorology, 113,* 1–41.

Finnigan, J. J, and Y. Brunet (1995), Turbulent airflow in forests on flat and hilly terrain, in *Wind and Trees,* edited by M. P. Coutts and J. Grace, pp. 3–40, Cambridge University Press, Cambridge.

Finnigan, J. J., et al (2003), A re-evaluation of long-term flux measurement techniques. Part I: averaging and coordinate rotation, *Boundary Layer Meteorology, 107,* 1–48.

Fishenden, M, and O. A. Saunders (1950), *An Introduction to Heat Transfer,* Clarendon Press, Oxford.

Fleischer, R. v. (1955), Der Jahresgang der Strahlungsbilanz sowie ihrer lang-und kurzwelligen Komponenten, in *Bericht des deutschen Wetterdienstes,* edited, pp. 32–40, Frankfurt.

Fowler, D., et al (1989), Deposition of atmospheric pollutants on forests, *Philosophical Transactions of the Royal Society of London, B, 324,* 247–265.

Fowler, D, and M. H. Unsworth (1979), Turbulent transfer of sulphur dioxide to a wheat crop, *Quarterly Journal of the Royal Meteorological Society, 105,* 767–783.

Frankland, B. (1981), Germination in shade, in *Plants and the Daylight Spectrum*, edited by H. Smith, Academic Press, London.

Fraser, A. I. (1962), Wind Tunnel Studies of the Forces Acting on the Crowns of Small Trees, *Report of Forest Research*.

Frohlich, C, and J. Lean (1998), The sun's total irradiance: cycles, trends and related climate change uncertainties since 1976, *Geophysical Research Letters*, *25*, 4377–4380.

Fuchs, M, and G. Stanhill (1980), Row structure and foliage geometry as determinants of the interception of light rays in a sorghum row canopy, *Plant, cell and environment*, *3*, 175–182.

Fuchs, N. A. (1964), *The Mechanics of Aerosols*, Pergamon Press, Oxford.

Funk, J. P. (1964), Direct measurement of radiative heat exchange of the human body, in *Nature*, edited, pp. 904–905.

Gale, J. (1972), Elevation and transpiration: some theoretical considerations with special reference to Mediterranean-type climate, *Journal of Applied Ecology*, *9*, 691–701.

Gardiner, B. A. (1995), The interactions of wind and tree movement in forest canopies, in *Wind and Trees*, edited by M. P. Coutts and J. Grace, pp. 41–59, Cambridge University Press, Cambridge.

Garland, J. A. (1977), The dry deposition of sulphur dioxide to land and water surfaces, *Proceedings of the Royal Society of London, A*, *354*, 245–268.

Garnier, B. J, and A. Ohmura (1968), A method of calculating the direct shorwave radiation income of slopes, *Journal of Applied Meteorology*, *7*, 796–800.

Garratt, J. R. (1978), Flux profile relations above tall vegetation, *Quarterly Journal of the Royal Meteorological Society*, *104*, 199–211.

Garratt, J. R. (1980), Surface influence upon vertical profiles in the atmospheric near-surface layer, *Quarterly Journal of the Royal Meteorological Society*, *106*, 803–819.

Gash, J. H. C. (1986), Observations of turbulence downwind of a forest-heath interface, *Boundary-Layer Meteorology*, *36*, 227–237.

Gatenby, R. M. (1977), Conduction of heat from sheep to ground, *Agricultural Meteorology*, *18*, 387–400.

Gatenby, R. M., et al (1983), Temperature and humidity gradients in the steady state, *Agricultural Meteorology*, *29*, 1–10.

Gates, D. M. (1980), *Biophysical Ecology*, Springer-Verlag, New York.

Gilby, A. R. (1980), Transpiration, temperature and lipids in insect cuticle, in *Advances in Insect Physiology*, edited by M. J. Berridge et al, pp. 1–33, Academic Press, New York.

Gloyne, R. W. (1972), The diurnal variation of global radiation on a horizontal surface—with special reference to Aberdeen, in *Meteorological Magazine*, edited, pp. 44–51.

Goudriaan, J. (1977), *Crop Micrometeorology: a Simulation Study*, Center for Agricultural Publishing and Documentation, Wageningen.

Goulden, M. L, et al (1996), Measurements of carbon sequestration by long-term eddy covariance: methods and critical evaluation of accuracy, *Global Change Biology*, *2*, 169–182.

Grace, J. (1978), The turbulent boundary layer over a flapping *Populus* leaf, *Plant, Cell and Environment*, *1*, 35–38.

Grace, J, and M. A. Collins (1976), Spore liberation from leaves by wind, in *Microbiology of Aerial Plant Surfaces*, edited by C. H. Dickinson and T. F. Preece, pp. 185–198, Academic Press, London.

Grace, J, and J. Wilson (1976), The boundary layer over a Populus leaf, *Journal of Experimental Botany*, *27*, 231–241.

Graser, E. A, and C. H. M. v. Bavel (1982), The effect of soil moisture upon soil albedo, *Agricultural Meteorology*, *27*, 17–26.

Greene, C. F. (1987), Nitrogen nutrition and wheat growth in relation to absorbed radiation, *Agricultural and Forest Meteorology*, *41*, 207–248.

Grigg, G. C., et al (1979), Time constants of heating and cooling in the Eastern Water Dragon, *Physignathus lesueruii* and some generalizations about heating and cooling in reptiles, *Journal of Thermal Biology*, *4*, 95–103.

Gu, L., et al (2003), Response of a deciduous forest to the Mt. Pinatubo eruption: enhanced photosynthesis, *Science*, *299*.

Guenther, A., et al (1996), Isoprene fluxes measured by enclosure, relaxed eddy accumulation, surface layer gradient, mixed layer gradient, and mixed layer mass balance techniques, *Journal of Geophysical Research*, *101*, 18555–18567.

Hamer, P. J. C. (1986), The heat balance of apple buds and blossoms Part III. The water requirements for evaporative cooling by overhead sprinkler irrigation, *Agricultural and Forest Meteorology*, *37*, 175–188.

Hammel, H. T. (1955), Thermal properties of fur, *American Journal of Physiology*, *182*, 369–376.

Hansen, J., et al (2004), Carbonaceous aerosols in the industrial era, *EOS, Transactions, American Geophysical Union*, *85*, 241–244.

Hansen, J. E., et al (1996), Global surface air temperature in 1995: return to pre-Pinatubo level, *Geophysical Research Letters*, *23*, 1665–1668.

Hargreaves, B. R. (2003), Water column optics and penetration of UVR, in *UV effects in aquatic organisms and ecosystems*, edited by E. W. Helbling and H. E. Zagarese, pp. 59–105, Royal Society of Chemistry, London.

Haseba, T. (1973), Water vapour transfer from leaf-like surfaces within canopy models, *The Journal of Agricultural Meteorology*, *29*, 25–33.

Hatfield, J. L. (1983), Comparison of long-wave radiation calculation methods over the United States, *Water Resources Research*, *19*, 285–288.

Hayhoe, K., et al (2004), Emissions pathways, climate change, and impacts on California, *PNAS*, *101*, 12422–12427.

375

Heagle, A. S., et al (1973), An open-top field chamber to access the impact of air pollution on plants, *Journal of Environmental Quality*, *2*, 365–368.

Hemmingsen, A. M. (1960), Energy metabolism as related to body size and respiratory surfaces, and its evolution, Report of the Steno Memorial Hospital, Niels Steensens Hospital, Copenhagen. Reports of the Steno Memorial Hospital and the Nordisk Insulinlaboratorium, Copenhagen.

Henderson, S. T. (1977), *Daylight and its spectrum*, Adam Hilger, Bristol.

Herrington, L. P. (1969), On Temperature and Heat Flow in Tree Stems, School of Forestry Bulletin, 79 pp, Yale University, New Haven.

Hickey, J. R., et al (1982), Extraterrestrial solar irradiance variability. Two and one-half years of measurements from Nimbus 7, *Solar Energy*, *29*, 127.

Hinds, W. C. (1999), *Aerosol technology*, 2nd ed, 483 pp, John Wiley & Sons, New York.

Howell, T. A., et al (1983), Relationship of photosynthetically active radiation to short-wave radiation in the San Joaquin Valley, *Agricultural Meteorology*, *28*, 157–175.

Hsieh, C.-I., et al (2003), Predicting scalar source-sink and flux distributions within a forest canopy using a 2-D Lagrangian stochastic dispersion model, *Boundary-Layer Meteorology*, *109*, 113–138.

Huete, A., et al (2002), Overview of the radiometric and biophysical performance of the MODIS vegetation indices, *Remote Sensing of Environment*, *83*, 195–213.

Hutchinson, J. C. D., et al (1975), Measurements of the reflectances for solar radiation of the coats of live animals, *Comparative Biochemistry and Physiology*, *52A*, 343–349.

Idso, S. B., et al (1975), The dependence of bare soil albedo on soil water content, *Journal of Applied Meteorology*, *14*, 109–113.

Impens, I. (1965), *Experimentele Studie van de Physische en Biologische Aspektera van de Transpiratie*, Ryklandbouwhogeschool, Ghent.

IPCC (2001), *Climate Change: the Scientific Basis*, Cambridge University Press, Cambridge.

Jarosz, N., et al (2003), Field measurements of airborne concentration and deposition rate of maize pollen, *Agricultural and Forest Meteorology*, *119*, 37–51.

Jarvis, P. G., et al (1976), Coniferous forest, in *Vegetation and the Atmosphere*, edited by J. L. Monteith, pp. 171–240, Academic Press, London.

Jarvis, P. G, and K. G. McNaughton (1986), Stomatal control of transpiration, in *Advances in Ecological Research*, edited, pp. 1–49, Academic Press, New York.

Jarvis, P. G., et al (1985), Modelling canopy exchanges of water vapor and carbon dioxide in coniferous forest plantations, in *The Forest-Atmosphere Interaction*, edited by B. A. Hutchinson and B. B. Hicks, pp. 521–542, D. Reidel, New York.

Johnson, G. T, and I. D. Watson (1985), Modelling longwave radiation exchange between complex shapes, *Boundary-Layer Meteorology*, *33*, 363–378.

Jones, H. G. (1992), *Plants and microclimate: a quantitative approach to environmental plant physiology*, 2 ed, Cambridge University Press, Great Britain.

Kaimal, J. C, and J. J. Finnigan (1994), *Atmospheric boundary layer flows: their structure and measurements*, Oxford University Press, Inc, New York.

Kaminsky, K. Z, and R. Dubayah (1997), Estimation of surface net radiation in the boreal forest and northern prairiefrom shortwave flux measurements, *Journal of Geophysical Research, 102*, 29707–29716.

Katul, G., et al (1995), Estimation of surface heat and momentum fluxes using the flux-variance method above uniform and non-uniform terrain, *Boundary-Layer Meteorology, 74*, 237–260.

Katul, G. G, and J. D. Albertson (1999), Modeling CO2 sources, sinks, and fluxes within a forest canopy, *Journal of Geophysical Research, 104*, 6081–6091.

Katul, G. G., et al (2001), Estimating CO2 source/sink distributions within a rice canopy using higher-order closure models, *Boundary-Layer Meteorology, 98*, 103–125.

Kaufman, Y. J., et al (2002), A satellite view of aerosols in the climate system, *Nature, 419*, 215–223.

Kelliher, F. M., et al (1995), Maximum conductances for evaporation from global vegetation types, *Agricultural and Forest Meteorology, 73*, 1–16.

Kerslake, D. M. (1972), *The stress of hot environments*, Cambridge University Press, Cambridge.

Kleiber, M. (1965), Metabolic body size, in *Energy Metabolism*, edited by K. L. Blaxter, pp. 427–435, Academic Press, London.

Kondratyev, K. J, and M. P. Manolova (1960), The radiation balance of slopes, *Solar Energy, 4*, 14–19.

Landsberg, J. J, and R. H. Waring (1997), A generalised model of forest productivity using simplified concepts of radiation-use efficiency, carbon balance and partitioning, *Forest Ecology and Management, 95*, 209–228.

Lang, A. R. G., et al (1983), Inequality of eddy transfer coefficients for vertical transport of sensible and latent heats during advective inversions, *Boundary-Layer Meteorology, 25*, 25–41.

Law, B. E., et al (2001), Estimation of leaf area index in open-canopy ponderosa pine forests at different successional stages and managment regimes in Oregon, *Agricultural and Forest Meteorology, 108*, 1–14.

Lee, X., et al (Eds.) (2004), *Handbook of micrometeorology*, 250 pp, Springer Verlag, New York.

Legg, B. J. (1983), Movement of plant pathogens in the crop canopy, *Philosophical Transactions of the Royal Society of London, B, 302*, 559–574.

Legg, B. J., et al (1981), Aerodynamic properties of field bean and potato crops, *Agricultural Meteorology, 23*, 21–43.

Lettau, H. (1969), Note on the aerodynamic roughness parameter estimation on the basis of roughness-element description, *Journal of Applied Meteorology, 8*, 828–832.

Leuning, R. (1983), Transport of gases into leaves, *Plant, Cell and Environment, 6*, 181–194.

Leuning, R. (2000), Estimation of scalar source/sink distributions in plant canopies using Lagrangian dispersion analysis: corrections for atmospheric stabilitiy and comparison with a multilayer canopy model, *Boundary-Layer Meteorology, 96*, 293–314.

Leuning, R., et al (1982), Effects of heat and water vapor transport on eddy covariance measurement of CO_2 fluxes, *Boundary-Layer Meteorology, 23*, 209–222.

Leuning, R, and M. J. Judd (1996), The relative merits of open- and closed-path analysers for the measurement of eddy fluxes, *Archives of Insect Biochemistry and Physiology, 2*, 241–253.

Lewis, H. E., et al (1969), Aerodynamics of the human microenvironment, *Lancet, 1*, 1273–1277.

Leyton, L. (1975), *Fluid behaviour in biological systems*, 235 pp, Clarendon Press, Oxford.

Linacre, E. T. (1972), Leaf temperature, diffusion resistances, and transpiration, *Agricultural Meteorology, 10*, 365–382.

Link, T. E., et al (2004), The dynamics of rainfall interception by a seasonal temperate rainforest, *Agricultural and Forest Meteorology, 124*, 171–191.

List, R. J. e. (1966), *Smithsonian Meteorological Tables*, Smithsonian Institution, Washington DC.

Little, P, and R. D. Whiffen (1977), Emission and deposition of petrol engine exhaust lead. I. Deposition of exhaust lead to plant and soil surfaces, *Atmospheric Environment, 11*, 437–447.

Liu, B. Y, and R. C. Jordan (1960), The interrelationship and characteristic distribution of direct, diffuse and total solar radiation, *Solar Energy, 4*, 1–19.

Lumb, F. E. (1964), The influence of cloud on hourly amounts of total solar radiation at the sea surface, *Quarterly Journal of the Royal Meteorological Society, 90*, 43–56.

Mahoney, S. A, and J. R. King (1977), The use of the equivalent black-body temperature in the thermal energetics of small birds, *Journal of Thermal Biology, 2*, 115–120.

Mahrt, L, and D. Vickers (2003), Formulation of Turbulent Fluxes in the Stable Boundary Layer, *Journal of the Atmospheric Sciences, 60*, 2538–2548.

Mahrt, L., et al (2003), Sea-surface aerodynamic roughness, *Journal of Geophysical Research, 108*, 1–9.

Marks, D, and J. Dozier (1992), Climate and Energy Exchange at the Snow Surface in the Alpine Region of the Sierra Nevada 2. Snow Cover Energy Balance, *Water Resources Research, 28*, 3043–3054.

Massman, W. J. (1999), A model study of kBH-1 for vegetated surfaces using "localized near-field" Lagrangian theory, *Journal of Hydrology, 223*, 27–43.

Mayhead, G. J. (1973), Some drag coefficients for British forest trees derived from wind tunnel studies, *Agricultural Meteorology, 12*, 123–130.

McAdams, W. H. (1954), *Heat transmission*, 3rd ed, McGraw Hill, New York.

McArthur, A. J. (1987), Thermal interaction between animal and microclimate—a comprehensive model, *Journal of Theoretical Biology, 126*, 203–238.

McArthur, A. J. (1990), An accurate solution to the Penman equation, *Agricultural and Forest Meteorology, 51*, 87–92.

McArthur, A. J, and J. A. Clark (1988), Body temperature of homeotherms and the conservation of energy and water, *Journal of Thermal Biology, 13*, 9–13.

McArthur, A. J, and J. L. Monteith (1980a), Air movement and heat loss from sheep. I. Boundary layer insulation of a model sheep with and without fleece, *Proceedings of the Royal Society of London, B, 209*, 187–208.

McArthur, A. J, and J. L. Monteith (1980b), Air Movement and heat loss from sheep. II. Thermal insulation of fleece in wind, *Proceedings of the Royal Society of London, B, 209*, 209–217.

McCartney, H. A. (1978), Spectral distribution of solar radiation. II. Global and diffuse, *Quarterly Journal of the Royal Meteorological Society, 104*, 911–926.

McCartney, H. A, and B. D. L. Fitt (1998), Dispersal of foliar fungal plant pathogens: mechanisms, gradients and spatial patterns, *Plant Disease Epidemiology. Kluwer Publishers, London*, 138–160.

McCartney, H. A, and M. H. Unsworth (1978), Spectral distribution of solar radiation. I. Direct radiation, *Quarterly Journal of the Royal Meteorological Society, 104*, 699–718.

McCree, K. J. (1972), The action spectrum, absorptance and quantum yield of photosynthesis in crop plants, *Agricultural Meteorology, 9*, 191–216.

McCulloch, J. S. G, and H. L. Penman (1956), Heat flow in the soil, *Report of the 6th International Soil Science Congress, B*, 275–280.

McFarland, W. N, and F. W. Munz (1975), The visible spectrum during twilight and its implications to vision, in *Light as an Ecological Factor*, edited by G. C. Evans et al, pp. 249–270, Cambridge University Press, Cambridge.

McNaughton, K. G. (1989), Regional interactions between canopies and the atmosphere, in *Plant canopies: their growth, form and function*, edited by G. Russell et al, pp. 63–82, Cambridge University Press, Cambridge.

McNaughton, K. G, and P. G. Jarvis (1983), Predicting effects of vegetation changes on transpiration and evaporation, in *Water Deficits and Plant Growth*, edited by T. T. Kozlowski, pp. 1–47, Academic Press, New York.

McNaughton, K. G, and P. G. Jarvis (1986), Stomatal Control of Transpiration: Scaling Up from Leaf to Region, *Advances in Ecological Research, 15*, 1–49.

McNaughton, K. G, and P. G. Jarvis (1991), Effects of spatial scale on stomatal control of transpiration, *Agricultural and Forest Meteorology, 54*, 279–301.

McNaughton, K. G, and T. W. Spriggs (1986), A mixed-layer model for regional evaporation, *Boundary-Layer Meteorology*, *34*, 243–262.

McNaughton, K. G, and B. J. van den Hurk (1995), A 'Lagrangian' revision of the resistors in the two-layer model for calculating the energy budget of a plant canopy, *Boundary-Layer Meteorology*, *74*, 261–288.

Meidner, H, and T. A. Mansfield (1968), *Physiology of Stomata*, McGraw Hill, London.

Meinzer, F, and G. Goldstein (1985), Some Consequences of Leaf Pubescence in the Andean Giant Rosette Plant Espeletia Timotensis, *Ecology*, *66*, 512–520.

Mellor, R. S., et al (1964), Leaf temperatures in controlled environments, *Planta*, *61*, 56–72.

Mercer, T. T. (1973), *Aerosol Technology in Hazard Evaluation*, Academic Press, New York.

Milne, R. (1991), Dynamics of swaying of *Picea sichensis*, *Tree Physiology*, *9*, 383–399.

Milthorpe, F. L, and H. L. Penman (1967), The diffusive conductivity of the stomata of wheat leaves, *Journal of Experimental Botany*, *18*, 422–457.

Mitchell, J. W. (1976), Heat transfer from spheres and other animal forms, *Biophysical Journal*, *16*, 561–569.

Monteith, J. L. (1957), Dew, *Quarterly Journal of the Royal Meteorological Society*, *83*, 322–341.

Monteith, J. L. (1962), Measurement and interpretation of carbon dioxide fluxes in the field, *Netherlands Journal of Agricultural Science*, *10*, 334–346.

Monteith, J. L. (1963a), Calculating evaporation from diffusive resistances, in *Investigations of energy and mass transfers near the ground, including the influences of the soil-plant-atmosphere system. Report number DA-36-039-SC-80334*, edited, pp. 177–189, University of California, Davis, California.

Monteith, J. L. (1963b), Calculating evaporation from diffusive resistances. (Chapter 10 in a report 'Investigation of energy and mass transfers near the ground including influences of the soil–plant–atmosphere system), 177–189 pp, Water Resources Center, University of California, Davis, Davis, California.

Monteith, J. L. (1965), Evaporation and Environment, paper presented at Symposium of the Society for Experimental Biology.

Monteith, J. L. (1969), Light interception and radiative exchange in crop stands, in *Physiological Aspects of Crop Yield*, edited by J. D. Eastin, American Society of Agronomy, Madison Wisconsin.

Monteith, J. L. (1972), Latent heat of vaporization in thermal physiology, *Nature*, *236*, 96.

Monteith, J. L. (1973), *Principles of Environmental Physics*, 1st ed, Edward Arnold, London.

Monteith, J. L. (1981a), Evaporation and surface temperature, *Quarterly Journal of the Royal Meteorological Society*, *107*, 1–27.

Monteith, J. L. (1981b), Coupling of plants to the atmosphere, in *Plants and Their Atmospheric Environment*, edited by J. Grace., et al, pp. 1–29, Blackwell Scientific Publications, Oxford.

Monteith, J. L. (1994), Fifty years of potential evaporation.

Monteith, J. L. (1995), A reinterpretation of stomatal responses to humidity, *Plant, Cell and Environment*, *18*, 357–364.

Monteith, J. L, and D. Butler (1979), Dew and thermal lag: a model for cocoa pods, *Quarterly Journal of the Royal Meteorological Society*, *105*, 207–215.

Monteith, J. L, and G. S. Campbell (1980), Diffusion of water vapour through integuments–potential confusion, *Journal of Thermal Biology*, *5*, 7–9.

Monteith, J. L, and J. Elston (1983), Performance and productivity of foliage in the field, in *The Growth and Functioning of Leaves*, edited by J. E. Dale and F. L. Milthorpe, pp. 499–518, Cambridge Unviersity Press.

Moon, P. (1940), Proposed standard solar radiation curves for engineering use, *Journal of the Franklin Institute*, *230*, 583–617.

Mount, L. E. (1967), Heat loss from new-born pigs to the floor, *Research in Veterinary Science*, *8*, 175–186.

Mount, L. E. (1968), *The Climatic Physiology of the Pig*, Edward Arnold, London.

Mount, L. E. (1979), *Adaptation of Thermal Environment of Man and his Productive Animals*, Edward Arnold, London.

Mulholland, B. J., et al (1998), Growth, light interception and yield responses of spring wheat (Triticum aestivum L.) grown under elevated CO2 and O3 in open-top chambers, *Global change biology*, *4*, 121–130.

Munro, D. S, and T. R. Oke (1975), Aerodynamic boundary-layer adjustment over a crop in neutral stability, *Boundary-Layer Meteorology*, *9*, 53–61.

Murray, F. W. (1967), On the computation of saturation vapour pressure, *Journal of Applied Meteorology*, *6*, 203–204.

Nakai, T., et al (2006), Correction of sonic anemometer angle of attack errors, *Agricultural and Forest Meteorology*, *136*, 19–30.

Nielsen, B. (1996), Olympics in Atlanta: a fight against physics, *Medicine & Science in Sports & Exercise*, *28*, 665–668.

Niemela, S. P., et al (2001), Comparison of surface radiative flux parameterizations. Part 1: Longwave radiation, *Atmospheric Research*, *58*.

Nobel, P. S. (1974), Boundary layers of air adjacent to cylinders, *Plant Physiology*, *54*, 177–181.

Nobel, P. S. (1975), Effective thickness and resistance of the air boundary layer adjacent to spherical plant parts, *Journal of Experimental Botany*, *26*, 120–130.

Norman, J. M. (1992), Scaling processes between leaf and canopy levels, in *Scaling physiological processes: leaf to globe*, edited by J. R. Ehleringer and C. B. Field, pp. 41–76, Academic Press, San Diego.

Offerle, B., et al (2003), Parameterization of net all-wave radiation for urban areas, *Journal of Applied Meteorology, 42*, 1157–1173.

Ohmura, A, and M. Wild (2002), Is the Hydrological Cycle Accelerating? *Science, 298*, 1345–1346.

Ohtaki, E. (1984), Application of an infrared carbon dioxide and humidity instrument to studies of turbulent transport, *Boundary-Layer Meteorology, 29*, 85–107.

Oosthuizen, P. H, and S. Madan (1970), Combined convective heat transfer from horizontal cylinders in air. Transactions of the American Society of Mechanical Engineers, Series C, *Journal of Heat Transfer, 92*, 194–196.

Papastamati, K., et al (2004), Modelling leaf wetness duration during rosette stage in oilseed rape, *Agricultural and Forest Meteorology, 123*, 69–78.

Parkhurst, D. F. (1976), Effects of Verbascum thapsus leaf hairs on heat and mass transfer: a reassessment, *New phytologist, 76*, 453–457.

Parkhurst, D. F., et al (1968), Wind-tunnel modelling of convection of heat between air and broad leaves of plants, *Agricultural Meteorology, 5*, 33–47.

Parlange, J. Y., et al (1971), Boundary layer resistance and temperature distribution on still and flapping leaves, *Plant Physiology, 48*, 437–442.

Parlange, M. B, and W. Brutsaert (1998), Hydrologic cycle explains the evaporative paradox, *Nature, 396*, 30.

Parlange, M. B., et al (1995), Regional scale evaporation and the atmospheric boundary layer, *Reviews of Geophysics, 33*, 99–124.

Paulson, C. A. (1970), The mathematical representation of wind speed and temperature profiles in the unstable atmospheric surface layer, *Journal of Applied Meteorology, 9*, 857–861.

Peat, J., et al (2005), Effects of climate on intra- and interspecific size variation in bumble-bees.

Penman, H. L. (1948), Natural evaporation form open water, bare soil and grass, *Proceedings of the Royal Society of London, A, 194*, 120–145.

Penman, H. L, and R. K. Schofield (1951), Some Physical aspects of assimilation and transpiration, *Symposium of the Society of Experimental Biology, 5*, 115–129.

Phillips, P. K, and J. E. Heath (1992), Heat exchange by the pinna of the African elephant (Loxodonta africana). *Comparative Biochemistry and Physiology, 101*, 693–699.

Phillips, P. K, and J. E. Heath (2001), Heat loss in Dumbo: a theoretical approach, *Journal of Thermal Biology, 26*, 117–120.

Pohlhausen, E. (1921), Der Warmestausch zwischen festen Korpen und Flussigkeiten mit Reibung und kleiner Warmeleitung, *Zeitschrift für angewandte Mathematik und Mechanik, 1*, 115–121.

Porter, W. P., et al (1973), Behavioural implications of mechanistic ecology. Thermal and behavioural modelling of desert ectotherms and their microenvironment, *Oecologia*, *13*, 1–54.

Powell, R. W. (1940), Further experiments on the evaporation of water from saturated surfaces, *Transactions of the Institutions of Chemical Engineers*, *18*, 36–55.

Prata, A. J. (1996), A new longwave formula for estimating downward clear-sky radiation at the surface, *Quarterly Journal of the Royal Meteorological Society*, *122*, 1127–1151.

Priestley, C. H. B. (1957), The heat balance of sheep standing in the sun, *Australian Journal of Agricultural Research*, *8*, 271–280.

Priestley, C. H. B, and R. J. Taylor (1972), On the assessment of surface heat flux and evaporation using large scale parameters, *Monthly Weather Review*, *100*, 81–92.

Prothero, J. (1984), Scaling of standard energy metabolism in mammals: I. Neglect of circadian rhythms, *Journal of Theoretical Biology*, *106*, 1–8.

Pypker, T. G., et al (2006), The role of epiphytes in the interception of rainfall in old-growth Douglas-fir/ western hemlock forests of the Pacific Northwest: Part II—Field measurements at the branch and canopy scale, *Canadian Journal of Forest Research*, *36*, 819–832.

Rainey, R. C., et al (1957), *The Behaviour of the Red Locust*, Anti-locust Research Centre, London.

Rapp, G. M. (1970), Convective mass transfer and the coefficient of evaporative heat loss from human skin, in *Physiological and Behavioural Temperature Regulation*, edited by J. D. Hardy., et al, C. C. Thomas, Illinois.

Raupach, M. (1989a), Applying Lagrangian fluid mechanics to infer scalar source distributions from concentration profiles in plant canopies, *Agricultural and Forest Meteorology*, *47*, 85–108.

Raupach, M. (1989b), A practical Lagrangian method for relating scalar concentrations to source distributions in vegetation canopies, *Quarterly Journal of the Royal Meteorological Society*, *115*, 609–632.

Raupach, M. (1995), Vegetation-atmosphere interaction and surface conductance at leaf, canopy and regional scales, *Agricultural and Forest Meteorology*, *73*, 151–179.

Raupach, M. R, and B. J. Legg (1984), The uses and limitations of flux-gradient relationships in micrometeorology, *Agricultural Water Management*, *8*, 119–131.

Raupach, M. R, and A. S. Thom (1981), Turbulence in and above plant canopies, *Annual Review of Fluid Mechanics*, *13*, 97–129.

Raynor, G. S., et al (1972), Dispersion and deposition of corn pollen from experimental sources, *Agronomy Journal*, *64*, 420–427.

Reeve, J. E. (1960), Appendix to 'Inclined Point Quadrats' by J. Warren Wilson, *The New Phytologist*, *59*, 1–8.

Reynolds, A. M. (2000), Prediction of particle deposition onto rough surfaces, *Agricultural and Forest Meteorology, 104*, 107–118.

Rich, P. M., et al (1993), Long-term study of solar radiation regimes in a tropical wet forest using quantum sensors and hemispherical photographs, *Agricultural and Forest Meteorology, 65*, 107–127.

Roderick, M. L, and G. D. Farquhar (2002), The Cause of Decreased Pan Evaporation over the Past 50 Years, *Science, 298*, 1410–1411.

Roderick, M. L., et al (2001), On the direct effect of clouds and atmospheric particles on the productivity and structure of vegetation, *Oecologia, 129*, 121–130.

Ross, J. (1975), Radiative transfer in plant communities, in *Vegetation and the Atmosphere*, edited by J. L. Monteith, pp. 13–55, Academic Press, London.

Ross, J. (1981), *The radiation regime and architecture of plant stands*, 391 pp, Dr. W. Junk, The Hague.

Roth-Nebelsick, A. (2001), Computer-based analysis of steady-state and transient heat transfer of small-sized leaves by free and mixed convection, *Plant Cell and Environment, 24*, 631–640.

Roy, J. C., et al (2002), Convective and ventilation transfers in greenhouses. 1: The greenhouse considered as a perfectly stirred tank, *Biosystems Engineering, 83*, 1–20.

Rudnicki, M., et al (2004), Wind tunnel measurements of crown streamlining and drag relationships for three conifer species, *Canadian Journal of Forest Research, 34*, 666–676.

Russell, G., et al (1989), Absorption of radiation by canopies and stand growth, in *Plant Canopies: their growth, form and function*, edited by G. Russell., et al, pp. 21–39, Cambridge University Press, Cambridge.

Sakai, R. K., et al (2001), Importance of low-frequency contributions to eddy fluxes observed over rough surfaces, *Journal of Applied Meteorology, 40*, 2178–2192.

Schmidt-Nielsen, K. (1965), *Desert Animals*, Oxford University Press, Oxford.

Schmidt-Nielsen, K., et al (1956), Body temperature of the camel and its relation to water economy, *American Journal of Physiology, 188*, 103–112.

Schmugge, T. (1998), Applications of passive microwave observations of surface soil moisture, *Journal of Hydrology, 212/213*, 188–197.

Scholander, P. F., et al (1950), Body insulation of some arctic and tropical mammals and birds, *Biological Bulletin, 99*, 225–234.

Schuepp, P. H. (1993), Leaf boundary layers, *New phytologist, 125*, 477–507.

Shaw, R. H, and A. R. Pereira (1982), Aerodynamic roughness of a plant canopy: a numerical experiment, *Agricultural Meteorology, 26*, 51–65.

Shuttleworth, W. J. (1988), Evaporation from Amazonian rainforest, *Proceedings of the Royal Society of London, B, 233*, 321–340.

Shuttleworth, W. J. (1989), Micrometeorology of temperate and tropical forest, *Phil. Trans. R. Soc. Lond, B 324*, 299–334.

Shuttleworth, W. J. e. a. (1984), Eddy correlation measurements of energy partition for Amazonian forest, *Quarterly Journal of the Royal Meteorological Society, 110,* 1143–1162.

Sinokrot, B. A, and H. G. Stefan (1993), Stream temperature dynamics: measurement and modeling, *Water Resources Research, 29,* 2299–2312.

Slatyer, R. O, and I. C. McIlroy (1961), *Practical Microclimatology,* CSIRO Australia and UNESCO.

Smith, H, and D. C. Morgan (1981), The spectral characteristics of the visible radiation incident upon the surface of the earth, in *Plants and the Daylight Spectrum,* edited by H. Smith, Academic Press, London.

Spence, D. H. N. (1976), Light and plant response in fresh water, in *Light as an Ecological Factor,* edited by G. C. Evans., et al, pp. 93–133, Blackwell Scientific Publications, Oxford.

Sperry, J. S. (1995), Limitations on stem water transport and their consequences, in *Plant Stems,* edited by B. Gardner, pp. 105–124, Academic Press, San Diego.

Stacey, G. R., et al (1994), Wind flows and forces in a model spruce forest, *Boundary-Layer Meteorology, 69,* 311–334.

Stanhill, G. (1969), A simple instrument for the field measurement of turbulent diffusion flux, *Journal of Applied Meteorology, 8,* 509–513.

Stanhill, G. (1970), Some results of helicopter measurements of albedo, *Solar Energy, 13,* 59.

Stannard, D. I. (1997), Theoretically based determinations of Bowen Ratio fetch requirements, *Boundary-Layer Meteorology, 83,* 375–406.

Steven, M. D. (1977), Standard distributions of clear sky radiance, *Quarterly Journal of the Royal Meteorological Society, 103,* 457–465.

Steven, M. D. (1983), *The Physical and Physiological Interpretation of Infrared/Red Spectral Ratios Over Crops,* 12 pp, Royal Chemical Society, London.

Steven, M. D., et al (1983), Estimation of sugar beet productivity from reflection in the red and infrared spectral bands, *International Journal of Remote Sensing, 4,* 325–334.

Steven, M. D., et al (1984), Atmospheric attenuation and scattering determined form multiheight multispectral scanner imagery, *International Journal of Remote Sensing, 5,* 733–747.

Steven, M. D, and M. H. Unsworth (1979), The diffuse solar irradiance of slopes under cloudless skies, *Quarterly Journal of the Royal Meteorological Society, 105,* 593–602.

Stewart, J. B, and A. S. Thom (1973), Energy budgets in pine forest, *Quarterly Journal of the Royal Meteorological Society, 99,* 154–170.

Stigter, C. J, and V. M. M. Musabilha (1982), The conservative ratio of photosynthetically active to total radiation in the tropics, *Journal of Applied Ecology, 19,* 853–858.

Styles, J. M., et al (2002), Soil and canopy CO2, 13CO2, H2O and sensible heat flux partitions in a forest canopy inferred from concentration measurements, *Tellus*, *54B*, 655–676.

Su, Z., et al (2001), An evaluation of two models for estimation of the roughness heaight for heat transfer between the land surface and the atmosphere, *Journal of Applied Meteorology*, *40*, 1933–1951.

Sunderland, R. A. (1968), Experiments on Momentum and Heat Transfer with Artificial Leaves, B. Sc. dissertation thesis, University of Nottingham.

Sutton, M., et al (1993a), The exchange of atmospheric ammonia with vegetated surfaces. 1: unfertilized vegetation, *Quarterly Journal of the Royal Meteorological Society*, *119*, 1023–1045.

Sutton, M., et al (1993b), The exchange of atmospheric ammonia with vegetated surfaces. 2: fertilized vegetation, *Quarterly Journal of the Royal Meteorological Society*, *119*, 1047–1070.

Sutton, M. A., et al (1995), Plant-atmosphere exchange of ammonia, *Philosophical Transactions of the Royal Society of London, A*, *351*, 261–278.

Swinbank, W. C. (1963), Long-wave radiation from clear skies, *Quarterly Journal of the Royal Meteorological Society*, *89*, 339–348.

Szeicz, G. (1974), Solar radiation for plant growth, *Journal of Applied Ecology*, *11*, 617–636.

Tageeva, S. V, and A. B. Brandt (1961), Optical properties of leaves, in *Progress in Photobiology*, edited by B. C. Cristenson, Elsevier, Amsterdam.

Tani, N. (1963), The wind over the cultivated field, *Bulletin of the National Institute of Agricultural Science, A.10*, 99.

Taylor, C. R, and C. P. Lyman (1972), Heat storage in running antelopes: independence of brain and body temperatures, *American Journal of Physiology*, *222*, 114–117.

Tetens, O. (1930), Uber einige meteorologische Begriffe, *Zeitschrift Geophysic*, *6*, 297–309.

Thom, A. S. (1968), The exchange of momentum, mass and heat between an artificial leaf and the airflow in a wind-tunnel, *Quarterly Journal of the Royal Meteorological Society*, *94*, 44–55.

Thom, A. S. (1971), Momentum absorption by vegetation, *Quarterly Journal of the Royal Meteorological Society*, *97*, 414–428.

Thom, A. S. (1972), Momentum, mass and heat exchange of vegetation, *Quarterly Journal of the Royal Meteorological Society*, *98*, 124–134.

Thom, A. S. (1975), Momentum, mass and heat exchange of plant communities, in *Vegetation and the Atmosphere*, edited by J. L. Monteith, pp. 57–109, Academic Press, London.

Thom, A. S, and H. R. Oliver (1977), On Penman's equation for estimating regional evaporation, *J. R. Met. Soc*, *103*, 345–357.

Thom, A. S., et al (1975), Comparison of aerodynamic and energy budget estimates of fluxes over a pine forest, *Quarterly Journal of the Royal Meteorological Society*, *101*, 93–105.

Thomas, S. C, and W. E. Winner (2000), Leaf area index of an old-growth Douglas-fir forest estimated from direct structural measurements in the canopy, *Canadian Journal of Forest Research*, *30*, 1922–1930.

Thornthwaite, C. W, and B. Holzman (1942), Measurement of Evaporation from Land and Water Surfaces, US Department of Agriculture Technical Bulletin, 75 pp.

Thorpe, M. R. (1978), Net radiation and transpiration of apple trees in rows, *Agricultural Meteorology*, *19*, 41–57.

Thorpe, M. R, and D. R. Butler (1977), Heat transfer coefficients for leaves on orchard apple trees, *Boundary-Layer Meteorology*, *12*, 61–73.

Tibbals, E. C., et al (1964), Radiation and convection in conifers, *American Journal of Botany*, *51*, 529–538.

Tooming, H. G, and B. E. Gulyaev (1967), *Methods of Measuring Photosynthetically Active Radiation (in Russian)*. Nauka, Moscow.

Tucker, C. J., et al (1985), African land-cover classification using satellite data, *Science*, *227*, 369–375.

Tucker, V. A. (1969), The energetics of bird flight, *Scientific American*, *220*, 70–78.

Tullett, S. G. (1984), The porosity of avian eggshells, *Comparative Biochemistry and Physiology*, *78A*, 5–13.

Turnpenny, J. R., et al (2000), Thermal balance of livestock. 2. Applications of a parsimonious model, *Agricultural and Forest Meteorology*, *101*.

Tuzet, A., et al (2003), Stomatal control of photosynthesis and transpiration: results from a soil–plant–atmosphere continuum model, *Plant, Cell and Environment*, *26*, 1097–1116.

Underwood, C. R, and E. J. Ward (1966), The solar radiation area of man, *Ergonomics*, *9*, 155–168.

Unsworth, M. H. (1975), Geometry of long-wave radiation at the ground. II. Interception by slopes and solids, *Quarterly Journal of the Royal Meteorological Society*, *101*, 25–34.

Unsworth, M. H., et al (1984a), Gas exchange in open-top field chambers. I. Measurement and analysis of atmospheric resistances to gas exchange, *Atmospheric Environment*, *18*, 373–380.

Unsworth, M. H., et al (1984b), Gas exchange in open-top field chambers. II. Resistance to ozone uptake by soybeans, *Atmospheric Environment*, *18*, 381–385.

Unsworth, M. H., et al (1984), Radiation interception and the growth of soybeans exposed to ozone in open-top field chambers, *Journal of Applied Ecology*, *21*, 1059–1079.

Unsworth, M. H, and J. L. Monteith (1972), Aerosol and solar radiation in Britain, *Quarterly Journal of the Royal Meteorological Society*, *99*, 778–797.

Unsworth, M. H, and J. L. Monteith (1975), Geometry of long-wave radiation at the ground. I. Angular distribution of incoming radiation, *Quarterly Journal of the Royal Meteorological Society, 101*, 13–24.

Unsworth, M. H., et al (2004), Components and controls of water flux in an old-growth Douglas fir-Western hemlock ecosystem, *Ecosystems, 7*, 468–481.

Van Eimern, J. (1964), *Untersuchungen uber das Klima in Pflanzengestanden*, Offenbach.

Van Wijk, W. R, and D. A. De Vries (1963), Periodic temperature variations, in *Physics of Plant Environment*, edited by W. R. Van Wijk, North-Holland Publishing Company, Amsterdam.

Vickers, D, and L. Mahrt (2002), The Cospectral Gap and Turbulent Flux Calculations, *Journal of Atmospheric and Oceanic Technology, 20*, 660–672.

Vogel, S. (1970), Convective cooling at low airspeeds and the shapes of broad leaves, *Journal of Experimental Botany, 21*, 91–101.

Vogel, S. (1993), When leaves save the tree, *Natural History, 102*, 58–62.

Vogel, S. (1994), *Life in moving fluids:the physical biology of flow*, 2nd ed, 467 pp, Princeton University Press, Princeton.

Vong, R. J., et al (2004), Eddy correlation measurements of aerosol deposition to grass, *Tellus, 56B*, 105–117.

Waggoner, P. E, and W. E. Reifsnyder (1968), Simulation of the temperature, humidity and evaporation profiles in a leaf canopy, *Journal of Applied Meteorology, 7*, 400–409.

Walsberg, G. E., et al (1978), Animal coat colour and radiative heat gain, *Journal of Comparative Physiology, 126*, 211–212.

Ward, J. M., et al (2001), Thermal resistance of chicken (*Gallus domesticus*) plumage: a comparison between broiler and free-range birds, *British Poultry Science, 42*, 558–563.

Waring, R. H., et al (1998), Net primary production of forests: a constant fraction of gross primary production? *Tree Physiology, 18*, 129–134.

Waring, R. H, and S. W. Running (1998), *Forest Ecosystems: Analysis at Multiple Scales*, 2nd ed, 370 pp, Academic Press, New York.

Wathes, C, and J. A. Clark (1981a), Sensible heat transfer from the fowl: I. Boundary layer resistance of model fowl, *British Poultry Science, 22*, 161–173.

Wathes, C, and J. A. Clark (1981b), Sensible heat transfer from the fowl: II. Thermal resistance of the pelt, *British Poultry Science, 22*, 175–183.

Webb, E. K. (1970), Profile relationships: the log-linear range, and extension to strong stability, *Quarterly Journal of the Royal Meteorological Society, 96*, 67–90.

Webb, E. K., et al (1980), Correction of flux measurements for density effects due to heat and water vapour transfer, *Quarterly Journal of the Royal Meteorological Society, 106*, 85–100.

Webster, M. D., et al (1985), Cutaneous resistance to water vapour diffusion in pigeons, and the role of plumage, *Physiological Zoology, 58*, 58–70.

Weiss, A, and J. M. Norman (1985), Partitioning solar radiation into direct and diffuse, visible and near-infrared components, *Agricultural and Forest Meteorology, 34*, 205–213.

Welles, J. M. (1990), Some indirect methods of estimating canopy structure, *Remote Sensing Reviews, 5*, 31–43.

Wesely, M. L., et al (1978), Daytime variations of ozone eddy fluxes to maize, *Boundary-Layer Meteorology, 15*, 361–373.

Wesely, M. L, and B. B. Hicks (1977), Some factors that affect the deposition of sulfur dioxide and similar gases on vegetation, *Journal of the Air Pollution Control Association, 27*, 1110–1116.

Wheldon, A. E, and N. Rutter (1982), The heat balance of small babies nursed in incubators and under radiant warmers, *Early Human Development, 6*, 131–143.

Wiersma, F, and G. L. Nelson (1967), Nonevaporative convective heat transfer from the surface of a bovine, *Transactions of the American Society of Agricultural Engineers, 10*, 733–737.

Wigley, G, and J. A. Clark (1974), Heat transfer coefficients for constant energy flux models of broad leaves, *Boundary-Layer Meteorology, 7*, 139–150.

Williams, M., et al (2001), Evaluation different soil and plant hydraulic constraints on tree function using a model and sap flow data from ponderosa pine, *Plant, Cell and Environment, 24*, 679–690.

Willmer, P., et al (2005), *Environmental Physiology of Animals*, Second ed, 754 pp, Blackwell Science Ltd, Oxford.

Wilson, J. D., et al (1981), Numerical simulation of particle trajectories in inhomogeneous turbulence. II. Systems with variable turbulent velocity scale, *Boundary-Layer Meteorology, 21*, 423–441.

Wilson, N. R, and R. H. Shaw (1977), A higher-order closure model for canopy flow, *Journal of Applied Meteorology, 16*, 1198–1205.

Wood, C. J. (1995), Understanding wind forces on trees, in *Wind and Trees*, edited by M. P. Coutts and J. Grace, pp. 133–164, Cambridge University Press, Cambridge.

Wylie, R. G. (1979), Psychrometric wet-elements as a basis for precise physico-chemical measurements, *Journal of Research of the National Bureau of Standards (US), 84*, 161–177.

Yuge, T. (1960), Experiments on heat transfer from spheres including combined natural and forced convection. Transactions of the American Society of Mechanical Engineers, Series C, *Journal of Heat Transfer, 82*, 214–220.

Zhou, L., et al (2001), Variations in northern vegetation activity inferred from satellite data of vegetation index during 1981 to 1999, *Journal of Geophysical Research, 106*, 20069–20083.

►

Bibliography

GENERAL INTRODUCTORY TEXT BOOKS

The Climate near the Ground, R. Geiger. Harvard University Press, Cambridge, Mass, 1965. The classic review and discussion of the early literature of microclimatology.

Vegetation and the Atmosphere. Volume I. Principles, ed. J.L. Monteith. Academic Press, London, 1975. Radiation, transfer processes and general concepts.

An Introduction to Environmental Biophysics, 2nd edition, G.S. Campbell and J.M. Norman. Springer-Verlag, New York, 1998. Physics of soil, plant and animal microclimates. Remote sensing. Numerical examples and problems. Uses molar units.

Boundary Layer Climates, T.R. Oke. Methuen, London, Second edition, 1987. General review of microclimatology with geographical emphasis. Covers scales from those of insects and leaves to cities.

Plants and Microclimate, 2nd edition. H.G. Jones. Cambridge University Press, Cambridge, 1992. Microclimate and environmental physiology, strong on physiological responses to environment.

Air Pollution, an introduction. J. Colls. Chapman and Hall, London, 1997. Sources, dispersion and biological and physical impacts of a wide range of pollutants.

Climate Change 2007, the Fourth Assessment Report of the Intergovernmental Panel on Climate Change (IPCC). Available in several languages online at http://www.ipcc.ch/ Physical basis of climate change, revised every five years by an international panel of research scientists. Includes excellent summaries for general readership.

MORE SPECIALIZED BOOKS
Radiation

Solar and Terrestrial Radiation, K.L. Coulson. Academic Press. New York, 1975. Reviews short- and long-wave radiation with emphasis on instrumentation.

Radiative Processes in Meteorology and Climatology, G.W. Paltridge and C.M.R. Platt. Elsevier, Amsterdam, 1976. Principles and meteorological implications.

Plants and the Daylight Spectrum, ed. H. Smith. Academic Press, London, 1981. Conference proceedings reviewing many aspects of plant-light relations.

Transfer processes

(i) Physical principles

Introduction to micrometeorology, S.P. Arya. Academic Press, New York, 1988. Principles and measurement techniques for micrometeorology.

Atmospheric boundary layer flows: their structure and measurements, J.C. Kaimal and J.J. Finnigan. Oxford University Press, Oxford, 1994. Mathematical treatment of boundary layer theory and its measurement.

An Introduction to Boundary Layer Meteorology, R.B. Stull. Kluwer Academic, Dordrecht, 1988. Statistical and mathematical basis of geophysical boundary layers.

Evaporation into the Atmosphere, W.H. Brutsaert. D. Reidel, Dordrecht, Holland, 1982. Theoretical concepts and their applications.

Atmospheric Diffusion, F. Pasquill and F.B. Smith. Ellis Horwood, Chichester, 1982. Physics and mathematics of diffusion, and examples of dispersion of particles and gases.

(ii) Biophysical connections

Life in moving fluids:the physical biology of flow, S. Vogel. Princeton University Press, Princeton, 1994. Plant and animal interactions in air and water.

Fluid Behaviour in Biological Systems, L. Leyton. Oxford University Press, 1975. Comprehensive review of flow and transfer in air and liquids. Deals with plants and animals.

Windborne Pests and Diseases, D. Pedgeley. Ellis Horwood, Chichester, 1982. Reviews role of weather in dispersion and transport.

Particles

Relevant chapters will be found in the general texts: *Plants and the Atmospheric Environment: Vegetation and the Atmosphere Volume I*. Particles are also considered in *Atmospheric Diffusion*, and *Windborne Pests and Diseases* (Transfer processes).

The Mechanics of Aerosols, N.A. Fuchs. Pergamon Press, Oxford, 1964. Classic book translated from Russian covering all aspects of aerosol physics.

Aerosol Technology: properties, behavior and measurement of airborne particles, 2nd edition, W.C. Hinds,John Wiley 1998. Principles and measurement methods with good sections on inhalation and atmospheric aerosols.

Plant environment

Plant Physiological Ecology, H. Lambers, F. Stuart Chapin III and T.L. Pons, Springer, New York 1998. Comprehensive text on physiological and physical interactions.

Physicochemical and environmental plant physiology, P. S. Nobel, Elsevier, Boston and Amsterdam, 2005 Substantial text integrating biology, chemistry and physics.

Physiological Ecology of Forest Production, J.J. Landsberg. Academic Press, London, 1986. Short but comprehensive review of forest environment and tree physiology.

Forest Ecosystems: analysis at multiple scales, 2nd edition, R.H. Waring and S.W. Running. Academic Press, San Diego, 1998. Forest ecology, emphasizing quantitative modelling across large spatial and time scales.

Vegetation and the Atmosphere, Volume II, Case Studies, ed. J.L. Monteith. Academic Press, London, 1976. Summarizes field experiments on specific crops, types of forests and natural ecosystems.

Resource Capture by Crops, eds. J.L. Monteith, R.K. Scott and M.H. Unsworth. Nottingham University Press, Nottingham, 1994. Conference proceedings with chapters on capture of light, water and nutrients, and applications to arable, greenhouse and grass crops.

Plant Response to Wind, J. Grace. Academic Press, London, 1977. Basic physics and physiological processes.

Wind and Trees, ed. M.P. Coutts and J. Grace. Cambridge University Press, Cambridge, 1995. Conference proceedings with excellent reviews of airflow in forests and mechanics of trees in the wind.

Soil physics

Introduction to Environmental Soil Physics, D. Hillel. Elsevier Academic Press, New York, 2004. Principles of heat and mass transport in soils, practical aspects of the field water cycle, and plant water relations.

Principles of Soil and Plant Water Relations, M.B. Kirkham. Elsevier Academic Press, New York, 2005. Principles of soil physics, plant anatomy and physiology, instrumentation and measurement methods

Soil Physics with Basic, G.S. Campbell. Elsevier, Amsterdam, 1986. Simulation of transport processes in soil-plant systems with BASIC computer programs.

Animal environment

Environmental Physiology of Animals, 2nd edition, P. Willmer, G. Stone, and I Johnston. Blackwell Science, Oxford, 2005 Wide-ranging text with emphasis on the environmental context of the physiology and adaptation of animals.

Environmental Aspects of Housing for Animal Production, ed. J.A. Clark. Butterworths, London, 1981. Conference proceedings with reviews of principles of heat balance and practical details of housing.

Adaptation to Thermal Environment, L.E. Mount. Edward Arnold, London, 1979. Heat exchanges between animals (including man) and their environments.

Heat Loss from Animals and Man, eds J.L. Monteith and L.E. Mount. Butterworths, London, 1974. Conference proceedings with coverage of principles and measurements.

Man and his Thermal Environment, R.P. Clark and O.G. Edholm. Edward Arnold, London, 1985. Heat exchange and thermal physiology of man.

Instrumentation

Measuring the Natural Environment, 2nd edition. I. Strangeways. Cambridge University Press, Cambridge, 2003. Measurement techniques, operation of instruments, and discussion of accuracy and data-recording.

Remote sensing

Introduction to Remote Sensing, 4th edition. J.B. Campbell. Guilford Press, New York, 2006. Comprehensive basic text.

Manual of Remote Sensing (3rd edition) Volume I, Earth observing platforms and sensors, ed. R.A. Ryerson, American Society for Photogrammetry and Remote Sensing, Falls Church, Virginia, 1998. Multi-volume series useful for specific principles and practice.

Appendix A

Table A1 Système International (SI) units with c.g.s. equivalents

Quantity	Dimensions	SI	c.g.s.
Length	L	1 m	$= 10^2$ cm
Area	L^2	1 m²	$= 10^4$ cm²
Volume	L^3	1 m³	$= 10^6$ cm³
Mass	M	1 kg	$= 10^3$ g
Density	$M\,L^{-3}$	1 kg m⁻³	$= 10^{-3}$ g cm⁻³
Time	T	1 s (or min, h, etc.)	$= 1$ s
Velocity	$L\,T^{-1}$	1 m s⁻¹	$= 10^2$ cm s⁻¹
Acceleration	$L\,T^{-2}$	1 m s⁻²	$= 10^2$ cm s⁻²
Force	$M\,L\,T^{-2}$	1 kg m s⁻² = 1 N (Newton)	$= 10^5$ g cm s⁻² $= 10^5$ dynes
Pressure	$M\,L^{-1}\,T^{-2}$	1 kg m⁻¹ s⁻² = 1 N m⁻² (Pascal)	$= 10$ g cm⁻¹ s⁻² $= 10^{-2}$ mbar
Work, energy	$M\,L^2\,T^{-2}$	1 kg m² s⁻² = 1 J (Joule)	$= 10^7$ g cm² s⁻² $= 10^7$ ergs
Power	$M\,L^2\,T^{-3}$	1 kg m² s⁻³ = 1 W (Watt)	$= 10^7$ g cm² s⁻³ $= 10^7$ ergs s⁻¹
Dynamic viscosity	$M\,L^{-1}\,T^{-1}$	1 N s m⁻²	$= 10$ dynes s cm⁻² $= 10$ Poise
Kinematic viscosity	$L^2\,T^{-1}$	1 m² s⁻¹	$= 10^4$ cm² s⁻¹ $= 10^4$ Stokes
Temperature	H (or $M\,L^2\,T^{-2}$)	1 °C (or 1 K)	$= 1$ °C (or 1 K)
Heat energy	$H\,T^{-1}$	1 J	$= 0.2388$ cal
Heat or radiation flux	$H\,L^{-2}\,T^{-1}$	1 W	$= 0.2388$ cal s⁻¹
Heat flux density	$H\,M^{-1}$	1 W m⁻²	$= 2.388 \times 10^{-5}$ cal cm⁻² s⁻¹
Latent heat	$H\,M^{-1}\,\theta^{-1}$	1 J kg⁻¹	$= 2.388 \times 10^{-4}$ cal g⁻¹
Specific heat	$H\,M^{-1}\,\theta^{-1}$	1 J kg⁻¹ K⁻¹	$= 2.388 \times 10^{-4}$ cal g⁻¹ K⁻¹
Thermal conductivity	$H\,L^{-1}\,\theta^{-1}\,T^{-1}$	1 W m⁻¹ K⁻¹	$= 2.388 \times 10^{-3}$ cal cm⁻¹ s⁻¹ K⁻¹
Thermal diffusivity (and other diffusion coefficients)	$L^2\,T^{-1}$	1 m² s⁻¹	$= 10^4$ cm² s⁻¹

Table A.2 Properties of air, water vapor and CO_2 (treated as constant between -5 and $45°C$)

		Air	Water vapor	Carbon dioxide
Specific heat	$(J\ g^{-1}\ K^{-1})$	1.01	1.88	0.85
Prandtl number	$Pr = (\nu/\kappa)$	0.70_5	—	—
	$Pr^{0.67}$	0.79	—	—
	$Pr^{0.33}$	0.89	—	—
	$Pr^{0.25}$	0.92	—	—
Schmidt number	$Sc = (\nu/D)$	—	0.63	1.04
	$Sc^{0.67}$	—	0.74	1.02
	$Sc^{0.33}$	—	0.86	1.01
	$Sc^{0.25}$	—	0.89	1.01
Lewis number	$Le = (\kappa/D)$	—	0.89	1.48
	$Le^{0.67}$	—	0.93	1.32
	$Le^{0.33}$	—	0.96	1.14
	$Le^{0.25}$	—	0.97	1.11

Table A.3 Properties of air, water vapor and CO₂ (changing by less than 1% per K)

Temperature		Densities of air		Virtual temperature of air	Latent heat of vaporization of water		Thermal conductivity of air	Molecular diffusion coefficients of air			
T		ρ_a	$\rho_{as}(T)$	T_v	λ	γ	k	κ	ν	D_v	D_c
K		kg m⁻³		°C	J g⁻¹	Pa K⁻¹	mW m⁻¹ K⁻¹	10⁻⁶ m² s⁻¹			
°C (Unit)											
-5	268.2	1.316	1.314	-4.57	2513	64.3	24.0	18.3	12.9	20.5	12.4
0	273.2	1.292	1.286	0.64	2501	64.6	24.3	18.9	13.3	21.2	12.9
5	278.2	1.269	1.265	5.92	2489	64.9	24.6	19.5	13.7	22.0	13.3
10	283.2	1.246	1.240	11.32	2477	65.2	25.0	20.2	14.2	22.7	13.8
15	288.2	1.225	1.217	16.87	2465	65.5	25.3	20.8	14.6	23.4	14.2
20	293.2	1.204	1.194	22.62	2452	65.8	25.7	21.5	15.1	24.2	14.7
25	298.2	1.183	1.169	28.62	2442	66.2	26.0	22.2	15.5	24.9	15.1
30	303.2	1.164	1.145	34.97	2430	66.5	26.4	22.8	16.0	25.7	15.6
35	308.2	1.146	1.121	41.73	2418	66.8	26.7	23.5	16.4	26.4	16.0
40	313.2	1.128	1.096	49.03	2406	67.1	27.0	24.2	16.9	27.2	16.5
45	318.2	1.110	1.068	57.02	2394	67.5	27.4	24.9	17.4	28.0	17.0

ρ_a density of dry air
$\rho_{as}(T)$ density of air saturated with water vapor at temperature
T_v virtual temperature of saturated air
λ latent heat of vaporization of water
γ $c_p\rho/\lambda\varepsilon$—'psychrometer constant'

k thermal conductivity of dry air
κ thermal diffusivity of dry air
ν kinematic viscosity of dry air
D_v diffusion coefficient of water vapor in air
D_c diffusion coefficient of CO₂ in air

Table A.4 Quantities changing by more than 1% per K. $e_s(T)$ saturation vapor pressure at temperature T (°C); Δ change of saturation vapor pressure per K, i.e. $\partial e_s/\partial T$; σT^4 full radiation at temperature T (K); $4\sigma T^3$ change of full radiation per K. Note that the quantities Δ and $4\sigma T^3$ can be used as mean differences to interpolate between the tabulated values of e_s and σT^4 respectively

T (°C)	(K)	$e_s(T)$ (kPa)	$\Delta(T)$ Pa K^{-1}	σT^4 W m^{-2}	$4\sigma T^3$ W m^{-2} K^{-1}
− 5	268.2	0.421	32	293.4	4.4
− 4	269.2	0.455	34	297.8	4.4
− 3	270.2	0.490	37	302.2	4.5
− 2	271.2	0.528	39	306.7	4.5
− 1	272.2	0.568	42	311.3	4.6
0	273.2	0.611	45	315.9	4.6
1	274.2	0.657	48	320.5	4.7
2	275.2	0.705	51	325.2	4.7
3	276.2	0.758	54	330.0	4.8
4	277.2	0.813	57	334.8	4.8
5	278.2	0.872	61	339.6	4.9
6	279.2	0.935	65	344.5	5.0
7	280.2	1.001	69	349.5	5.0
8	281.2	1.072	73	354.5	5.1
9	282.2	1.147	78	359.6	5.1
10	283.2	1.227	83	364.7	5.2
11	284.2	1.312	88	369.9	5.2
12	285.2	1.402	93	375.1	5.3
13	286.2	1.497	98	380.4	5.3
14	287.2	1.598	104	385.8	5.4
15	288.2	1.704	110	391.2	5.4
16	289.2	1.817	117	396.6	5.5
17	290.2	1.937	123	402.1	5.6
18	291.2	2.063	130	407.7	5.6
19	292.2	2.196	137	413.3	5.7
20	293.2	2.337	145	419.0	5.7
21	294.2	2.486	153	424.8	5.8
22	295.2	2.643	162	430.6	5.8
23	296.2	2.809	170	436.4	5.9
24	297.2	2.983	179	442.4	6.0
25	298.2	3.167	189	448.3	6.0
26	299.2	3.361	199	454.4	6.1
27	300.2	3.565	210	460.5	6.2
28	301.2	3.780	221	466.7	6.2
29	302.2	4.006	232	472.9	6.3
30	303.2	4.243	244	479.2	6.3
31	304.2	4.493	257	485.5	6.4
32	305.2	4.755	269	492.0	6.5
33	306.2	5.031	283	498.4	6.5
34	307.2	5.320	297	505.0	6.6
35	308.2	5.624	312	511.6	6.7
36	309.2	5.942	327	518.3	6.7
37	310.2	6.276	343	525.0	6.8
38	311.2	6.262	357	531.8	6.9
39	312.2	6.993	376	538.7	6.9
40	313.2	7.378	394	545.6	7.0
41	314.2	7.780	413	552.6	7.1
42	315.2	8.202	432	559.7	7.1
43	316.2	8.642	452	566.8	7.2
44	317.2	9.103	473	574.0	7.3
45	318.2	9.586	494	581.3	7.3

Table A.5 Nusselt numbers for air

(a) Forced convection

Shape	Case	Range of Re	Nu
(1) Flat plates			
	Streamline flow	$<2 \times 10^4$	$0.60\ Re^{0.5}$
	Turbulent flow	$>2 \times 10^4$	$0.032\ Re^{0.8}$
(2) Cylinders			
	Narrow range of Reynolds numbers	$1-4$	$0.89\ Re^{0.33}$
		$4-40$	$0.82\ Re^{0.39}$
		$40-4 \times 10^3$	$0.62\ Re^{0.47}$
		$4 \times 10^3 - 4 \times 10^4$	$0.17\ Re^{0.62}$
		$4 \times 10^4 - 4 \times 10^5$	$0.024\ Re^{0.81}$
		or	
	Wide range of Reynolds numbers	$10^{-1} - 10^3$	$0.32 + 0.51\ Re^{0.52}$
		$10^3 - 5 > 10^4$	$0.24\ Re^{0.60}$
(3) Spheres			
		$0-300$	$2 + 0.54\ Re^{0.5}$
		$50 - 1.5 \times 10^5$	$0.34\ Re^{0.6}$

Notes (i) Arrows show direction of airflow

(ii) d is characteristic dimension; take width of a long crosswind strut as shown or mean side for a rectangle whose width and length are comparable

(iii) To find corresponding Sherwood numbers multiply Nu by $Le^{0.33}$ (see values in Table A.1)

(iv) Sources—Ede (1967), Fishenden and Saunders (1950), Bird, Stewart and Lightfoot (1960)

Table A.5—(*continued*)(b) Free convection

Shape and relative temperature	Range Laminar flow	Turbulent flow	Nu
(1) Horizontal flat plates or cylinders			
(i) Hot ──d── or ──Cold──	$Gr < 10^5$		$0.50\,Gr^{0.25}$
		$Gr > 10^5$	$0.13\,Gr^{0.33}$
(ii) Hot or Cold			$0.23\,Gr^{0.25}$
		Arrangement not conducive to turbulence	
(iii) Hot or cold ⦂ d	$10^4 < Gr < 10^9$		$0.48\,Gr^{0.25}$
		$Gr > 10^9$	$0.09\,Gr^{0.33}$
(2) Vertical flat plates or cylinders Hot or cold d	$10^4 < Gr < 10^9$		$0.58\,Gr^{0.25}$
		$10^9 < Gr < 10^{12}$	$0.11\,Gr^{0.33}$
(3) Spheres Hot or Cold d	$Gr^{0.25} < 220$		$2 + 0.54\,Gr^{0.25}$

Notes
(i) Arrows indicate direction of air circulation
(ii) d is characteristic dimensions for calculation of Gr: take height for vertical plate and average chord for horizontal plate
(iii) To find corresponding Sherwood numbers, multiply Nu by $Le^{0.25}$ for laminar flow or tubulent flow (see values in Table A.1)
(iv) Sources—Ede (1967), Fishenden and Saunders (1950), Bird, Stewart and Lightfoot (1960)

Table A.6 Characteristic quantities for particle transfer in air*: D diffusion coefficient ($m^2 s^{-1}$); τ Relaxation time (s); $\overline{\Delta x_B}$ Mean displacement in 1 s in a given direction $2(Dt/\pi)^{0.5}$ (μm); $\overline{\Delta x_s}$ Distance fallen in 1 s under gravity (μm)

Radius r (μm)	D ($10^{-9} m^2 s^{-1}$)	τ (10^{-6} s)	$\overline{\Delta x_B}$ (μm)	$\overline{\Delta x_s}$ (μm)
1.0×10^{-3}	1.28×10^3	1.33×10^{-3}	1.28×10^3	1.31×10^{-2}
5.0×10^{-3}	5.24×10^1	6.76×10^{-3}	2.58×10^2	6.63×10^{-2}
1.0×10^{-2}	1.35×10^1	1.40×10^{-2}	1.31×10^2	1.37×10^{-1}
5.0×10^{-2}	6.82×10^{-1}	8.81×10^{-2}	2.95×10^1	8.64×10^{-1}
0.1	2.21×10^{-1}	0.23	1.68×10^1	2.24
0.5	2.7×10^{-2}	3.54	5.90	3.47×10^1
1.0	1.3×10^{-2}	1.31×10^1	4.02	1.28×10^2
5.0	2.4×10^{-3}	3.08×10^2	1.74	3.0×10^3
10.0	1.4×10^{-3}	1.23×10^3	1.23	1.2×10^4

* From Fuchs (1964), calculated at 23°C, standard atmospheric pressure; particle density 1 g cm^{-3}.

Solutions to Selected Problems

CHAPTER 2

2.1 Assuming that g is unchanged, $\Gamma = g/c_p$, and $c_p = \frac{7}{2}\frac{R}{M} = \frac{7}{2} \times \frac{8.31}{10}$. Hence $\Gamma = \frac{9.81 \times 10 \times 2}{7 \times 8.31 \times 1000} = 0.0034$ K m^{-1}.

CHAPTER 3

3.1 $r = \int \frac{dz}{D} = \frac{1 \times 10^{-3}}{15.1 \times 10^{-6}} = 66$ s m^{-1}. If pressure is reduced from 101.3 to 70 kPa, $r = 66 \times 70/101.3 = 46$ s m^{-1}

To convert resistances from s m^{-1} to molar units of m^2 s mol^{-1}, divide by the molar volume of air, which is $273/(293 \times 0.0224) = 41.6$ mol m^{-3} at 101.3 kPa and 20°C, and $70.0 \times 273/(293 \times 101.3 \times 0.0224) = 28.7$ mol m^{-3} at 70.0 kPa and 20°C. Hence the resistance in molar units is independent of pressure, and is $66/41.6 = 46/28.7 = 1.59$ m^2 s mol^{-1}. This conservative property of molar resistances makes them preferable for some calculations.

CHAPTER 4

4.1 Energy per photon is $6.63 \times 10^{-34} \times 3.0 \times 10^8/300 \times 10^{-6} = 6.63 \times 10^{-22}$ J. The photon flux is therefore $20/6.63 \times 10^{-22} = 3.0 \times 10^{22}$ photons m^{-2} s^{-1} or $3.0 \times 10^{22}/6.02 \times 10^{23} = 0.050$ mol m^{-2} s^{-1}

4.2 $\lambda_{max} = 2897/3000 = 0.97$ μm.

CHAPTER 5

5.1 Civil twilight is when $\psi = 96°$ and sunset is when $\psi = 90°$. The small correction in the following calculations for the equation of time is ignored. To find the hour angle θ at civil twilight, solve the equation $\cos 96 = \sin 51.5 \sin 23.5 + \cos 51.5 \cos 23.5 \cos \theta$, giving $\cos \theta = -0.730$ and $\theta = 136.9°$. Hence astronomical twilight is $136.9/15 = 9.13$ hours after solar noon. Similarly, (or using the simplification of Equation 5.4) θ at sunset is $123.1°$, so sunset is $123.1/15 = 8.21$ hours after solar noon. The difference is therefore 0.92 hours, or 55 minutes.

5.2 At solar noon $\psi = \phi - \delta = 45.0 - 23.5 = 21.5°$. Hence $m = \sec 21.5 = 1.075$. Then

$$\mathbf{S}_p = 1366 \exp(-0.30 \times 1.075)(\exp -(1.075 \times \tau_a))$$
$$= 990 \exp(-1.075 \times \tau_a),$$

and $\mathbf{S}_s = \mathbf{S}_p \cos \psi$.
Diffuse radiation is calculated as $\mathbf{S}_d = 0.3 \times 1366 \times \cos 21.5(1 - \exp(-(\tau_m + \tau_a) \times 1.075))$, and $\mathbf{S}_t = \mathbf{S}_s + \mathbf{S}_d$. The table shows results of the calculations in W m^{-2} for the three values of τ_a.

τ_a	$\mathbf{S}_p$	$\mathbf{S}_s$	$\mathbf{S}_d$	$\mathbf{S}_t$
0.05	941	875	118	993
0.20	802	746	160	906
0.40	644	599	202	801

5.3 The daylength $n = 24\pi^{-1} \cos^{-1}(-\tan 45.0 \tan 23.5) = 15.4$ hours. Insolation is $2n\pi^{-1}\mathbf{S}_{tm}$ where $\mathbf{S}_{tm}$ is the irradiance at noon. Consequently the insolation at $\tau_a = 0.05, 0.20$ and 0.40 is $35.0, 32.0$ and 28.3 MJ m^{-2}.

5.4 $\mathbf{L}_d = 356$ W m^{-2}; $\varepsilon_a = 0.79$. When $c = 0.5$, $\varepsilon_a(c) = 0.88$, and $\mathbf{L}_d(c) = 0.88 \times 448 = 393$ W m^{-2}

5.5 Corrected irradiances are A $= 916$; B $= 120$; C $= 465$; D $= 25$ W m^{-2}. Then

   i. $120/916 = 0.13$;

   ii. $(120 - 25)/120 = 0.79$;

   iii. $(916 - 465 - 120 + 25)/(916 - 120) = 356/796 = 0.45$

5.6 Assuming that the soil and soot are perfect black-body radiators, $\mathbf{L}_u = 545$ W m^{-2}, so $T_{soot} = 40°C$.

CHAPTER 6

6.1 $k_{red} = 0.23$; $k_{bluegreen} = 0.014$. Fractional transmission at 100 m is $1.03 \times 10^{-10}/0.246 = 4.2 \times 10^{-10}$.

6.2 Assume that PAR is 50% of the total irradiance. Then $\rho_t = 0.5\rho_{PAR} + 0.5\rho_{NIR}$; $\rho_{NIR} = 0.3/0.5 = 0.6$. Benefits could be efficient use of PAR but weak absrption of NIR, thus reducing overheating.

CHAPTER 7

7.2 $x = 5$; $A_h/A = \frac{(2 \times 5 \times \cot 10°)/\pi + 1}{(2 \times 5) + 2} = \frac{19.1}{12} = 1.59$

$$
\begin{aligned}
\overline{S}_b &= 1.59 S_b = 1.59 \times \sin 10° \times S_p \\
&= 1.59 \times 0.174 \times 800 \\
&= 221 \text{ W m}^{-2}
\end{aligned}
$$

7.3 Short-wave: Upper surface

$$0.933 \times 150 + 0.067 \times 0.15 \times 150 = 142 \text{ W m}^{-2}$$

Lower surface
$$0.067 \times 150 + 0.933 \times 22.5 = 31 \text{ W m}^{-2}$$
Long-wave: Upper surface

$$
\begin{aligned}
0.933 \times 290 + 0.067 \times 391 &= 297 \text{ W m}^{-2} \\
0.067 \times 290 + 0.933 \times 391 &= 384 \text{ W m}^{-2}
\end{aligned}
$$

CHAPTER 8

8.1 $\mathcal{K}_s = 2(\cot \beta)/\pi = 0.637$

  i. $S_b(L) = 500 \exp(-0.637 \times 4) = 39.1 \text{ W m}^{-2}$

  ii. Fractional area of sunflecks is $S_b(L)/S_b(0) = 39.1/500 = 0.078$

  iii. Mean irradiance per unit leaf area is $\mathcal{K}_s S_b = 0.637 \times 39.1 = 24.9 \text{ W m}^{-2}$

8.2 When $\beta = 60°$, $\mathcal{K}_s$ would be $0.5/\sin 60°$ ($= 0.577$) if the leaf distribution was spherical, and $2(\cot 60°)/\pi$ ($= 0.368$) if the distribution was cylindrical. The observed $S(L)/S(0) = 1 - 0.8 = 0.2 = \exp(-\mathcal{K}_s \times 3)$. Hence observed transmission corresponds to $\mathcal{K}_s = -\ln 0.2/3 = 0.54$, so the distribution most closely corresponds to spherical.

8.3 Treat the radiation as diffuse when observed over a full day. Then Figure 8.3 gives $\mathcal{K} = 0.67$.

i. $\tau = \exp(-0.67 \times 3.0) = 0.13$

ii. Interception $= 1 - \tau = 0.87$

iii. $(1 - 0.95) = \exp(-0.67L); \ L = 4.5$

8.4 Given: $\rho = \tau = 0.15$ for PAR

    a) absorption coefficient

$$\alpha_p = 1 - \tau - \rho = 1 - 2(0.15) = 0.70$$

    b) reflection coefficient of the canopy (PAR)

$$\rho_c^* = (1 - \alpha_p^{0.5})/(1 + \alpha_p^{0.5}) = 0.09$$

    c) Repeat (a) and (b) for NIR when $\rho = \tau = 0.40$

$$\alpha_p = 1 - 0.40 - 0.40 = 0.20$$
$$\rho_c^* = (1 - \alpha_p^{0.5})/(1 + \alpha_p^{0.5}) = 0.38$$

    d) Calculate ρ_c^* for the whole solar spectrum

$$\rho_{ctotal}^* = 0.5 \times (\rho_{cPAR}^*) + 0.5 \times (\rho_{cNIR}^*) = 0.24$$

8.5

    a)

$$\alpha_p = 1 - \tau_p - \rho_p = 1 - 0.10 - 0.10 = 0.80$$
$$\mathcal{K} = \alpha_p^{0.5}\mathcal{K}_b = (0.80^{0.5}) \times 1 = 0.89$$

    b) neglecting second order terms,

$$\rho_c = \rho_c^* - (\rho_c^* - \rho_s) \exp(-2\mathcal{K}L)$$

and

$$\rho_c^* = (1 - \alpha_p^{0.5})/(1 + \alpha_p^{0.5}) = 0.056$$

so

$$\rho_c = 0.056 - (0.056 - 0.15) \exp(-2 \times 0.89 \times 1) = 0.072$$
$$\tau_c = \exp(-\mathcal{K}L) = \exp(-0.89 \times 1) = 0.41$$

c)

$$\alpha_c = 1 - \rho_c - \tau_c(1 - \rho_s) = 1 - 0.072 - 0.41(1 - 0.15)$$
$$= 0.58$$

d) Interception is defined as $(1 - \tau_c)$, so

$$\alpha_c/(1 - \tau_c) = 0.98.$$

Thus 98% of the PAR radiation intercepted by the sparse canopy is absorbed. If the problem had been set up for total solar radiation, the ratio would have been closer to 0.75.

CHAPTER 9

9.1 As $\mathrm{Re_p}$ is small ($\mathrm{Re_p} = 0.5 \times 10^{-3} \times 4.2 \times 10^{-6}/15 \times 10^{-6} = 0.14 \times 10^{-3}$), use the relation

$$c_d = 24/\mathrm{Re_p} = 171$$

CHAPTER 10

10.1 For the flat plate,

$$\mathrm{Re} = 2.0 \times 50 \times 10^{-3}/15 \times 10^{-6} = 6.6 \times 10^{-3}$$

$\mathrm{Nu} = 0.60\,\mathrm{Re}^{0.5} = 49$. For the leaf, $\mathrm{Nu} = 2 \times 49 = 98$.
Then convective heat transfer is

$$\mathbf{C} = 98 \times 26 \times 10^{-3} \times 1.5/50 \times 10^{-3} = 76 \ \mathrm{W\,m^{-2}}$$

10.2 $\mathrm{Re} = 8 \times 10^{-3} \times 0.05/15 \times 10^{-6} = 2.7$

$$\mathrm{Gr} = 158 \times 0.8^3 \times 5 = 404$$

$\mathrm{Gr/Re}^2 = 55$, so heat transfer is dominated by free convection. Assume that $\mathrm{Nu} = 0.58\mathrm{Gr}^{0.25}$. Hence $\mathrm{Nu} = 4.5$ and

$$\mathbf{C} = 4.5 \times 26 \times 10^{-3} \times 5/8 \times 10^{-3} = 73 \ \mathrm{W\,m^{-2}}$$

10.3 In equilibrium, the net radiation must balance convective heat loss. The net radiation is $\mathbf{R}_n = (1 - 0.4)300 + \sigma T_a^4 - \sigma T_t^4$. Writing the difference between air and thermometer temperature $(T_a - T_t)$ as ΔT, it follows that, for small temperature differences, $\mathbf{R}_n = (1 - 0.4)300 + 4\sigma T_a^3 \Delta T$. The rate of convective heat loss is given by $\mathbf{C} = -\rho c_p \Delta T/80$. Hence, equating the fluxes, $-\rho c_p \Delta T/80 = 180 + 4\sigma T_a^3 \Delta T$. Solving for ΔT gives $\Delta T = -8.7°\mathrm{C}$, i.e, $T_t = 28.7°\mathrm{C}$.

i. Increasing the reflection coefficient gives $\Delta T = -1.4°C$, i.e, $T_t = 21.4°C$

ii. Adding a radiation shield that reduced ventilation increases r_H to 113 s m^{-1} and gives $\Delta T = -5.5°C$, i.e, $T_t = 25.5°C$

10.4 The net radiation balance of the bud, temperature T_b, is $\mathbf{R}_n = 0.5 \times 230 + 0.5 \times \sigma \times 273^4 - \sigma T_b^4$. If T_b is to be maintained at 273K, it follows that

$$\mathbf{R}_n = 115 + 158 - 316 = -43 \text{ W m}^{-2}.$$

This radiative loss must be balanced by heat gain of 43 W m^{-2} if the bud is to maintain thermal equilibrium. The heat gain could be from convection (fans warming the air near the ground) or from latent heat (spraying the bud with water).

10.5 $\text{Re} = 0.30 \times 0.15/16 \times 10^{-6} = 2.8 \times 10^3$

$$\text{Gr} = 158 \times 30^3 \times 40 = 1.7 \times 10^8$$

$\text{Gr}/\text{Re}^2 = 22$, so heat transfer is dominated by free convection and the flow is assumed laminar. Using the relation $\text{Nu} = 0.48\text{Gr}^{0.25}$ gives $\text{Nu} = 55$, and $\mathbf{C} = 55 \times 27 \times 10^{-3} \times 40/0.30 = 198 \text{ W m}^{-2}$.

10.6 Using Equation 10.23, $\mathbf{G} = k'(T_1 - T_2)/(r_2 \ln(r_2/r_1))$. Then $\mathbf{G} = 0.60 \times 7/(0.10 \ln(0.10/0.$
188 W m^{-2}.

CHAPTER 11

11.1 Saturation vapour pressure at 25°C is 3167 Pa. Vapour pressure at 60% relative humidity is $0.60 \times 3167 = 1900$ Pa.

i. Absolute humidity in the leaf is $\chi_l = 2.17 \times 3167/298 = 23.1 \text{ g m}^{-3}$. Absolute humidity in the air is $\chi_a = 0.60 \times 23.1 = 13.9 \text{ g m}^{-3}$.

ii. a. $F_w = D_w(\chi_l - \chi_a)/l = 25.3 \times 10^{-6} \times (23.1 - 13.9)/10 \times 10^{-6} = 23.5$ $\text{g m}^{-2}\text{ s}^{-1}$.

b. Resistance of one stoma is $l/D_w = 10 \times 10^{-6}/25.3 \times 10^{-6} = 0.40 \text{ s}$ m^{-1}.

iii.

$$r_l = 4[l + (\pi d/8)]/\pi n d^2 D_w$$
$$= 4[10 \times 10^{-6}$$
$$+(\pi \times 5 \times 10^{-6}/8)]/(\pi \times 200 \times 10^6 \times (5 \times 10^{-6})^2 \times 25.3 \times 10^{-6}$$

$r_l = 120 \text{ s m}^{-1}$. This value is typical for herbaceous plants when freely supplied with water.

11.2 For a single pore, $100\mu m$ long and $6\mu m$ diameter, the resistance is

$$r_p = (l + \pi d/8)/D_w$$
$$= (100 \times 10^{-6} + (\pi \times 6 \times 10^{-6}/8))/25 \times 10^{-6}$$
$$= 4.0 \text{ s m}^{-1}.$$

The evaporation rate per unit area of membrane is $\mathbf{E} = (n\pi d^2/4)\Delta\chi/r_p$ where $\Delta\chi$ is the difference in absolute humidiy across the pore. Since $\chi_{\text{inside}} = 2.17 \times 4243/303 = 30.4 \text{ g m}^{-3}$, and $\chi_{\text{outside}} = 2.17 \times 0.30 \times 421/268 = 1.02 \text{ g m}^{-3}$, $\Delta\chi = 29.4 \text{ g m}^{-3}$. Then $\mathbf{E} = 10^9 \times \pi \times (6 \times 10^{-6})^2 \times 29.4/4.0 = 0.21 \text{ g m}^{-2}$ s^{-1}. Hence $\lambda\mathbf{E} = 525 \text{ W m}^{-2}$.

11.3 Since r_H for the upper and lower leaf surfaces in parallel is 40 s m^{-1}, the value of r_H for each side is 80 s m^{-1}. Then, combing total resistances for each side in parallel gives

$$(r_t)^{-1} = (80 + 100)^{-1} + (80 + 200)^{-1}.$$

Hence $r_t = 110 \text{ s m}^{-1}$.

11.4 $\mathbf{E} = (\chi_l - \chi_a)/r_t$. The value of χ_l is $2.17 \times 2337/293 = 17.3 \text{ g m}^{-3}$, and χ_a is $0.5 \times 17.3 = 8.65 \text{ g m}^{-3}$. Rearranging the flux equation gives $r_t = 8.65/(10.0 \times 10^{-6} \times 10^4) = 87 \text{ s m}^{-1}$. The flux of carbon dioxide from the ambient air to the leaf is $\mathbf{F}_c = 100 \times 10^{-6} \times 1.87 \times 10^3/r_c$ where r_c is the combined stomatal and boundary layer resistance for CO_2 transfer. Assuming that the transfer is taking place by forced convection, the ratio of the boundary layer resistances for CO_2 and water vapour transfer r_c/r_V is $1.32/0.93 = 1.42$ and the ratio of the stomatal resistances is $1.14/0.96 = 1.19$ (see Equation 11.5). Since the boundary layer and stomatal resistances are not known individually, the mean value for r_c/r_V , 1.30 may be used to estimate r_t for carbon dioxide. Hence $\mathbf{F}_c = 100 \times 10^{-6} \times 1.87 \times 10^3/(87 \times 1.30) = 1.7 \text{ mg } CO_2 \text{ m}^{-2} \text{ s}^{-1}$.

11.5 To determine whether the boundary layer flow is likely to be laminar or turbulent, calculate Re, which is $\text{Re} = 1.0 \times 50 \times 10^{-3}/15.8 \times 10^{-6} = 3165$. Hence flow is in the margin between laminar and turbulent for the leaf. If flow is laminar, then $\text{Nu} = 0.60 \text{Re}^{0.5} = 34$, and $r_H = l/(\kappa Nu) = 50 \times 10^{-3}/(22 \times 10^{-6} \times 34) = 67$ s m^{-1}. Adopting an empirical correction factor of 1.5 to allow for boundary layer turbulence, the value of r_H becomes $67/1.5 = 45 \text{ s m}^{-1}$, and $r_V = 0.93 r_H = 41 \text{ s}$ m^{-1}.

If the leaf surface is wet, absolute humidity on the surface is $\chi_l = 2.17 \times 3167/298 = 23.1 \text{ g m}^{-3}$, and $\chi_a = 0.6 \times 23.1 = 13.8 \text{ g m}^{-3}$. Then $\mathbf{E} = (23.1 - 13.8)/41 = 0.23$ $\text{g m}^{-2} \text{ s}^{-1}$ and $\lambda\mathbf{E} = 2436 \times 0.23 = 553 \text{ W m}^{-2}$.

11.6 The evaporation rate per unit leaf area is $\mathbf{E} = 0.70/(100 \times 10^{-4} \times 600) = 0.117 \text{ g}$ $\text{m}^{-2} \text{ s}^{-1}$. As in problem 11.5, $\chi_l = 2.17 \times 3167/298 = 23.1 \text{ g m}^{-3}$, but this time

$\chi_a = 0.75 \times 23.1 = 17.3$ g m^{-3}. Then the boundary layer resistance is given by $r_b = (\chi_l - \chi_a)/\mathbf{E} = (23.1 - 17.3)/0.117 = 49.3$ s m^{-1}.

11.7 As this is a free convection problem, begin by calculating the Grashof number and then the Nusselt number.

    i. Ignoring the humidity gradient.

$$\text{Gr} = 158d^3(T_s - T_a) = 158 \times 50^3 \times (30 - 25) = 99 \times 10^6.$$

Table A.5 indicates that the flow will be laminar and Nu $= 0.48\text{Gr}^{0.25} = 48$. Then $r_H = d/\kappa\text{Nu} = 0.50/(23 \times 10^{-6} \times 48) = 453$ s m^{-1}, and, assuming $r_V/r_H = 0.93$, $r_V = 421$ s m^{-1}.

    ii. Considering the humidity gradient it is necessary to find the difference in virtual temperature between the surface and the air. This is

$$\begin{aligned}
T_{vs} - T_{va} &= T_s(1 + 0.38e_s/p) - T_a(1 + 0.38e_a/p) \\
&= (T_s - T_a) + 0.38p^{-1}(e_s T_s - e_a T_a) \\
&= (T_s - T_a) + 0.38 \times 10^{-5}(4243 \times 303 - (0.30 \times 3167 \times 298).
\end{aligned}$$

(Note the use of temperature in Kelvin). Hence $T_{vs} - T_{va} = (T_s - T_a) + 3.8 = 8.8$ K. Then Gr $= 158d^3(T_s - T_a) = 158 \times 50^3 \times (8.8) = 174 \times 10^6$, and Nu $= 55$. The resistance to sensible heat transfer is $r_H = d/\kappa\text{Nu} = 0.50/(23 \times 10^{-6} \times 55) = 395$ s m^{-1} and $r_V = 368$ s m^{-1}.
The evaporation rate from the wet mud is $\mathbf{E} = \rho c_p \delta e/\lambda \gamma r_V$. Taking humidity gradients into account, this yields

$$\begin{aligned}
\mathbf{E} &= 1.2 \times 10^3 \times (4243 - 0.30 \times 3167)/(2430 \times 66.5 \times 368) \\
&= 66 \times 10^{-3} \text{ g m}^{-2} \text{ s}^{-1}.
\end{aligned}$$

Ignoring the buoyancy created by the humidity gradient gives $\mathbf{E} = 58 \times 10^{-3}$ g m^{-2} s^{-1}, i.e. a 12% underestimation in the flux.

CHAPTER 12

12.1   i. If the pollen grain obeys Stokes' Law, then $V_s = 2\rho g r^2/9\rho_a v$. Hence

$$\begin{aligned}
V_s &= 2 \times 0.8 \times 10^6 \times 9.81 \times (5 \times 10^{-6})^2/9 \times 1.29 \times 10^3 \times 15 \times 10^{-6} \\
&= 2.4 \text{ mm s}^{-1}.
\end{aligned}$$

To determine whether using Stokes' Law is appropriate, calculate the particle Reynolds number Re$_p$.

$$\text{Re}_p = 2.4 \times 10^{-3} \times 10 \times 10^{-6}/(15 \times 10^{-6}) = 1.6 \times 10^{-3}$$

So Stokes' Law is valid.

ii. In this case, Stokes' Law predicts that $V_s = 545$ m s^{-1}, but then $\text{Re}_p = 109 \times 10^3$, so Stokes' Law clearly is not adequate for calculating V_s. Using the iterative method described in the text, with a first guess for drag coefficient of 0.44 eventually yields $V_s = 13$ m s^{-1}. There is an interesting discussion of drag coefficients at http://exploration.grc.nasa.gov/education/rocket/termvr.html

12.2 $\tau = m/6\pi v \rho_a r = 2r^2 \rho_p / 9 v \rho_a$

i.

$$\tau = 2 \times (10 \times 10^{-6})^2 \times 1 \times 10^6 / (9 \times 15 \times 10^{-6} \times 1.2 \times 10^3$$
$$= 1.23 \times 10^{-3} \text{ s.}$$

$S = \tau V_0 = 1.23 \times 10^{-3} \times 2.0 = 2.5$ mm. Hence the probability of deposition from turbulent flow in a bronchus of diameter 4mm would be high.

ii. $\tau = 3 \times 10^{-6}$ s, and $S = 6$ μm. The probability of deposition would be low.

12.4 (This problem and solution were written by Dr. A.C Chamberlain, who was a special professor in the Environmental Physics group at the University of Nottingham.)

i. The drops do not obey Stokes' Law, so their terminal velocities must be calculated by trial and error, seeking to balance the gravitional force by the drag force.

Radius of drop r (μm)	100	1000
Projected area A (m^2)	3.14×10^{-8}	3.14×10^{-6}
Gravitational force $F_g = mg$ (N)	4.10×10^{-8}	4.10×10^{-5}

The drag force on a drop with cross-section A, falling at velocity V, is given by $F_d = 0.5\rho_a A c_d V^2$. The dependence of the drag coefficient c_d on drop Reynolds' number is shown in Figure 9.6 or may be calculated from Equation 9.XX. Drag forces for a range of fall speeds for the two drop sizes are calcualated in the table below

Radius of drop r (μm)	100		
Velocity V (m s^{-1})	0.5	0.6	0.7
Re_p	6.7	8.0	9.3
c_d	5.5	5.2	5.0
Drag force (N)	2.65×10^{-8}	3.6×10^{-8}	4.7×10^{-8}

Radius of drop r (μm)	1000		
Velocity V (m s^{-1})	5.0	6.0	7.0
Re_p	670	800	930
c_d	0.53	0.50	0.48
Drag force (N)	2.6×10^{-5}	3.4×10^{-5}	4.5×10^{-5}

By interpolation, the gravitational force and drag force are equal when $V = 0.65$ m s^{-1} (100μm drops) and $V = 6.7$ m s^{-1} (1000μm drops). These are the terminal velocities of the drops.

ii. Knowing the terminal velocities, the stopping distances S_0 and Stokes numbers Stk ($= S_0/r$) for 10μm diameter particles impacting on falling drops can be found using Table A.6 ($S_0 = 200\mu$m, Stk $= 2.0$ (for 100μm drops); $S_0 = 1700\mu$m, Stk $= 1.7$ (for 1000μm drops))

iii. The impaction efficiency c_p can then be read from Figure 12.3, $c_p = 0.58$ (100μm drops) and 0.54 (1000μm drops). Note that, because c_p increases with increasing relative velocity between the drop and the particle, it decreases with increasing drop radius, so the net variation in c_p with drop size is small.

iv. Now calculate the number of possible impacts on a drop per second. A raindrop has volume $4\pi r^3/3$ and projected area πr^2. Hence each drop, considered as sweeping out a cylinder that just fits it, is eqivalent to $4r/3$ m of rain. A rainfall rate of 1 mm h^{-1}, or 0.28×10^{-6} m s^{-1} is therefore equivalent to $0.28 \times 10^{-6}/(4r/3) = 2.1 \times 10^{-7}r^{-1}$ drops s^{-1} through each point in the atmosphere. Hence the number of drops per second (n) for 1mm h^{-1} rainfall is 2×10^{-3} with 100μm drops and 2×10^{-4} with 1000μm drops. The washout coefficient Λ (s^{-1}) $= nc_p = 1.2 \times 10^{-3}$ (s^{-1}) for 100μm drops and 1.1×10^{-4} (s^{-1}) for 1000μm drops. Hence, for a given rate of rainfall, small drops are more efficient than large drops in removing particles by impaction.

v. The fraction of aerosol remaining after a time t is $f_w = \exp -\Lambda t$. Hence, after one hour ($t = 3600$ s), f_w is 0.013 (1.3%) for the 100μm rain drops and 0.49 (49%) for the 1000μm rain drops.

CHAPTER 13

13.1 The resistance of the thermometer to sensible heat transfer is $r_H = d/\kappa$Nu. The Nusselt number is given by Nu $= 0.24\,\mathrm{Re}^{0.60} = 0.24 \times (3 \times 10^{-3})^{0.6} \times V^{0.6} = 5.65V^{0.6}$. Hence $r_H = 3 \times 10^{-3}/(22.2 \times 10^{-6} \times 5.65 \times V^{0.6}) = 23.9V^{-0.6}$ s m^{-1}. The radiative resistance to heat transfer is $r_R = \rho c_p/4\sigma T^3 = 210$ s m^{-1}. From Equation 13.5, $T_t = (r_H T_{sh} + r_R T_a)/(r_R + r_H)$. Rearranging, $r_H = r_R(T_t - T_a)/(T_{sh} - T_t)$, which must be $0.1 \times 210/4.9 = 4.3$ s m^{-1}. Hence, $V^{-0.6} = 4.3/23.9$, and $V = 17.6$ m s^{-1}.

13.4 When air temperature $T_a = 22^\circ$C, $\Delta = 162$ Pa K^{-1} and $\gamma = 66.3$ Pa K^{-1}. The saturation deficit δ is $2643 - 1000 = 1643$ Pa. Substituting values into the Penman–Monteith equation gives

$$\lambda E = [162 \times 300 + (1.2 \times 10^3 \times 1643/40)]/[162 + 66.3(110/40)]$$
$$= 284 \text{ W m}^{-2}.$$

Then $\mathbf{C} = \mathbf{R}_n - \lambda E = 16$ W m^{-2}. Since $\mathbf{C} = \rho c_p(T_l - T_a)/r_H$, $(T_l - T_a) = 40 \times 16/(1.2 \times 10^3) = 0.53$, so $T_l = 22.5^\circ$C.

13.6 The runner's velocity, 19 km h^{-1}, is 5.3 m s^{-1}. Nusselt number Nu= $0.24\,\text{Re}^{0.6}$ = $0.24 \times (0.33 \times 5.3/16 \times 10^{-6})^{0.6}$ = 253. Reistance to sensible heat transfer is $r_H = d/\kappa\text{Nu} = 0.33/(22.8 \times 10^{-6} \times 253) = 57$ s m^{-1}. Assume that $r_V/r_H = 0.93$. Saturation deficit δ is $(0.75 \times 4243 - 2400) = 782$ Pa. Other parameters are $\Delta = 244$ Pa K^{-1} and $\gamma = 66.5$ Pa K^{-1}. Applying the Penman-Monteith equation gives

$$\lambda E = [(0.75 \times 244 \times (300 + 600)) + (1.2 \times 10^3 \times 782/57)]/[(0.75 \times 244) + (66.5 \times 0.93)]$$
$$= 739 \text{ W m}^{-2}.$$

If the salt was washed off, so that the relative humidity at the surface was 100%, but all other terms remained the same, $\lambda E = 843$ W m^{-2}.

CHAPTER 14

14.1 The net isothermal radiation is $\mathbf{R}_{ni} = (1 - 0.40)300 + 4\sigma T^3 \Delta T = 180 + 6.0 \times 9.6 = 240$ W m^{-2}. Then, applying Equation 14.3, $\mathbf{M} + \mathbf{R}_{ni} - \lambda\mathbf{E}_r - \lambda\mathbf{E}_s = 140 + 240 - 11 - 98 = \rho c_p(T_c - T_a)/r_{HR} = 1.2 \times 10^3(31.6 - 22)/r_{HR}$, and solving for r_{HR} gives $r_{HR} = 43$ s m^{-1}. Assuming that the coat is not penetrated by radiation, the flux of sensible heat through the coat is $\mathbf{M} - \lambda\mathbf{E}_r - \lambda\mathbf{E}_s = \rho c_p(T_c - T_s)/r_c$, so, rearranging and substituting values gives the coat resistance $r_c = 1.2 \times 10^3(34.0 - 31.6)/(140 - 11 - 98) = 93$ s m^{-1}.

Subtracting the second algebriac equation from the first gives $\mathbf{R}_{ni} = \rho c_p[((T_c - T_a)/r_{HR}) - ((T_s - T_c)/r_c)]$, so if shading reduced $\mathbf{R}_{ni}$ to 100 W m^{-2}, solution of this equation gives $T_c = 28.2°$C. Denoting the new rate of latent heat loss from the skin as $\lambda\mathbf{E}'_s$, the heat balance can be written

$$140 + 100 - 11 - \lambda\mathbf{E}'_s = 1.2 \times 10^3(28.2 - 22)/43,$$

giving $\lambda\mathbf{E}'_s = 56$ W m^{-2}.

14.2   i. When the skin is dry, $\mathbf{M} + \mathbf{R}_n - \lambda\mathbf{E}_r = \rho c_p(T_s - T_a)/r_{HR}$. Rearranging to solve for T_a gives

$$T_a = [-r_{HR}(\mathbf{M} + \mathbf{R}_n - \lambda\mathbf{E}_r)/\rho c_p] + T_s$$
$$= [-80(60 + 240 - 10)/1.2 \times 10^3] + 33$$
$$= 13.7°\text{C}.$$

  ii. When the skin is covered in wet mud, surface temperature T_m. Heat flow through the mud is described by $\rho c_p(T_s - T_m)/r_m = \mathbf{M} - \lambda\mathbf{E}_r$. Solving for T_m gives $T_m = 33 - (8 \times (60 - 10)/1.2 \times 10^3) = 32.7°$C. Then, for the mud-covered skin

$$\mathbf{M} + \mathbf{R}_{ni} - \lambda\mathbf{E}_r = [\rho c_p(T_m - T_a)/r_{HR}] + [\rho c_p(e_{sm} - e_a)/\gamma r_v].$$

Assume that $r_V/r_{HR} = 0.93$. Solving the equation gives $T_a = 80°$C.

CHAPTER 15

15.1 i. $\tau = 80$ s

    ii. $\tau = 31$ minutes

15.2 630 W m^{-2}. Sources are net radiation absorption and sensible heat transfer.

15.3 i. (a) $\rho' = 1.03 \times 10^6$ g m^{-3}, $c' = 0.90$ J g^{-1} K^{-1}; (b) $\rho' = 1.38 \times 10^6$ g m^{-3}, $c' = 1.73$ J g^{-1} K^{-1}

    ii. (a) $\kappa' = 0.32 \times 10^{-6}$ m^2 s^{-1}; $D = 9.4$ cm; (b) $\kappa' = 0.67 \times 10^{-6}$ m^2 s^{-1}; $D = 18.4$ cm

15.5 i. $x = 0.40$;

    ii. $\rho' c' = 2.13$ MJ m^{-3} K^{-1}

15.6 Extrapolating to the surface, $T_{\text{surface}} = -3.0°$C. Soil heat flux $G = -k\frac{dT}{dz} = -120$ W m^{-2}, and this must be equated to the net radiation $\mathbf{R_n}$. Assuming that the surface radiates like a perfect black body, it follows that $\mathbf{L_d} = -120 + 302 = 182$ W m^{-2}.

CHAPTER 16

16.1 i. $z_0 = 0.25$ m; $u_* = 0.164$ m s^{-1}

    iii. $r_{aM} = 52$ s m^{-1}

16.2 i. $u_* = 0.89$ m s^{-1}

    ii. $u_{30} = 3.89$ m s^{-1}; $r_{aM} = 4.9$ s m^{-1}

16.3 i. $\beta = 0.50$; $\mathbf{C} = 143$ W m^{-2}; $\lambda\mathbf{E} = 287$ W m^{-2}

16.4 i. The zero plane displacement can be found by trial and error, seeking a value of d that produces the best straight-line relationship between $\ln(z - d)$ and u. This value is approximately $d = 0.56$ m. Then $u_* = 0.32$ m s^{-1} and $z_0 = 6.3$ cm

    ii. $\tau = 0.12$ N m^{-2}

    iii. $r_{aM} = 24$ s m^{-1}

    iv. $\mathbf{F_c} = -\frac{331.1 - 324.5}{2.65 - 1.68} \times 0.32^2 \times \frac{605}{330} = 1.28$ mg m^{-2} s^{-1} = 4.6 g m^{-2} h^{-1}

16.5 Using the trial and error method in 16.4,

    i. $d = 0.15$ m;

    ii. $z_0 = 0.03$ m;

    iii. $u_* = 0.20$ m s^{-1};

    iv. $\tau = 0.048$ N m^{-2}

    v. $\mathbf{F}_{O3} = 0.49$ μg m^{-2} s^{-1};

    vi. $v_g = 0.49/96 = 5 \times 10^{-3}$ m s^{-1} = 5 mm s^{-1}.

CHAPTER 17

17.3 $r_c = 228$ s m^{-1}. If the resistance remained constant, $\lambda\mathbf{E}$ would increase linearly with saturation deficit, but in reality many tree species increase their stomatal resistance as saturation deficit increases, adjusting transpiration to balance the rate at which water can be transported from the soil through roots, stem and foliage.

17.4 Resultant total resistance is $\frac{1}{r_t} = \frac{1}{228} + \frac{1}{300} = 130$ s m^{-1}. Then $\mathbf{F}_{SO_2} = 100/130 = 0.77$ μg m^{-2} s^{-1}. The flux into the plant is $100/228 = 0.44\mu$g m^{-2} s^{-1}, so the fraction entering the plant is 0.57.

Index

Abundance ratio, 24
Acid deposition, 18
Aerosol
    effect on plant productivity, 123
Air mass number, 60, 62
Albedo, *see* Reflection coefficient
Ammonia, 358
Amphistomatous, 244
Angle
    hour, 54
    Zenith, 63
    zenith, 54
Assmann psychrometer, *see* Psychrometer
Atmospheric transmissivity, 62
Atmospheric window, 74
Attenuation coefficient, 49
Avogadro's constant, 5

Basal metabolic rate, 259, 269
Beer's Law, 49
Bi-directional reflectance, 47
Big leaf model, 250
Black body radiation,
            see Radiation, black body
Boltzmann's constant, 5
Bounce-off, 216
Boundary layer
    convective, 252
    displacement, 144
    turbulent, 143, 301
Boundary layers
    laminar, 310
    turbulent, 310
Bowen ratio, 327
Boyle's Law, 2, 5
Broadening
    Collision, 39
    Doppler, 39

Bulk density, 291
Bushel basket experiment, 316

Calorimeter, 266
Carbon dioxide, 18
    Rate of increase, 20
Cavitation, 353
Charles' Law, 5
Clausius-Clapeyron equation, 12
Clo, 181, 279
Cloud water deposition,
            see Occult precipitation
Clumping index, 125
Cold limit, 268
Condensation, 289
Conductance, 34, 201
Conservative quantities, 10
Convection
    forced, 162, 191
    free, 163, 192
    mixed, 164
Convective boundary layer, 252
Convergence, 293
Cosine law, 44

Dalton's Law, 6
Damping depth, 295
Declination, 53
Decoupling coefficient, 255
Density
    Dry air, 6
Deposition velocity, 34, 217
Dew point, 14
Dew point temperature, 14
Diffusion
    Brownian, 226
    molecular, 190
    turbulent, 190

Diffusivity
  Thermal, 30
Divergence, 293
Drag coefficient, 147
Dry deposition, 356, 359

Earth's orbit
  eccentricity , 53
  obliquity , 53
  precession, 53
Eddy covariance, 304
Eddy diffusivity, 310
Emissivity
  animal coats, 87
  leaves, 87
  soils, 93
Emittance, 44
Enhanced vegetation index, 131
Entrainment, 252
Equilibrium evaporation, 250, 252, 255
Equivalent temperature,
        see Temperature, equivalent
Eulerian models, 360, 363
Evaporation, 11
  equilibrium, 255
  imposed, 255
  respiratory, 261
EVI, see Enhanced vegetation index

Fetch, 301, 328
Fick's Law of Diffusion, 33
Field capacity, 22
Fleece, 274
Flux density
  mole, 35
Form drag, 146, 338
Friction velocity, 311

Gap fraction, 124
Generalized stability factor, 322
Goretex, 209
Grashof number, 163
Grazing incidence, 47
Greenhouse effect, 75
Greenhouse gases, 18

Heat capacity, 285, 291
Hemispherical photography, 124

Humidity
  Absolute, 15
  increment, 276
  Relative, 16
  Specific, 15
Hydrostatic Equation, 7
Hyperthermia, 270
Hypostomatous, 244
Hypothermia, 268

Ideal Gas Equation, 5
Impaction, 216, 226
  efficiency, 216, 217
Inertial sublayer, 302
Insensible perspiration, 261
Insolation, 69
Insulation, 181
Intercepted rainfall, 247, 343, 344
Interception
  particles, 218, 227
Inverse methods, 362
Irradiance, 44
Isotropic scattering, 51

K-theory, 360, 363
kB^{-1}, see Resistance,
        additional aerodynamic
Kinetic theory, 4
Kirchhoff's Principle, 39, 78
Kubelka and Munk equations, 50

Lagrangian models, 360, 363
Lambert's Cosine Law, 46
Lambertian surface, 47
Laminar flow, 165
Laminar sublayer, 143
Lapse rate, 8
  Dry adiabatic, 9
  Saturated adiabatic, 9
Latent heat, 8
  of condensation, 8
  of fusion, 8
  of melting, 8
  of vaporization, 8
Leaf angle distribution
  conical, 120
  spherical and ellipsoidal, 118
  vertical, 118

Leaf area index, 96, 116, 123
Lewis number, 192
Lifetime
    of gases in the atmosphere, 19
Locust, 272
Lodging of crops, 155
Lower critical temperature,
        see Temperature, critical
Lungs
    alveoli, 225
    head airways, 224
    particle deposition in, 224
    pulmonary airways, 225
    tracheobronchial airways, 224

Methane, 18
Micrometeorology, 300
Milankovich Theory, 53
Minute volume, 198
Mixing length, 310
Mixing ratio, 14
Models, 2
Moisture characteristic, 21
Molar gas constant, 5
Mole flux density, 35
Mole fraction, 35
Mole unit, 43
Molecular weight
    Dry air, 6
Monin-Obhukov Similarity Theory,
        see Similarity theory

NDVI, *see* Normalized difference vegetation
        index
Net ecosystem exchange, NEE, 349
Net isothermal radiation,
        see Radiation, isothermal net
Net radiometer, 139
Nitrogen oxides, 18
Normalized difference vegetation index, 130
Nusselt number, 161

Occult precipitation, 218
Ohm's Law, 3
Open-top chambers, 198
Optical thickness, 62
Ozone, 61
    Tropospheric, 18

Partial pressure, 6
Penman equation, 240, 250
Penman-Monteith equation, 243, 250
Permanent wilting point, 22
Phase lag, 290
Photon, 43
Photosynthetically Active Radiation, 56
Planck's Law, 42
Potential Temperature, 10
Prandtl number, 162
Precipitable water, 60
Priestley-Taylor equation, 252
Productivity
    gross primary, 349
    net ecosystem, 349, 352
    net primary, 349
Profile, 310
Psychrometer, 233
    Assmannn, 237
Psychrometer constant, 234
    modified, 237

Quantum unit, 43

Radiance, 44
Radiant flux density, 44
Radiant intensity, 44
Radiation
    absorption, 58
    Absorption by gases, 60
    Black body, 40
    Diffuse, 62, 64
    Direct, 62
    Downward long-wave, 75
    Global, *see* Radiation, total
    increment, 270
    isothermal net, 241
    Net, 79
    Quantum content of, 64
    Terrestrial, 73
        from cloudy skies, 78
    Total, 62, 66
    transmission in soil, 92
Radiation interception
    black leaves
        diffuse radiation, 121
        direct radiation, 116
Radiative forcing, 19

416

Rayleigh number, 163
Reflection coefficient, 47, 87
    animal coats and plumage, 87
    canopies, 87, 95
    leaves, 87, 94
Reflectivity, 46, 87
    animal coats, 97
    dependence on water content of soil, 92
    leaves, 93
Relative humidity, 16, *see* Humidity
    Equilibrium, 23
Relaxation time, 215
Relaxed eddy accumulation , 309
Resistance, 3
    additional aerodynamic, 339, 344
    aerodynamic, 250, 318, 344
    animal coats, 208
    apparent canopy, 342
    canopy, 250, 337
    clothing, 208
    coat, 272
    combined for convection and radiation, 232
    dependence on pressure, 208
    eggshells, 202
    fungi, 202
    heat transfer, 161
    incursion, 199
    inequality of momentum and heat or mass transfer, 147
    insect integuments, 202
    leaf stomatal, 205
    mass transfer, 191
    molar units, 201
    radiative, 231
    stability, 340
    stomatal, 201
    stomatal pore, 203
    tissue, 269
Respiration
    autotrophic, 349
    heterotrophic, 349
    hetrotrophic, 350
Reynolds averaging, 304
Reynolds number, 143
Reynolds stress, 151
Richardson number, 321

Roughness length, 312
Roughness sublayer, 302

Saltation, 317
Saturation ratio, 222
Saturation vapor pressure deficit, 14
Scattering
    Mie, 58
    Multiple, 49, 65
    Rayleigh, 57, 63
    Single, 49
Schmidt number, 191
Sedimentation velocity, 37, 212
Sequestration of CO_2, 355
Shape factor
    cone, 107
    diffuse radiation
      of cone, 114
      of cylinders, 114
      on plane, 112
    direct radiation, 101
    ellipsoid, 102
    horizontal cylinder, 105
    inclined plane, 109
    sphere, 102
    vertical cylinder, 103
Shearing stress, 29,
        see Reynolds stress, 311
Sheep, 273
Shelter factor, 150
Sherwood number, 191
Similarity Theory, 319
Simple ratio, 130, 132
Skin friction, 145
Soil water content
    microwave sensing, 93
Solar Constant,
        see Total Solar Irradiance, 62
Solarimeter
    tube, 123
Specific heat
    at Constant Pressure, 7
    at Constant volume, 7
    bulk, 291
    volumetric, 291
Spectrum
    Line, 39
Spore dispersal, 154

Stability, 10, 11
Stability function, 320
Stable isotopes, 24
Standard Overcast Sky, 65
Stefan Boltzmann Law, 41
Stokes diameter, 213
Stokes number, 216
Stomata, 200
Stopping distance, 215
Sublayer
    inertial, 302
    roughness, 302
Sulfur dioxide, 18, 356
Sweat, 269, 277

Temperature
    apparent equivalent, 276
    critical, 269
    Dew point, 14
    effective, 270
    Equivalent, 236
    equivalent, 276, 327
    Potential, 10
    Thermodynamic wet-bulb, 17
    Virtual, 16
    Wet-bulb, 17
    wet-bulb, 237, 279
Thermal conductivity, 31, 291
Thermal diffusivity, 292
Thermo-neutral diagram, 266
Thermo-neutral zone, 269
Thermodynamic wet-bulb temperature, 234
Thermometer
    dry-bulb, 232
    wet-bulb, 234, 236
Time constant, 285
Tog, 181
Total Solar Irradiance, 52
Transmission coefficient, 117
Turbulence
    properties of, 303
Turbulent flow, 165
Turbulent intensity, 169, 197
Turbulent transfer coefficient, 152, 310
Twilight, 54

Ultraviolet radiation, 55
Uniform Overcast Sky, 65

Units
    alternatives for resistance
        and conductance, 35

Vapor pressure, 11
    Saturation, 12
Vapor pressure deficit,
        see Saturation vapor pressure deficit,
    353
Vasoconstriction, 183
Vasodilation, 183
Ventilation
    greenhouse, 197
    lungs, 198
Virtual temperature, 16
Viscosity
    Dynamic, 30
    Kinematic, 29
von Karman constant, 313

Washout coefficient, 228
Water
    content, 20
    potential, 20
        gravitational, 21
        matric, 21
        osmotic, 21
        presssure, 21
Water vapor, 11
Wet-bulb temperature, 17
Wilting point, see Permanent wilting point
Wind shear, 147
Windthrow, 155
Wood area index, 124

Zero plane displacement, 313

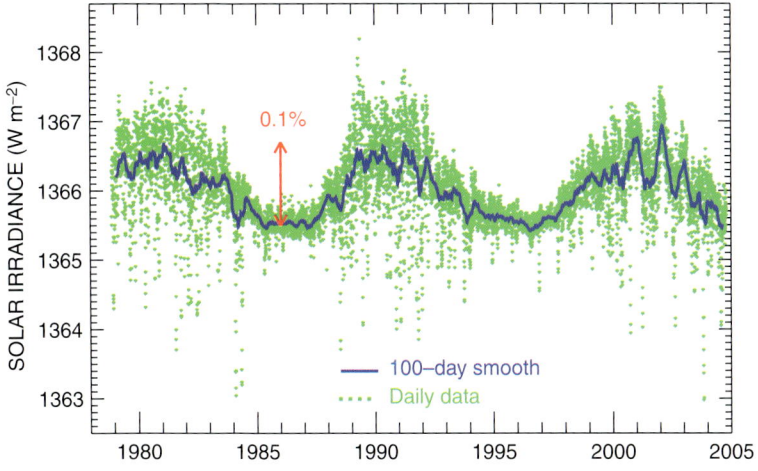

FIGURE 5.1 Composite record of Total Solar Irradiance (the 'Solar Constant') from 1978 to 2005 compiled from satellite radiometric measurements adjusted to a standard reference scale. (data courtesy of Dr. Judith Lean)

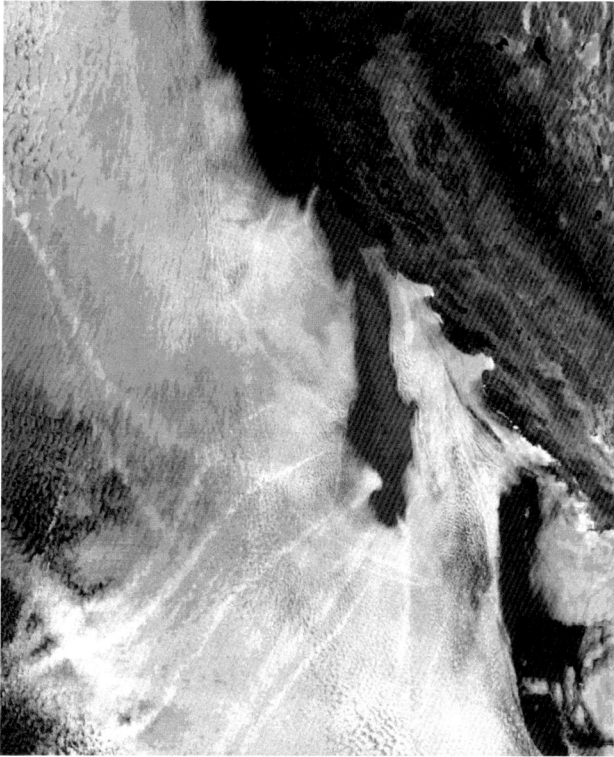

FIGURE 5.3 Infrared satellite image of reflected solar radiation at 2.1μm off the coast of California near San Francisco. Pollution from ship smoke emissions increases the reflectivity of marine stratus clouds by supporting the formation of larger numbers of smaller drops. This enables ship tracks to be seen clearly in the image. (Courtesy of Dr J. A. Coakley)

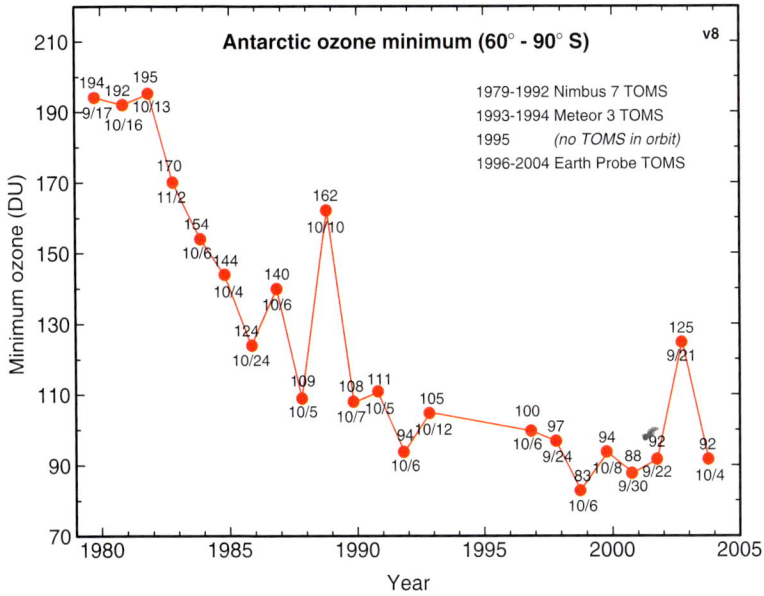

FIGURE 5.4 Variation since 1980 in observed minimum ozone column depth over Antarctica (in Dobson units) derived from satellite observations. The month and date when each minimum was observed are indicated below each point on the graph (1 Dobson unit, DU, corresponds to a layer of pure gaseous ozone 0.01mm thick at STP).

FIGURE 7.8 A conifer with a conical shape, casting a shadow that closely resembles the lower part of Figure 7.9. The amount of direct solar radiation intercepted by the tree could therefore be calculated from Equation 7.16.

A

FIGURE 8.7A Farmers at a cattle market in Andhra Pradesh. The bareheaded farmer needs the shade of an umbrella. The two with white turbans do not. (Would a white umbrella be more or less effective?)

B

FIGURE 8.7B Cattle in a field in Israel. The dark cattle have sought shade, but the white animals are apparently comfortable in full sunshine.

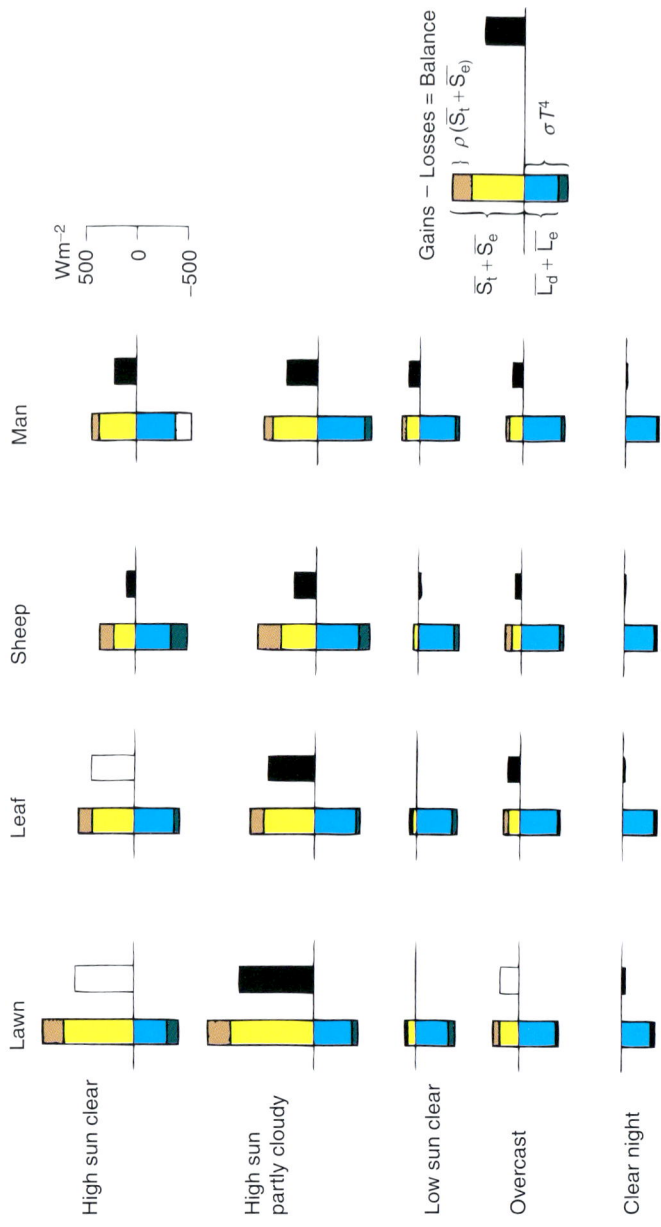

FIGURE 8.8

FIGURE 9.10A wind tunnel experiment using an array of model trees with realistically scaled properties. The researcher is holding an element of the array showing how the trees are attached to strain gauges to measure wind forces. (from Wood, 1995)

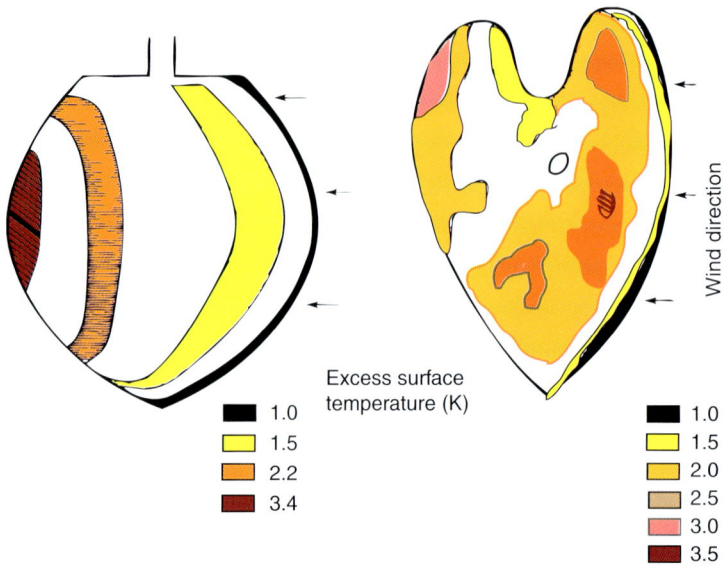

Excess surface temperature (K)

⬛	1.0
🟨	1.5
🟧	2.2
🟥	3.4

Wind direction

⬛	1.0
🟨	1.5
🟧	2.0
🟫	2.5
🟥	3.0
🟥	3.5

FIGURE 10.3

FIGURE 11.5 Open-top chambers in a field of beans at Sutton Bonington. The boxes adjacent to the chambers contain fans for ventilating the chambers, and some also contain charcoal for absorbing gaseous air pollutants. The "frustum" design at the top reduces the rate of incursion of unfiltered air.

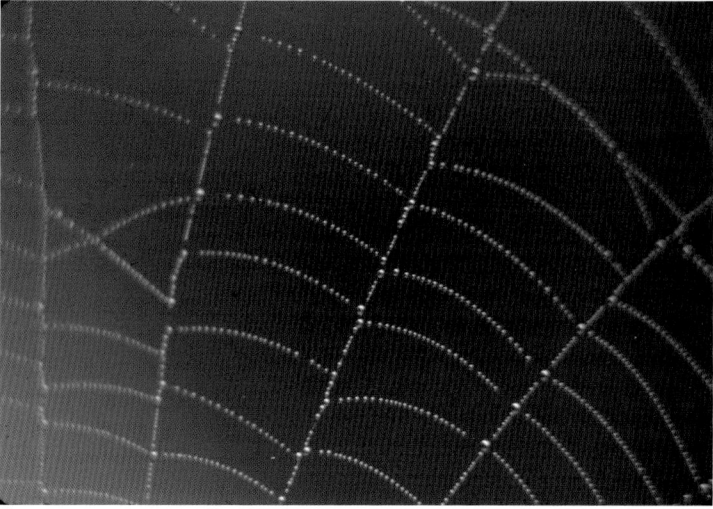

A

FIGURE 12.4A Drops of fog that have impacted on and been intercepted by threads of a spider's web. The threads of the web are at least an order of magnitude smaller than individual drops (typically 10 μm diameter), and so are efficient collectors by interception.

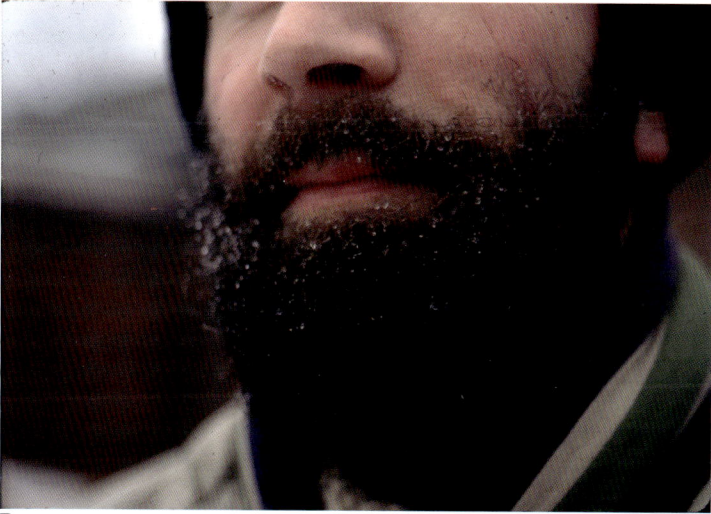

FIGURE 12.4B Drops of fog collected on an author's beard after cycling. Human hair is about 50 μm diameter, so the drops visible in this picture must have coalesced from many impacted fog drops.

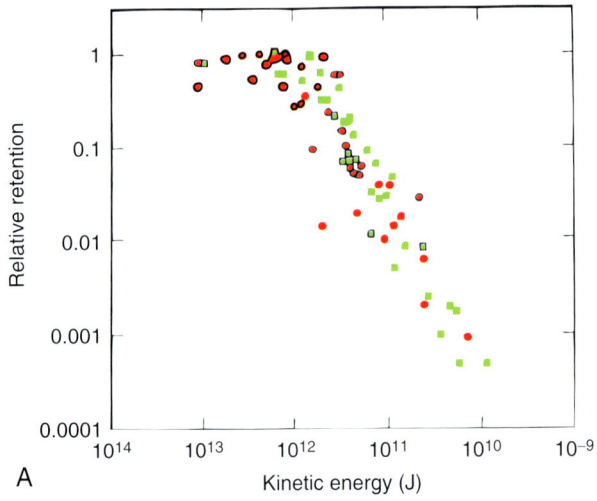

A

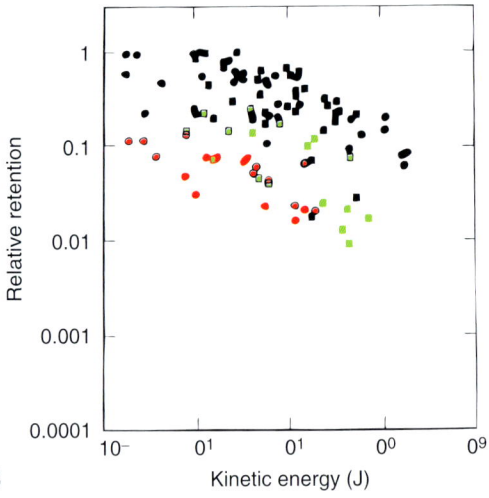

B

FIGURE 12.5

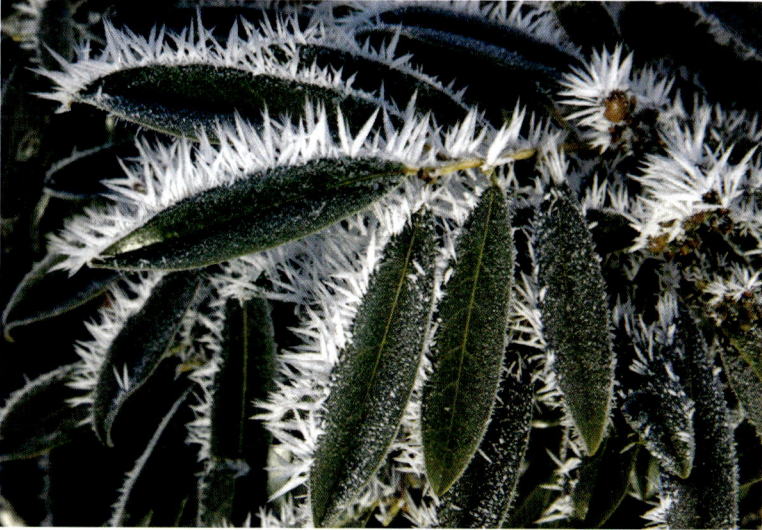

FIGURE 13.10   Hoarfrost on leaves. Note the preferential formation of ice on the spikes at the edges of the leaves. Exchange of heat and water vapor is faster round the edges than in the center of the laminae because the boundary layer is thinner at the edge. A faster rate of heat exchange implies that the spikes and leaf edges should be warmer than the rest of the leaf (i.e., closer to air temperature at night). A faster rate of mass exchange implies that the spikes should collect hoarfrost by condensation faster than the rest of the leaf when their temperature is below the frost-point of the air. (Courtesy of Randall L. Milstein)

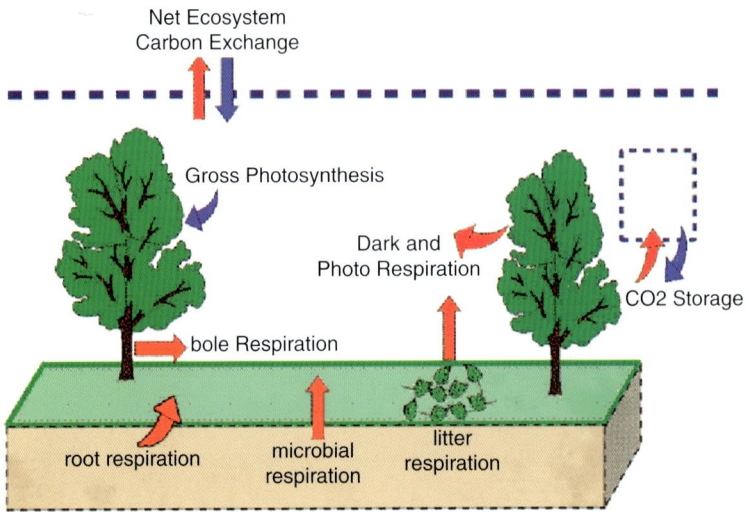

FIGURE 17.7 Schematic diagram of the main components of the carbon exchange between vegetation and the atmosphere. (Courtesy of Dr. D. Baldocchi)